T0348746

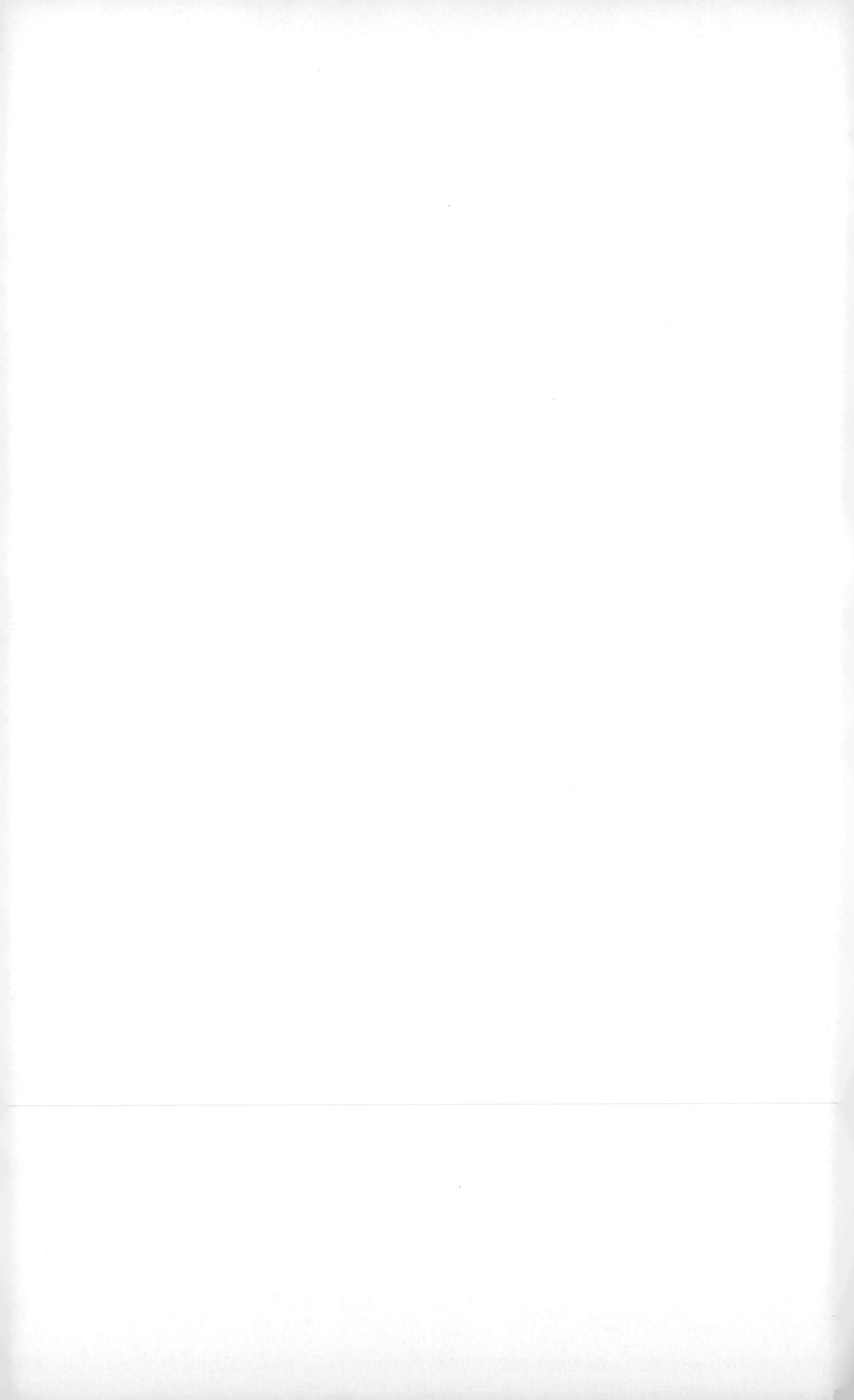

VOLUME FORTY FIVE

ADVANCES IN
INSECT PHYSIOLOGY

VOLUME FORTY FIVE

ADVANCES IN INSECT PHYSIOLOGY

Behaviour and Physiology of Root Herbivores

Edited by

SCOTT N. JOHNSON

Hawkesbury Institute for the Environment, University of Western Sydney, New South Wales, Australia

IVAN HILTPOLD

Division of Plant Sciences, University of Missouri, Columbia, Missouri, USA

TED C.J. TURLINGS

Institute of Biology, University of Neuchâtel, Switzerland

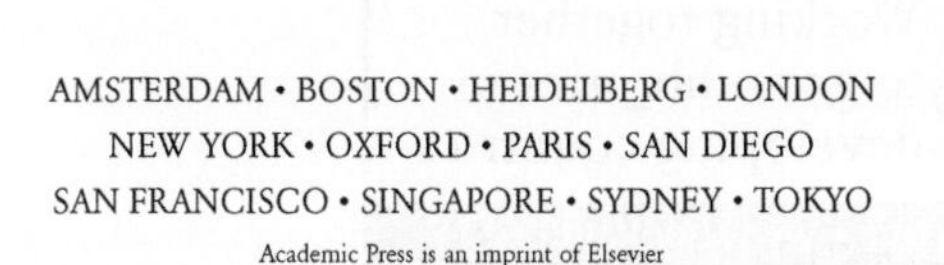
AMSTERDAM • BOSTON • HEIDELBERG • LONDON
NEW YORK • OXFORD • PARIS • SAN DIEGO
SAN FRANCISCO • SINGAPORE • SYDNEY • TOKYO
Academic Press is an imprint of Elsevier

Academic Press is an imprint of Elsevier
The Boulevard, Langford Lane, Kidlington, Oxford, OX5 1GB, UK
32 Jamestown Road, London NW1 7BY, UK
Radarweg 29, PO Box 211, 1000 AE Amsterdam, The Netherlands
225 Wyman Street, Waltham, MA 02451, USA
525 B Street, Suite 1800, San Diego, CA 92101-4495, USA

First edition 2013

ISBN: 978-0-12-417165-7
ISSN: 0065-2806

For information on all Academic Press publications
visit our website at store.elsevier.com

Printed and bound by CPI Group (UK) Ltd, Croydon, CR0 4YY

13 14 15 16 11 10 9 8 7 6 5 4 3 2 1

CONTENTS

CONTRIBUTORS

Nadir Alvarez
Department of Ecology and Evolution, University of Lausanne, Lausanne, Switzerland

Carla C.M. Arce
Root-Herbivore Interactions Group; Department of Molecular Ecology, Max Planck Institute for Chemical Ecology, Jena, Germany, and Departamento de Entomologia, Universidade Federal de Viçosa, Viçosa, Minas Gerais, Brazil

Kirk Barnett
Hawkesbury Institute for the Environment, University of Western Sydney, New South Wales, Australia

Carly M. Benefer
Centre for Agricultural and Rural Sustainability, Plymouth University, Plymouth, United Kingdom

Elisa Bernklau
Department of Bioagricultural Sciences and Pest Management, Colorado State University, Fort Collins, Colorado, USA

Louis B. Bjostad
Department of Bioagricultural Sciences and Pest Management, Colorado State University, Fort Collins, Colorado, USA

Rod P. Blackshaw
Centre for Agricultural and Rural Sustainability, Plymouth University, Plymouth, United Kingdom

Paul D. Cooper
Evolution, Ecology and Genetics, Research School of Biology, College of Medicine, Biology and Environment, Australian National University, Canberra, Australia

Matthias Erb
Root-Herbivore Interactions Group, and Department of Molecular Ecology, Max Planck Institute for Chemical Ecology, Jena, Germany

Abigail P. Ferrieri
Root-Herbivore Interactions Group, and Department of Molecular Ecology, Max Planck Institute for Chemical Ecology, Jena, Germany

Astrid Forneck
Department of Crop Sciences, Division of Viticulture and Pomology, University of Natural Resources and Applied Life Sciences, Vienna, Austria

Bruce E. Hibbard
USDA-ARS, Plant Genetic Research, University of Missouri, Columbia, Missouri, USA

Ivan Hiltpold
Division of Plant Sciences, University of Missouri, Columbia, Missouri, USA

Meret Huber
Root-Herbivore Interactions Group, and Department of Biochemistry, Max Planck Institute for Chemical Ecology, Jena, Germany

Scott N. Johnson
Hawkesbury Institute for the Environment, University of Western Sydney, New South Wales, Australia

Jonathan G. Lundgren
USDA-ARS, North Central Agricultural Research Laboratory, Brookings, South Dakota, USA

Ricardo A.R. Machado
Root-Herbivore Interactions Group, and Department of Molecular Ecology, Max Planck Institute for Chemical Ecology, Jena, Germany

Nicole E. Miller-Struttmann
Division of Biological Sciences, University of Missouri, Columbia, Missouri, USA

Kevin S. Powell
Biosciences Research Division, Department of Environment and Primary Industries, Rutherglen, Victoria, Australia

Christelle A.M. Robert
Root-Herbivore Interactions Group, and Department of Biochemistry, Max Planck Institute for Chemical Ecology, Jena, Germany

PREFACE

Root herbivore research has traditionally been the domain of applied entomology and pest management, with most research to date focussing on economically important pest species. The past two decades, however, have seen steadily burgeoning research in this area as scientists have realized that root herbivores significantly shape many ecosystem processes. The ecological role of root feeding insects has been particularly apparent in the field of aboveground–belowground ecology with researchers aiming to understand not just their impacts on plants and other trophic groups but also the underlying physiology and behaviour of root herbivores. All of the chapters in this volume comment on how our understanding of the behaviour and physiology of root herbivores lags way behind that for aboveground herbivores. This is undoubtedly true, but as the five chapters in this volume illustrate, we have made significant advances in the field. It is immediately apparent that living in the soil and feeding on roots present root herbivores with unique challenges, to which many herbivores have evolved behavioural and physiological mechanisms to inhabit this unfavourable environment.

Barnett and Johnson describe how the abiotic environment shapes herbivore behaviour and physiology. For example, root herbivores have evolved behavioural (e.g. cell construction and burrowing) and physiological (e.g. cuticular pumps and spiracle valves) mechanisms to prevent desiccation and suffocation. Erb and co-authors go on to discuss the role of primary and secondary metabolites in shaping the behavioural and physiological adaptations of root herbivores, pointing out that root herbivores display a high prevalence of tolerance mechanisms to toxic plant secondary compounds. Rhizopheric chemical ecology is discussed by Hiltpold et al. in an evolutionary context. Hiltpold et al. pay particularly close attention to the western corn rootworm, which remains one of the best studied species. A second case study is provided by Powell et al., this time focusing on grape phylloxera, one of the most infamous root herbivores. They comment on how recent genetic characterisation and partial characterisation of the internal anatomy of phylloxera have improved our knowledge of its life cycle, genetic diversity, physiology and geographic distribution of holocyclic and anholocyclic populations. Benefer and Blackshaw end the volume with an excellent appraisal of the molecular approaches being used to study root herbivores. These methodologies will play a pivotal role in gaining a better understanding of root herbivore behaviour and physiology and illustrate how researchers in the field should look to other disciplines for novel approaches to research challenges. Methods used by soil microbiologists, such as the recovery of whole-soil DNA for next-generation sequencing being a prime example.

While many studies focus on root herbivores as economic pests, root herbivores are also mooted as an effective means of controlling undesirable plants. Their beneficial effects as biocontrol agents merit further attention, but this will ultimately rely on a sound grasp of their biology if we are to predict their impacts on the wider ecosystem. Root herbivore biology is sometimes described as a black box, which refers to both their opaque environment and the uncertainties that surround their ecology. We are therefore grateful to Jérôme and Stephen for commissioning this volume and to the many authors who have produced remarkable and comprehensive reviews which we hope will go some way to unlocking the black box on this hidden group of insects.

Scott N. Johnson
University of Western Sydney, Australia

Ivan Hiltpold
University of Missouri, USA

Ted C.J. Turlings
Université de Neuchâtel, Switzerland

CHAPTER ONE

Living in the Soil Matrix: Abiotic Factors Affecting Root Herbivores

Kirk Barnett, Scott N. Johnson
Hawkesbury Institute for the Environment, University of Western Sydney, New South Wales, Australia

Contents

Abstract

Root herbivores live in the soil matrix where they are in constant physical contact with their environment and have become behaviourally and physiologically adapted to its abiotic properties. Soil moisture is amongst the most important factors influencing root herbivores, though this is tightly linked with soil temperature. Root herbivores have

Advances in Insect Physiology, Volume 45
ISSN 0065-2806
http://dx.doi.org/10.1016/B978-0-12-417165-7.00001-5

evolved behavioural (e.g. cell construction and burrowing) and physiological (e.g. cuticular pumps and spiracle valves) mechanisms to prevent desiccation and suffocation. Carbon dioxide (CO_2) is higher within the soil than in the air, and hence, root feeders are physiologically adapted to high concentrations. Indirectly, root herbivores can benefit from elevated atmospheric CO_2 through stimulation of root growth, though studies are scarce. Similarly, in an agricultural context, fertilisation usually affects root herbivores positively through improved nutritional quality of host plants. Soil texture and structure is critical for root herbivore movement, and compacted soils cause significant mortality. As a consequence, many maternal insects specifically oviposit in soils amenable to offspring movement.

1. INTRODUCTION

For the most part, root herbivores live in constant physical contact with the soil environment, and as such, they are intimately affected by the prevailing conditions in the soil matrix. Unlike aboveground herbivores, moreover, they cannot disperse rapidly from adverse conditions, and so, many have evolved behavioural and physiological mechanisms to cope with these conditions. Soil is a particularly heterogeneous environment, and so, the range of abiotic conditions experienced by root herbivores can be both diverse and complex. In this chapter, we review existing knowledge about how soil moisture, temperature, carbon dioxide, pH, texture and structure and, finally, soil nutrients affect the behaviour and physiology of root herbivores. We aim to identify trends and the prevailing concepts about how these dominant abiotic factors operate in the soil and what the implications of these are for root herbivores. In addition to characterising the proximate and direct effects of these factors on herbivores, we identify those indirect effects (often including other organisms in the rhizosphere) that likely shape the net effect on root herbivores overall.

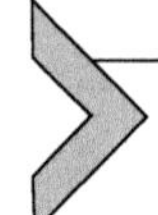

2. SOIL MOISTURE

2.1. Direct effects on root herbivores

The soil, as a substrate, is far from homogeneous. Rather, it contains much more organismal diversity than normally thought due to its complex, porous nature (Dindal, 1990). Within this matrix, soil moisture is one of the key elements for the growth and successful reproduction of a plant. At the same time, root herbivores that feed upon those plants are also affected by this same abiotic property (Veihmeyer and Hendrickson, 1950). On top of this, climate change is expected to affect these interactions through the alteration of soil moisture (Staley and Johnson, 2008). Therefore, it is important that scientists begin to focus attention on these interactions, tease them apart and possibly add this information to existing predictive climate models. To this end, it is useful to examine the basic interactions that soil moisture has with soil-dwelling arthropods, especially insects. These include the basic physiology of an insect, its movement through the soil, host plant location and feeding behaviours, all of which can be additionally altered through plant-, predator- and anthropogenic-mediated effects.

2.1.1 Physiology

Firstly, soil moisture can affect soil-dwelling arthropods in the most direct way possible: by changing the maintenance of osmotic pressure from one side of the cuticle to the other. Early on, researchers speculated that the flow of moisture through the insect cuticle, for example, was asymmetrical and passive. Beament (1961) hypothesised that there was an undiscovered cuticular valve that maintained water balance, but these had rarely been examined up to the date of that publication (but see Wigglesworth, 1956). However, since then, some mechanisms have been explored, including spiracles (Bursell, 1964) and epidermal cells that actively pump water against the gradient (Verhoef and Witteveen, 1980; Winston and Beament, 1969).

Beetles from the family Tenebrionidae, in particular, have been examined in the search for the mechanisms behind cuticular water relations because they can surprisingly be some of the most abundant organisms in arid environments (Ahearn, 1970; Buxton, 1930; Cooper, 1983; Hadley, 1978; Mellanby, 1932). For example, when parts of the Tenebrionid *Phrynocolus petrosus* were examined under altered humidity, researchers found that the majority of water loss, about 70%, occurred through the pronotal spiracles and not over either the abdomen or the subelytral cavity (Zachariassen, 1991).While the pathways of water movement through the cuticle for adult insects are numerous (Cooper, 1983), in larvae, it is largely assumed to be through thermo-osmosis. For example, the osmotic balance of the cuticle in the larvae of *Costelytra zealandica* was examined and determined to be the result of thermo-osmosis across the air space in the surrounding soil (Galbreath, 1975).

While water flow across the cuticle remains an important component of arthropod water relations, feeding, absorption of water vapour through the soil atmosphere, and metabolic waste could also contribute (Mellanby and French, 1958). The most studied pathway of water intake in arthropods is through the consumption of its food. This is because water content is an important component of plant biomass in relation to arthropod growth and reproduction. For instance, varying water content of plants can reduce growth of herbivorous lepidopterans by up to 40% (Scriber, 1984). This seems to be especially true for tree feeders, which are more sensitive to water content than forb feeders (Scriber, 1979). Because the host plant and feeding guild appear to have an effect on the importance of food-borne water, we should expect that root-feeding herbivores will experience this phenomenon differently. For example, when researchers measured insect biomass in the tallgrass prairies of Kansas in response to different range management practices, they found that feeding method was predictive of the biomass of different insect groups. Xylem feeders, such as cicadas, responded negatively to mowing, while white grubs showed increased biomass. They hypothesised that continual removal of the aboveground plant parts caused decreased productivity, similar to the response to drought which caused less transfer of fluid through the xylem compared to burned prairie (Seastedt et al., 1986). Lastly, some larvae have been shown to swell rapidly (more so than should be possible through either water movement through the cuticle or spiracles) when previously dry soil becomes saturated, suggesting that some larvae may even be able to 'drink' water when needed (Hassan and Hilditch, 1976). The relationship between moisture in the soil microclimate is more complex, however, because the soil atmosphere contains a range of soil moisture tensions in which temperature and humidity interact. When soil is below the soil moisture tension required for plant functioning (wilting point pF 4.2), larvae lose moisture proportional to the saturation deficit. For example, both the larvae of the scarabs *Sericesthis nigrolineata* and *Aphodius howitti* have been found to have water loss linearly related to the saturation deficit of soil (Davidson et al., 1972; Maelzer, 1961). Additionally, soil moisture tension can change with soil type; larvae of *S. nigrolineata* had a maximum survival between pF of 4.2 and 2.7, which is between 5% and 10% moisture in silt and 6% and 12% moisture in loam (Davidson et al., 1972).

Because soil moisture is inherently more stable than aboveground air water vapour, the focus on food-borne water may be moot, but at the same time, soil insects are adapted to high-relative-humidity environments and therefore could be more sensitive to changes in moisture than aboveground folivorous arthropods (Haile, 2001). In order to mitigate this, some arthropods, soil-dwelling scarab larvae, for example, construct an earthen chamber, which allows for the control of the microclimate (Villani and Wright, 1990; Villani et al., 1999; Fig. 1.1).

Figure 1.1 Chamber building of both the European chafer *Rhizotrogus majalis* (top) and green June beetle *Cotinis nitida* (bottom). *Reproduced from Villani et al. (1999).*

One of the mechanisms insects can utilise to prevent desiccation is the production of metabolic water, which is a by-product of the oxidative catabolism of fats and carbohydrates (Wharton, 1985). This can be especially useful when the atmospheric relative humidity is low enough to restrict the uptake of moisture through water vapour (Dautel, 1999). Its importance, however, will change depending on the environmental stresses, and in very dry conditions, metabolism could be the main source of water for an arthropod. This coupled with the fact that most metabolically derived water is from enhanced energy loss (increased movement) (Jindra and Sehnal, 1990; Wharton, 1985) and that movement patterns are altered by changing soil moisture environments (Villani and Wright, 1990) could mean that soil moisture changes due to climate change will make it hard to predict its effect on arthropod metabolically derived water relations.

2.1.2 Behaviour

Aside from its physiological impacts, soil moisture can also affect the movement of some larvae through substrate. This is seen as a short-term response to varying rainfall and soil drying (Villani and Wright, 1990). Movement, however, can be both increased and hindered in differing soil moisture conditions. For example, early researchers found that larvae of the Coleopteran genus *Agriotes* responded in two ways to soil moisture; the first was to increase activity and migrate out of dry sand into wet sand, but conversely, if soil moisture conditions change too rapidly, they became trapped and burrowing activity desisted (Lees, 1943b).

While the gravity of a soil can have an effect, the aeration or porousness can also help or hinder movement and can additionally be altered by soil moisture. The first instars of the clover root weevil have been shown to prefer slightly moist silty clay-loam and coarse sand over extremely moist soil of any type (Pacchioli and Hower, 2004). This, like in another study involving neonate root-feeding leaf beetle larvae, was thought to have been due to large cracks and pores filling up in wet soils, limiting larval movement and preventing host plant location (Simelane, 2007). Additionally, the movement of larval *Diabrotica virgifera virgifera* has been determined to be important for host location, and the distance an individual can move within the soil depends on soil type, moisture, porosity and bulk density (Krysan, 1999). Bulk soil density or gravity is highly dependent on the type of soil, a determination that describes its constituents. For example, larvae and adults of the invasive scarab *Heteronychus arator* have significantly different survival in various soil types, a response that interacts with soil moisture (Matthiessen, 1999; Fig. 1.2). Finally, soil texture and moisture are related because texture determines the rate at which soil moisture changes. This will have significant impacts on soil invertebrates because rapidly drying soils cause faster desiccation when compared to those that dry less rapidly (Brust and House, 1990).

Soil-dwelling organisms can normally migrate out of unfavourable environments or modify them to their needs, but quick changes could be detrimental (Villani and Wright, 1990). However, anhydrobiosis, dehydration-induced hibernation, can occur in some soil-dwelling arthropods, and physical adaptations, such as spiracle plastrons, could spare them from rapid and large changes in soil moisture (Belgnaoui and Barra, 1988; Hinton, 1960; Villani and Wright, 1990; Fig. 1.3). In the end, it is clear that arthropods are directly affected by soil moisture. With an increase in the frequency of extreme weather patterns expected due to climate change, we can

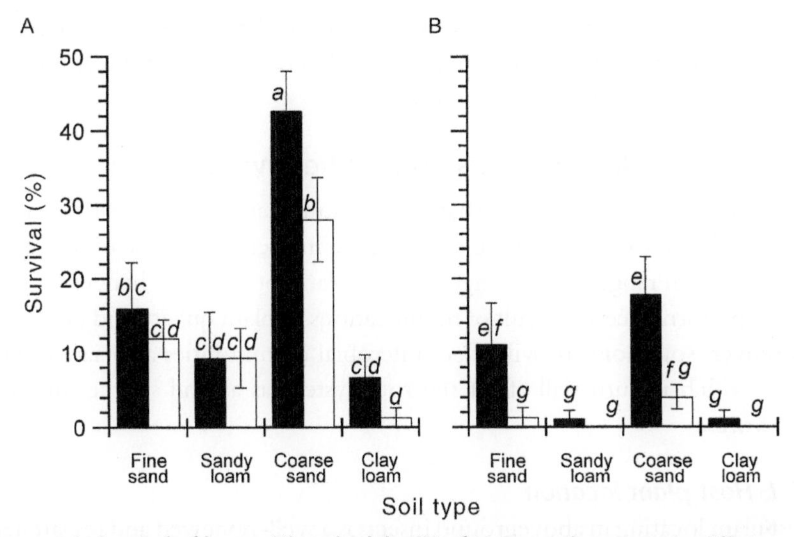

Figure 1.2 Survival of larvae (A) and adults (B) of *Heteronychus arator* in differing soil types with two different moisture treatments: watered regularly and allowed to dry normally. *Reproduced from Matthiessen (1999).*

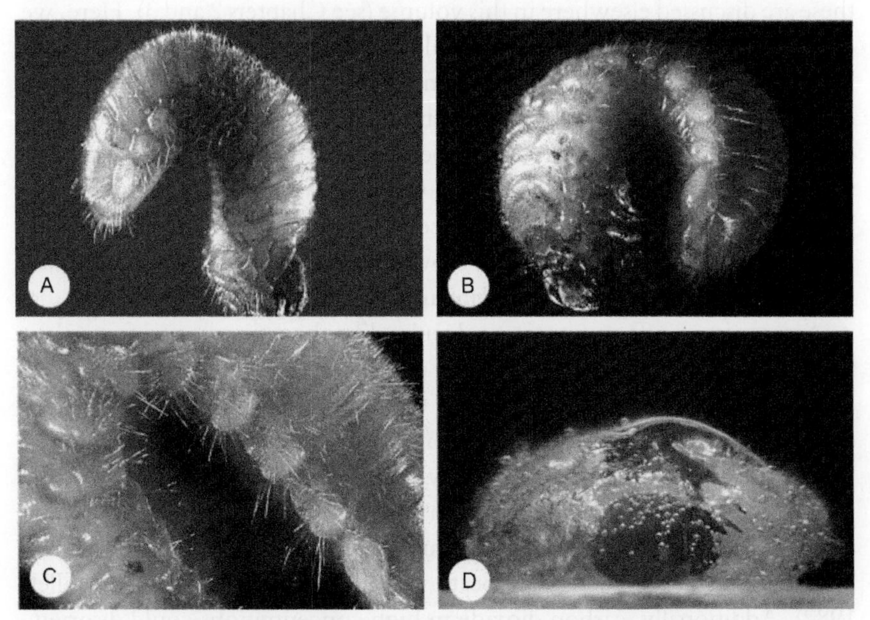

Figure 1.3 (A) Larvae of the cranberry root grub *Lichnanthe vulpine*. (B) The exterior surface of the grub promotes the formation of plastrons. (C) Hydrofugic hairs. (D) Air bubble formed around a fully submerged grub. *Reproduced from Villani et al. (1999).*

only predict that the result of these interactions will become more stochastic and, for most species, increased mortality will result when rapid wetting and drying occurs in highly porous soils.

2.2. Plant-mediated effects on root herbivores

While there are obvious ways that soil moisture can have a direct effect on arthropods, for example, physiologically, the indirect effects are much more subtle and can range from changes in host plant location to changes in herbivore performance as a result of modifications to plant quality and quantity. Moreover, soil moisture will affect microbial communities associated with roots, which in turn will affect the root system as a food source for root herbivores.

2.2.1 Host plant location

Host plant locating in aboveground insects is a well-reviewed and researched topic (Bernays and Chapman, 1994; Elkinton and Cardé, 1984; Mayhew, 1997). How this occurs in the soil environment is becoming better understood (Johnson and Gregory, 2006; Johnson and Nielsen, 2012). Chemicals released by, and contained within, the roots underpin host plant location, and these are discussed elsewhere in this volume (see Chapters 2 and 3). Here, we consider this in the context of how soil moisture could alter the connection between host and herbivore by changing the concentration and transmission of these chemical cues. To start, early studies utilising radiography show that some root-feeding larvae can discern benign sources of moisture from moisture emanating from their host plants; when larvae were allowed to choose between sponges or irrigated sod placed on sides of an arena, root feeders were more often found in the sod. However, in some situations, the movement was species-dependent (Villani and Wright, 1990).

Soil moisture affects gaseous molecules, like the ones released from plant roots and decaying matter, because it can change the diffusion rates when compared to standard atmospheric air (Payne and Gregory, 1988). Carbon dioxide has been shown to be a major player when it comes to plant host location (Klingler, 1965), and many soil insects are attracted to CO_2 (Bernklau and Bjostad, 1998b; Johnson and Gregory, 2006). Carbon dioxide dissolves more quickly in water compared to plant volatiles and could make it more difficult for insects to locate plants within saturated soils (Hillel, 1982). Additionally, carbon dioxide in high concentrations could disorientate or even anaesthetise insects (Bernklau et al., 2004), to the point where they would be unable to escape dramatic changes in soil moisture levels.

In the end, both gases and plant exudates interact with each other and influence soil-dwelling herbivore behaviour at different scales and concentrations (Reinecke et al., 2008).

2.2.2 Host plant quality

It is obvious that changes in root growth in response to soil moisture will affect the herbivores that feed upon them. For example, the absolute rooting depth of plants in moisture-limited environments has a tendency to increase with increasing levels of precipitation (Schenk and Jackson, 2002). But, at the same time, some plant species have been shown to actually increase the size of their rooting systems when soils become drier (Gregory, 2006). Relatively few studies, however, have examined the consequences of the sudden wetting or drying of soils on root system growth and size (Gregory, 2006), which can also have an effect on root herbivores. For example, some plant systems increased lateral root growth from 300% to 500% when soils dry but then extended further when soil is rewetted (Jupp and Newman, 1987). Soil arthropods should be able track and take advantage of these changes (Jones, 1979).

While drought is thought by some to increase the uptake of nutrients in plants from root systems through the aggregation of organic and inorganic materials by capillary forces (Gregory, 2006), it also has the potential to decrease nutrient uptake, especially in crop plants (Pessarakli, 1999), by means of shifts in stoichiometry and nutrient availability. For example, in an evergreen Mediterranean forest in northeast Spain, a reduction of 15% in soil moisture decreased the nutrient uptake by *Quercus ilex* up to 40% by reducing the availability of phosphorus and increasing the recalcitrance of minerals in the soil (Sardans and Peñuelas, 2007). On the other hand, the nutrient content of annual grasses per percent of dry matter increased under reducing water availability in the grasslands of California (Gerakis et al., 1975). While this will certainly have an effect on root herbivores, the magnitude and direction on performance will depend on the herbivore guild. This is because moisture stress might lead to an increase in mobilisation of nitrogen in the phloem (Huberty and Denno, 2004), which is beneficial to sucking herbivores but could also be detrimental to chewing herbivores (Koricheva et al., 1998). Mineral uptake, as well, can be altered by drought because as soil temperature increases and soil moisture decreases, root growth and ion movement are reduced and roots have a tendency to become tougher (Kramer, 1983).

Along with root growth, the actual quality of the plant biomass can change because different parts of a plant have the potential to hold different levels of antiherbivore compounds. For example, the fruits of wild parsnip contain 4 × higher furanocoumarin than its leaves and 800 × more than its roots (Zangerl and Bazzaz, 1992). When water stress is added, changes in allelopathic and antiherbivore compounds can shift and increase in different plant parts. This has been shown to be true with Mexican sunflowers grown in moisture-limited soils. The water from new leaves of these stressed plants showed higher phytotoxic activity than mature leaves, roots or stems (Tongma et al., 2001). Additionally, concentrations of terpenoids and phenolics can increase in *Cyperus rotundus* when grown in water-stressed soils (Tang et al., 1995). Lastly, water stress can be a large component determining plant quality changes in response to root herbivory, for example, in the case of *D. virgifera* and *Zea mays* (Erb et al., 2011). Under this framework, we should expect the same changes in root growth and quality belowground, followed by similar responses of soil-dwelling herbivores.

Soluble nitrogen increases in plants during times when water is limited, but the effect on total plant nitrogen is tissue- and stress-dependent (Chaves et al., 2003). When water stress is mild to moderate, total nitrogen declines in roots and older aboveground tissues and increases in newer tissue (Mattson and Haack, 1987). Prairie grasses, in particular, seem to reduce the nitrogen content of their roots in response to drought (Heckathorn and DeLucia, 1994). To complicate matters, nitrogen and water contents can covary and have been shown to change throughout a growing season for many plant species (Schoonhoven et al., 2005; Slansky, 1993). Therefore, belowground nitrogen content in plants, which can sometimes be the highest nitrogen-containing portion of a plant (Mattson, 1980), might be an indicator of root water content.

2.2.3 Plant microbes and host plant quality

Microbial activity connected to the roots of plants might also be affected by soil moisture, which could in turn affect root herbivores. For example, soil moisture is crucial for the infectivity of arbuscular mycorrhizae in plant roots. Previous research has shown that periods of soil wetting and drying can have both negative and positive effects (Braunberger et al., 1996), the result of which can have a large effect on root herbivores (Gange et al., 1994). For example, root-feeding weevil larvae incurred a reduction in survival by 41% when they fed on arbuscular mycorrhizae-infected plants (Gange et al., 1994). This relationship can also operate in reverse, where root herbivores can influence the infection of plants with arbuscular mycorrhizae. In a study examining the effect of leatherjacket root feeding on the infectivity of *Agrostis capillaris*, researchers found that plants that were exposed to previous root herbivory had significantly more infection by arbuscular mycorrhizae than those plants that had not (Currie et al., 2006). The presence or absence of arbuscular mycorrhizae can significantly affect both the soil moisture in the rhizosphere through changes in hydraulic conductivity and, hence, a plant's ability to survive rapidly changing moisture conditions (Hardie and Leyton, 1981).

Other microbes besides arbuscular mycorrhizae are also affected by soil moisture and can have effects on carbon cycling that could change plant–root herbivore interactions. For example, when soil microbes experience increased precipitation, soil respiration is increased (Liu et al., 2009). Certain microbes were found to reduce plant vigour and root length when experimentally removed with targeted biocides (Ingham and Coleman, 1984). Hence, the interaction of moisture and microbes could alter root herbivore performance either by making plant hosts more vulnerable to death or simply by reducing the amount of food available belowground.

Some plants (Fabaceae and some rosids) contain nitrogen-fixing root nodules that, during times of limiting soil nitrogen, can utilise atmospheric fixing bacteria to increase intraplant nitrogen (Doyle and Luckow, 2003). This activity can be altered by several abiotic and biotic factors and is important to arthropods because nitrogen is often a limiting resource. Drought, in particular, has been shown to cause extreme reductions in nitrogen-fixing activity by bacteria (Serraj et al., 1998; Streeter, 2003). In addition, drought can cause reductions in overall root and stolon production and inhibit nodule formation (Allen and Allen, 1981; Sanderson and Elwinger, 2002; Whitehead, 1983). This should translate into effects on the feeding and performance of belowground herbivores and elicit a subsequent response from a plant. For example, some plants respond to nodule herbivory by increasing nodule growth, but, as a result of drought, this response can be reduced making it harder for the plant to overcome the effects of herbivory (Quinn and Hall, 1992).

2.3. Soil moisture and other abiotic soil properties

2.3.1 Oxygen and carbon dioxide

Soil moisture can affect the soil atmosphere by altering its composition through evaporation and condensation. Oxygen levels within the soil are lower, and carbon dioxide levels are higher compared to the aboveground habitat (Villani et al., 1999), and the concentration of these gases depends on soil moisture (Hillel, 1982). Because the exchange of oxygen and carbon dioxide occurs at the soil surface, they can be affected by several abiotic factors (Marshall and Holmes, 1979; Russell, 1973). One of these factors, water, can infiltrate the soil, causing a reduction in diffusion and both lower oxygen and increase carbon dioxide levels significantly. While this is the case in heavy soils, the top portion (the first 30 cm) in some light soils can contain gas concentrations close to those at the surface (Boynton and Compton, 1944). Additional effects of carbon dioxide on root feeders are discussed in Section 4.

2.3.2 Soil pH

Soil moisture not only can affect root-feeding herbivores directly through physiology but can also affect them through changes in pH. For example, soil moisture content could change the concentration of minerals or substances in the water and cause shifts in pH as soil wets and dries. Additionally, some heavy metals can change chemically in reaction to water content, increasing or decreasing the ion holding capacity of the molecules, eventually leading to a change in pH. The bioavailability and toxicity of some heavy metals is partially dependent on soil pH because they become more available to organisms as pH decreases, via soil pore water (Sijm et al., 2000). Direct effects of pH on soil-dwelling arthropods are considered in Section 5.

2.4. Soil moisture and root herbivore antagonists

While on one hand, root herbivores eat plant roots and thus cause damage to other organisms, they are also targeted and killed by predators. Within this framework, root herbivores are controlled from the bottom-up (availability and quality of plant material) and top-down (predators and human activities). The top trophic level is affected by the same abiotic forces as the root herbivores themselves (Hunter and Price, 1992) and, hence, might also alter the interactions between arthropod root herbivores and their predators (Hilborn, 1975; Murdoch, 1977) by reducing enemy dispersal rates (Brockhurst et al., 2006) or through the development of enemy-free space (Schrag and Mittler, 1996).

2.4.1 Entomopathogenic nematodes

Firstly, one of the most researched areas of root herbivore predation has been on the attack by entomopathogenic nematodes. Like root herbivores, these organisms are also attracted by plant exudates (Laznik and Trdan, 2013). For example, Rasmann and colleagues (2005) found that the volatile organic compounds that were produced as a result of root herbivore attack attracted nematodes. Soil moisture, too, can alter this relationship because the population increases of nematode predators in some areas follow the southern oscillation cycles, which influence rainfall patterns (Preisser et al., 2006; Strong et al., 1995, 1996, 1999). At a fine scale, soil surrounding lupines tends to remain moist in between rainfalls in the summer and increases the survival of nematodes (Preisser et al., 2006). Some nematodes are obligate parasites on insects and have a symbiotic relationship with bacteria. In this case, there is only one stage of the life cycle that exists outside of the host and that is the infective juvenile stage. This stage does not feed or reproduce and its only function is to find an insect host. Once located, the nematode will infect its host and then resume its life cycle, releasing endosymbiotic bacteria, which induce host mortality via septicaemia. The type of feeding behaviours appears to play some role in their distribution throughout the soil, which is normally patchy and the degree to which depends on the species. In general, nematode distribution is more patchy than their hosts, indicating that soil factors could play a bigger role (Campbell et al., 1996). Moreover, they do not have phoresy with their hosts, so all dispersal is self-generated. Another explanation for patchy distributions is linked to the fact that infected insects seem to attract other nematodes and invite continued attack (Grewal et al., 1997), leading to larger aggregations.

In general, many species of nematode are highly sensitive to soil moisture (Staley and Johnson, 2008). This could be due to several reasons, such as the inducibility of volatile signals in plants being linked to water status (Gouinguené and Turlings, 2002) or the diffusion of volatiles through soil being affected by soil moisture (Hiltpold and Turlings, 2008) or movement and activity within the soil (Grant and Villani, 2003). On the whole, it is thought that the best soil environment for entomopathogenic nematodes is one that is humid and well structured (Erb and Lu, 2013).

2.4.2 Carnivorous insects

In addition to nematodes, the larvae of click beetles (wireworms) (Coleoptera: Elateridae), while traditionally thought as herbivores and detritivores, can be effective predators of root feeders. The microorganisms present in soil are highly affected by the abiotic conditions of that soil such as moisture, nutrient content and pH (Fierer and Jackson, 2006), and some species of click beetles have a preference for feeding on root-feeding insects (Robertson, 1987). While the number of studies on the behaviour of carnivorous and omnivorous wireworm larvae remains small, we might be able to use the same patterns described for herbivorous wireworms and apply them to these species. This could mean predatory wireworms might also avoid drying soil but, on the other hand, cease activity in fully saturated soil (Campbell, 1937). The ceasing of activity might be more related to the trapping of the insect within the soil, rather than a metabolic slow down, however (Lees, 1943a,b). Some insects also create earthen cells, similar to white grubs, in order to modify the soil environment until conditions become more favourable (Villani and Gould, 1986). Of all the predators discussed so far, the organism predicted to be the most adapted to the changing soil moisture conditions might actually be the wireworm because the heavier sclerotisation of their cuticle should translate into less intrabody moisture changes (Evans, 1943; Jones, 1951; Villani et al., 1999).

2.4.3 Insect pathogens

Entomopathogenicity is a lifestyle that has commonly reoccurred in many lines of fungi (Humber, 2008; Rehner and Buckley, 2005; Roberts and Humber, 1981). Water retention might be one of the most important factors in microbial activity (Griffin, 1963, 1981). Soils with higher water retention in general have reduced fungal persistence (Li and Holdom, 1993). The soil moisture conditions surrounding fungal infection might also be important for determining retention and effectiveness (Bruck, 2005). Severe cycles of wetting and drying can be detrimental to microbial populations because the dramatic increases in water potential cause microbial plasmoptysis, which results in the release of intracellular nutrients (Kieft, 1987).

Hypocreales contains the largest number of entomopathogenic fungi, and its species have been used extensively when eradicating soil- and plant-borne pests (Kaya and Lacey, 2007). They have been proposed to be used instead of chemical pesticides, but are not adopted widely because of a lack of understanding of their place in the ecosystem (Jaronski, 2007; Vega et al., 2009). Several successes in the application of fungal insect pathogens can be noted, such as the use of *Beauveria bassiana* to control the European corn borer (Lewis and Bing, 1991). In plant pathology, the basic principle that disease is the result of an interaction between a host, a pathogen and the environment is called a 'disease triangle' (Agrios, 2005; Cooke et al., 2006; McNew, 1960). Hence, the effectiveness of any entomopathogenic fungus depends heavily on the environment. For example, the pathogen *Metarhizium anisopliae* was evaluated for its effectiveness in field environments and how it performed in relation to the rhizosphere effect (organisms that show enhanced growth in response to developing roots) (Schmidt, 1979). Researchers found that the rhizosphere effect was most pronounced in the top 3 cm of soil because that is where roots were most numerous and where the fungal pathogen resided due to application at the soil's surface (Hu and St. Leger, 2002). However, it has also been recovered 10 cm deep within the soil, which might suggest either vertical movement or percolation through the soil by water (Bruck, 2005; Hu and St. Leger, 2002). Additionally, the pathogen is easily washed from roots, indicating that quickly changing water conditions might make it less infective (Hu and St. Leger, 2002).

Bacteria in relation to the infection of root feeders are a well-studied subject because they are often used as a bulwark against infestations. This is especially true for the agents of milky disease, *Paenibacillus popilliae* and *Paenibacillus lentimorbus*, and its effects on the Japanese beetle (*Popillia japonica* (Coleoptera: Scarabaeidae)) (Klein, 1992; Klein and Bedding, 1993). Grubs ingest the spores with root and other materials in the soil. The spores then germinate and invade the haemocoel, where it causes fat depletion and eventually bacteraemia (Régnière and Brooks, 1978). The final stage of infection brings about a proliferation of spores and spore bodies in the haemolymph, which turns it to a milky white colour. Under drought and flooding conditions, grass species may experience large amounts of herbivores due to aggregated oviposition and enhanced root growth, which could lead to larval combat (Allsopp, 1992; Potter et al., 1996; Régnière et al., 1979). Due to starvation, grubs injure and bite each other, which create lacerations through which bacteria enter and cause mortality (Jackson and Klein, 2006). Some bacteria toxins have been integrated into plants or applied onto the soil in order to reduce soil-dwelling arthropod numbers. Such toxins are synthesised by *Bacillus thuringiensis*, and its genes are commonly spliced into crop plants. There are, in reality, several endotoxic proteins synthesised by *B. thuringiensis*, and the genes to produce several of those toxins are even spliced into other bacteria, such as *Pseudomonas fluorescens*, in order to incubate *B. thuringiensis* proteins. This is especially helpful because as *P. fluorescens* grows and reproduces, it yields even greater amounts of endotoxin. Soil moisture can affect the application of these bacterial-borne insecticides because percolating water is one of the most important dispersion agents of applied bacteria in soil (Gerba and Bitton, 1994; Harvey et al., 1989; Madsen and Alexander, 1982; van Elsas et al., 1991). The movement of the endotoxins can be both vertical and horizontal through the soil (Guimaraes et al., 1997) and hence should be altered by patterns of wetting and drying, in addition to the initial soil moisture status at the application. Patterns of heavy rainfalls are also predicted to change their effectiveness in both temperate and tropical agricultural environments (Guimaraes et al., 1997).

2.5. Soil moisture and anthropogenic pest control

On a related note, the effectiveness of synthetic insecticides has been known for decades to be highly dependent on the moisture status of the soil (Harris, 1972). On the whole, in clay soils, the efficacy of insecticides and pesticides is positively correlated with soil moisture. They are also less volatile in dry soils, which results in reduced fumigant toxicity (Ehlers et al., 1969a,b; Guenzi and Beard, 1970). Within the same types of soils, relative humidity also plays an important role in insecticide effectiveness (Hadaway and Barlow, 1951). This is because drying soil can inactivate insecticides such as DDT, which then shows marked reduced effectiveness when sorbed into wetted soil (Barlow and Hadaway, 1956, 1958). Researchers have also found that increasing the relative humidity of the soil can increase the exposure of the insect to the toxin (Harris, 1972). While much research has found that lower soil moisture can mean lower sorption, the opposite has also been found. For example, ethylene dibromide is strongly sorbed by dry soils and decreases with increasing relative humidity (Wade, 1954, 1955). Additionally, ethylene dibromide and similar pesticides can be easily displaced by even a small amount of soil moisture (Farmer and Jensen, 1970; Guenzi and Beard, 1970; Spencer et al., 1969) or even have reduced fumigant toxicity (Harris and Lichtenstein, 1961). In some cases, the vapour density of a pesticide was not affected until soil moisture content reached a critical level, after which the vapour densities fell dramatically (Spencer et al., 1969). All soil moisture effects observed, however, are highly dependent on soil type (Harris, 1966; Wolcott, 1970).

3. SOIL TEMPERATURE

Almost all of the thermal energy that soil contains is derived from incoming solar radiation. How that radiation affects the actual temperature of the soil, however, is more complex. This not only is due to the physical laws governing soil properties but also is the result of seasonality and vegetation structure. Additionally, soil depth and stratification complicates this relationship by altering temperature fluxes diurnally and seasonally. Before we can begin to examine how temperature affects soil arthropods, we must first link abiotic factors to changes in soil temperature.

3.1. Mechanisms of temperature change

3.1.1 Depth

The surface temperature of the soil will closely track the temperature of the air sitting above it, which is linked to the incoming radiation based on the time of day. In addition, it absorbs some radiation that is hit by it, which then can be radiated back into the surface air, heating the soil further. The amount of radiation the soil surface can absorb depends on its albedo (the reflectance of the surface). The soil surface temperature lags somewhat behind the air temperature during the day, but at night, it is essentially the same. The same can be said of the soil below the surface, which lags behind the soil surface and additionally experiences less diurnal variation. These lag time effects become greater with depth because of the physical dampening of heat conductance. For example, during August, soil at 20 cm depth reaches its maximum temperature at 19.00 hours. At a depth of 40 cm, however, diurnal variation is almost nonexistent (Russell, 1988). Soil temperatures at higher latitudes also track the season, naturally dropping towards winter and warming as summer approaches. Compared to the maximum air temperature, the maximum soil temperature is reached later in the year at these latitudes. This also depends on depth, however, with greater seasonal variation occurring further down (several metres) in temperate regions, while at the equator temperature variation can sometimes only reach one metre down (Keen, 1931; McCulloch, 1959).

3.1.2 *Physical characteristics of the soil*

At the simplest level, the temperature change within soil is the result of heat flow in and out. The exchange of heat in the soil is affected by temperature gradients and thermal conductivity, which is a product of the composition of the air, water and solid matter in the soil. This results in significant changes in temperature gradients because, for example, air has a thermal conductivity of $0.025 \text{ W m}^{-1} \text{ K}^{-1}$, while water has a much higher thermal conductivity of $0.6 \text{ W m}^{-1} \text{ K}^{-1}$. Hence, soil thermal heat flux can change considerably in a short amount of time due to moisture changes alone. Additionally, compaction can change the amount of pore space within the soil and depending on the degree of soil saturation and can either increase its conductivity through water conductivity or reduce it through air insulation (Russell, 1988).

To further complicate matters, heat flow through soil can alter thermal diffusivity, which describes the velocity and depth of diurnal and seasonal changes in soil temperature. This property increases with soil moisture until it reaches maximum saturation, at which point the diffusivity plateaus and eventually decreases (Russell, 1988). This is because when the soil surface becomes moist, the net radiation that is absorbed is mostly used to evaporate the water sitting at the surface and will not be conducted downward. When the soil is unsaturated, soil moisture is actually quite mobile, via air temperature gradients in the soil. Depending on the compaction, water moves from warmer areas to cooler areas through evaporation and eventual condensation in areas that contain that cooler air. This has the capacity to transfer heat and also increases the apparent conductivity of soil (de Vries, 1969).

The core temperature of soil is essentially determined by a few key factors. These include the quantity of incoming radiation, the soil's albedo, the moisture content, the air content and how much evaporation is occurring at the surface (Russell, 1988). The quantity of radiation is mostly determined by the latitude and aspect of the location. In the Northern Hemisphere, south-facing slopes are warmer than horizontal surfaces, which are, in turn, warmer than north-facing slopes. Additionally, west-facing slopes are warmer than east-facing slopes. This results in faster germination for some crops early in the spring, if they are planted on south-facing slopes (Ludwig et al., 1957).

The albedo of a soil surface can also have significant effects on its temperature. Most soils have an albedo between 0.10 and 0.15 (the higher the albedo, the higher its reflectance), and the soil moisture or soil colour interacts with this value. This is important because a soil's albedo will determine, in part, its variation in temperature fluxes. In order to reduce the damaging effects of temperature fluctuations, researchers have added chalk- or light-coloured soil to the soil surface and found that it can reduce the maximum daily temperature fluctuations from 5 to 10 °C at depths of 5–7 cm when compared to surface soils more darkly coloured (Ludwig and Harper, 1958; Stanhill, 1966). Other coverings, such as mulch or polythene, can alter the way heat is received or dissipated from the soil surface. In general, mulch reduces fluctuations in temperature but reduces warming in the spring and cooling in the autumn, which could have effects on seed germination and crop performance (Burrows and Larson, 1962; Harrison-Murray and Lal, 1979; van Wijk et al., 1959).

As mentioned previously, soil moisture is a big determining factor when it comes to soil temperature. Moist soil conducts heat better than dry soil, but the surface of dry soil heats up much quicker than wet soil; however, the heat does not travel as deeply (Russell, 1988). This is also important in terms of soil type, for example, sandy versus clay soils, because heavily textured soils are usually wetter in the spring compared to lighter soils and hence contain more moisture and are usually more compact. This means that the surface of sandy soils heats up much quicker than clay, but the heat does not travel down as deeply. Crop growers in the past have used irrigation to cool the surface of soil, but too much irrigation can actually cause the surface to warm. This is due to the increased thermal conductance from the addition of water to the soil surface (Hallett and Bengough, 2012; Smith et al., 1931).

3.1.3 Vegetation

The effects of vegetation cover are similar to the effect of mulch in that it reduces diurnal and seasonal fluctuations in temperature. The magnitude of the effect, however, depends on how much leaf area shades the ground underneath it; the effect is proportional to the area of soil shaded. Under a complete canopy, all incoming radiation is blocked, and the soil surface will only receive long-wave radiation coming from the canopy. Seasonal temperature fluctuations are lagged as a result of vegetation covering the soil surface. However, this will depend on the thickness, height and ultimately type of vegetation found in the canopy (Flerchinger and Pierson, 1991). For example, grass leaves trap air above the soil surface, thus insulating it from fluctuation. Forest canopy cover, on the other hand, has a deep column of air sitting on the soil surface and therefore has higher convection potential. Additionally, soil under forest canopy tends to have more aeration than grassland and a higher probability for increased soil moisture (Russell, 1988). The penetration of frost, as well, is modified by vegetation type and has significant effects on its depth within the soil (Dreibelbis and Post, 1942; Salisbury, 1939; Schofield, 1940). For example, expansive and densely rooted vegetation types such as grasses cause shallower frosts than plants growing patchily.

3.2. Effects of temperature on root herbivores

A lot of species of soil-dwelling herbivores show vertical movements in response to the seasons, especially in the temperate regions (Dowdy, 1944; Falconer, 1945; Fisher et al., 1975; Jones, 1979; Jones and Shirck, 1942; Villani and Gould, 1986; Villani and Wright, 1990; Villani et al., 1999). Depending on the soil conditions, thermal energy has the potential to be transferred vertically. As a result, the vast majority of soil herbivores move further down in the soil, thereby preventing the effects of frost and unfavourably cold temperatures during winter. These individuals will return to shallower depths once again during spring and summer (Dowdy, 1944). Herbivores have a tendency to also move deeper in the summer months to avoid high temperatures caused by increased moisture and incoming radiation. For example, when wireworm larvae (Elateridae) were surveyed in soils around the Quebec area, researchers found that they showed responses to extremes of both temperatures. Most larvae moved from lower depths to the top 25 cm in early May, when surface temperatures were around 2 °C and soil moisture was approaching 200%. When summer came, soils approached 13 °C and larvae moved downward, with most individuals residing deeper within the soil at 22 °C. However, in August, larvae returned once again to the surface (top 10 cm) and only descended again at first frost (Lafrance, 1968).

While this represented a field study, which had inherent variability, controlled laboratory studies involving root-feeding scarab larvae (Coleoptera: Scarabaeidae) have shown similar results. For example, when temperatures in controlled sandy loam soil arenas were slowly lowered, two of the species studied (*Popillia japonica* and *Anomala orientalis*) responded by moving deeper within the soil. When temperatures were then increased slowly, larvae returned to the surface, and at least one species (*A. orientalis*) showed downward movement once again when temperatures were continually increased (Villani and Wright, 1988). Moreover, most of these trends have been confirmed by field observations (Lafrance, 1968; Parker and Howard, 2001; Vittum et al., 1987).

While the general patterns in soil-dwelling herbivores might be understood, there are still discrepancies when it comes to specific behaviour between species. In the previous example, we saw that, for the most part, larvae responded the same way to temperature fluxes. However, in the same study of *Rhizotrogus majalis*, individuals showed no response to changes in temperature, while *P. japonica* and *A. orientalis* moved deeper in response to cooling. Further still, only *A. orientalis* moved downward again when temperatures were raised too high (Villani and Wright, 1988). Differences between species movement patterns appear to be true for wireworms as well (Benefer et al., 2012; Zacharuk, 1962).

The trigger for these behaviours, however, does not always occur at an exact temperature point because soil-dwelling organisms most likely respond to a range around a given temperature. Meaning, they will endure a range of temperatures, but anything outside of that range will spur movement (Davis et al., 1996; Fisher et al., 1975; Villani and Wright, 1988). This is evolutionarily advantageous as some ecosystems can experience oscillations in temperature during a season or time period and movement in anticipation of a further increase or decrease could be wasted energy. However, no researchers since Villani and Wright (1988) have conducted laboratory experiments manipulating soil temperature to measure herbivore movements using 'real-time' observations through radiography.

Sometimes, the previous experiences with temperature can affect how that organism anticipates temperature changes in the future. In particular, many cold-acclimated herbivores increase fat accumulation in anticipation of decreasing temperatures (Holmstrup et al., 2007; Kostal and Simek,

1998). However, even though one can predict that soil-dwelling arthropods would be able to anticipate soil temperature changes, no study to our knowledge has tested specific pre-emptive movement by invertebrate soil herbivores, although some researchers have hinted at the possibility of the behaviour existing (Falconer, 1945).

3.2.1 Movement

The temperature of the soil certainly has an effect on the movement of the larvae. When temperature is low, enzymes react slowly, and, as a consequence, larvae are predicted to have decreased movement compared to optimal temperatures (Edwards and Cresser, 1992). Basic research on wireworm behaviour shows that movement activity is consistent within a certain temperature range (8–25 °C). However, when temperatures dropped to 6 °C and below, activity increased, potentially as a means of relocating to a warmer environment (Falconer, 1945). Recent research by van Herk and Vernon (2013), however, has concluded that weight and temperature are concomitant and both must be taken into account when evaluating the walking speed of wireworm larvae. Additionally, speed is positively and linearly related to temperature in a range from 6 to 18 °C. From this trend, it can be predicted that movement will cease completely from 4.8 to 3 °C, depending on the weight of the larvae (van Herk and Vernon, 2013).

Larvae of scarab beetles (Coleoptera: Scarabaeidae) have also been evaluated in terms of how temperature affects their activity, partly because they can be a major contributing factor to turfgrass death (Jackson and Klein, 2006; Klein et al., 2007; Potter, 1998). When researchers used acoustic sounds to monitor white grub movement activity in soil, they found that species and weight of scarab did not alter their sound rates (activity); rather, it was temperature that caused the change. Specifically, second and third instar larvae of *Phyllophaga crinita* made more sound per unit of time (1 min), and the frequency of activity was greater as temperature increased. At or below 9 °C, the larvae only made an average of 0.32 sounds a minute, and only 25% of the 1 min intervals had sound. However, between 23 and 26 °C, the rate that was 10.7 and 71% of 1 min intervals had sound, and at 30–32 °C, the rate that increased to 13.6 and 85% of 1 min intervals had sound (Zhang et al., 2003).

3.2.2 Feeding

Changes in temperature can affect how an organism feeds within the soil either directly or indirectly (i.e. plant-mediated effects). In the latter case, changes in root quality and quantity occur due to changes in soil temperature

(BassiriRad et al., 1997; Edwards et al., 2004; King et al., 1996; Xu and Huang, 2000a,b), although not always (Fitter et al., 1999). As a result, soil herbivores could be impacted, from not only direct warming in the soil but also reductions in root biomass and quality. Most studies have examined soil warming effects on root dynamics but not, to our knowledge, how this affects root herbivores. We can predict, however, that, like in aboveground studies on herbivory, reductions in plant quality will lead to slower development time (Goverde et al., 1999; Kukal and Dawson, 1989), increased mortality (Bosio et al., 1990; Karban, 1989; Zalucki and Brower, 1992), more time searching for quality hosts (Kause et al., 1999; Mayhew, 1997), and lower fecundity (Dixon and Watson, 1970; Weibull, 1987; Wu and Li, 1992).

3.2.3 Reproduction

The majority of insect herbivores with root-feeding stages reproduce as adults aboveground, although there are examples of adult insects reproducing in the soil too (see Chapter 4). Even though reproductive adults may not be directly affected by soil temperature, changes in larval development can influence the number of generations per year and even alter reproductive behaviour by adults due to asynchronisation between the sexes (Potter, 1981). In fact, soil temperature increases due to global climate change have been predicted to affect emergence and mating patterns of soil-dwelling herbivorous insects (Collier et al., 1991). This study predicted that an increase in average soil temperature by 3 °C would result in extended and earlier emergence of *Delia radicum* (Diptera: Anthomyiidae), with a third larger generation emerging in the late summer. Additionally, an increase to 5 °C allows for the possibility of a fourth generation. More importantly, they propose that it could cause a disruption in normal mating behaviour because of a desynchronisation in emergence (Collier et al., 1991). Not only emergence but also the formation of sex cells (Giojalas, 1993; Nguyen et al., 2012), the ability to compete when mating (Colinet and Hance, 2009), and egg development and production (Howe, 1967; Piyaphongkul et al., 2012) have been shown to be altered by variation in temperature.

3.2.4 Growth rate

Warming and cooling effects in soil have the potential to affect the growth of root-feeding larvae (Söndgerath and Müller-Pietralla, 1996). For example, the egg development time for the root weevil *Sitona lepidus* was 3.5 × faster in soil at 25 °C when compared to eggs in soil at 10 °C (Johnson et al., 2010). However, nonlethal high temperatures do not always speed development.

Researchers studying the effects of temperature on cabbage root fly development found that higher temperatures actually slowed development (Johnsen et al., 1997). Some root-feeding insects require multiple years as larvae in order to fully develop into adults. In some situations, extended larval development, for instance, when scarab larvae undergo a second year of root herbivory (K.S. Powell, personal communication), could result in temporal mismatch between insects and their host plants. In plant communities that undergo rapid changes in composition and host plants become scarce, root feeders could have insufficient nutrition, which could lead to delayed development and decreased performance.

4. CARBON DIOXIDE IN THE SOIL

Carbon dioxide (CO_2) is present in relatively high concentrations in nearly all soils (Payne and Gregory, 1988), produced mainly from plant and microbial respiration, the latter often involving large emissions from decomposition processes (Payne and Gregory, 1988). Soil invertebrates also emit CO_2 during respiration, though this probably contributes comparatively less to the overall amount of CO_2 in the soil (Bardgett, 2005). Because they encounter higher concentrations of CO_2 than herbivores living aboveground, root herbivores are physiologically adapted to higher concentrations of CO_2 than aboveground herbivores (Staley and Johnson, 2008). The behaviour and physiology of root herbivores is affected by fluctuations in soil CO_2, in terms of both direct effects and indirect plant-mediated effects.

4.1. Direct effects

The most widely reported effect of CO_2 on root herbivores, as mentioned before, is as a means of host plant location (Johnson and Nielsen, 2012). Essentially, respiratory emissions of CO_2 from roots are thought to underpin attraction of root herbivores towards their host plants (Mortimer et al., 1999; Robert et al., 2012; Turlings et al., 2012). At least 16 species of root-feeding insects have been shown to be behaviourally responsive to CO_2 reported in Table 1.1 (Johnson and Nielsen, 2012). Some root herbivores are sensitive to very low concentrations of CO_2: 0.02 mmol mol^{-1} for the wireworm (*Ctenicera destructor*) (Doane et al., 1975) and 0.03 mmol mol^{-1} for vine weevil (*Otiorhynchus sulcatus*) (Klingler, 1958), respectively.

Orientated attraction to point sources of CO_2 may have been observed for many root herbivores in a range of experimental setups (Johnson and Nielsen, 2012), but more detailed observations of root herbivore behaviour

Table 1.1 Soil insect herbivores showing behavioural responses to CO_2

Insect order	Insect species	Plant specificity	Dose response	References
Diptera	Carrot root fly *Psila rosae*	s	dr nm	Jones and Coaker (1977, 1979) and von Städler (1971)
	Cabbage root fly *Delia brassicae*	s	na	Jones and James (unpublished) cited in Jones and Coaker (1978)
Lepidoptera	Lesser cornstalk borer *Elasmopalpus lignosellus*	s	dr	Huang and Mack (2001, 2002)
Coleoptera	Western corn rootworm *Diabrotica virgifera virgifera*	s	dr nm	Strnad et al. (1986), Strnad and Bergman (1987a), MacDonald and Ellis (1990), and Bernklau and Bjostad (1998a, 1998b)
	Grass grub *Costelytra zealandica*	s	nm	Galbreath (1988)
	Black vine weevil *Otiorhynchus sulcatus*	g	nm	Klingler (1957, 1958, 1965, 1966)
	Wireworms *Agriotes* spp.	g	nm	Klingler (1957, 1958, 1965, 1966) and Thorpe et al. (1946)
	Ctenicera destructor	g	nm	Doane et al. (1975)
	Agriotes obscurus and A. lineatus	g	nm	Doane et al. (1975)
	Limonius californicus	g	nm	Doane et al. (1975)
	Hypolithus bicolor	g	nm	Doane et al. (1975)
	Cockchafer *Melolontha vulgaris*	g	nm	Klingler (1957)
	European cockchafer *Melontha vulgaris*	g	na	Reinecke et al. (2008)

Table 1.1 Soil insect herbivores showing behavioural responses to CO_2—cont'd

Insect order	Insect species	Plant specificity	Dose response	References
	Southern corn rootworm *Diabrotica undecimpunctata*	s	nm	Jewett and Bjostad (1996)
	Ground beetle *Evarthrus sodalis*	g	dr	Hamilton (1917)
	Clover root weevil *Sitona lepidus*	s	dr	Johnson et al. (2006b)

Plant specificity refers to the host-plant range of the insects; g = generalist feeders (polyphagous), s = specialist feeders (mono/oligophagous). Dose response refers to whether insects showed a dose-dependent response to CO_2; dr = dose dependent, nm = not measured and na = information not available.
Adapted from Johnson and Nielsen (2012).

and movement patterns suggest that the situation may be more complex. In particular, Johnson et al. (2006b) found that the clover root weevil (*Sitona lepidus*) larvae were not directly attracted to emissions of CO_2, but would alter their searching behaviour when exposed to CO_2 concentrations resembling those found in the proximity of roots. Using fractal dimensions to characterise behavioural responses (see Dicke and Burrough 1988 for an explanation of this approach to different CO_2 concentrations), they demonstrated that *S. lepidus* displayed more convoluted movement patterns (values closer to 1, with a perfectly linear trajectory being 0) when moving in concentrations resembling those in the immediate rhizosphere of its host plant (ca. 1000 ppm) (Fig. 1.4). This more intensive search behaviour within a smaller area (see Fig. 1.5 for examples of movement patterns) was considered to be an adaptive behaviour for detecting other host plant chemical cues, potentially allowing *S. lepidus* to distinguish between host and nonhost plants (Johnson et al., 2004a, 2006b). Similar observations were made for wheat bulb fly (*Delia coarctata*) larvae, which were also induced to search areas more intensively in response to elevated CO_2 concentrations (Rogers, 2011). The fact that there are multiple sources of CO_2 in the soil, including nonplant sources, does bring into question the effectiveness of CO_2 as a host plant location cue by itself. Moreover, CO_2 gradients in the soil profile fluctuate vertically and horizontally, so perceiving and responding purely to CO_2 emissions could be difficult, particularly in mixed plant communities (Johnson and Nielsen, 2012). This might be a more effective strategy for root

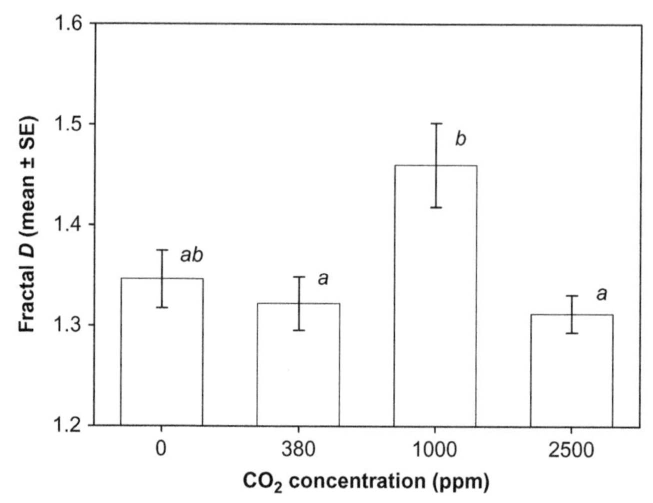

Figure 1.4 Mean fractal dimension (*D*) values ± SE for neonatal *Sitona lepidus* larvae moving in closed arenas containing different CO_2 conditions, showing statistically significantly higher values for larvae in the 1000 ppm arena ($F_{3,102} = 4.08$, $P = 0.009$). Lowercase superscripts indicate significant differences between CO_2 concentrations ($P < 0.05$). The difference between 1000 and 0 ppm approached statistical significance ($t = 2.39$, $P = 0.086$). *Reproduced from Johnson et al. (2006b).*

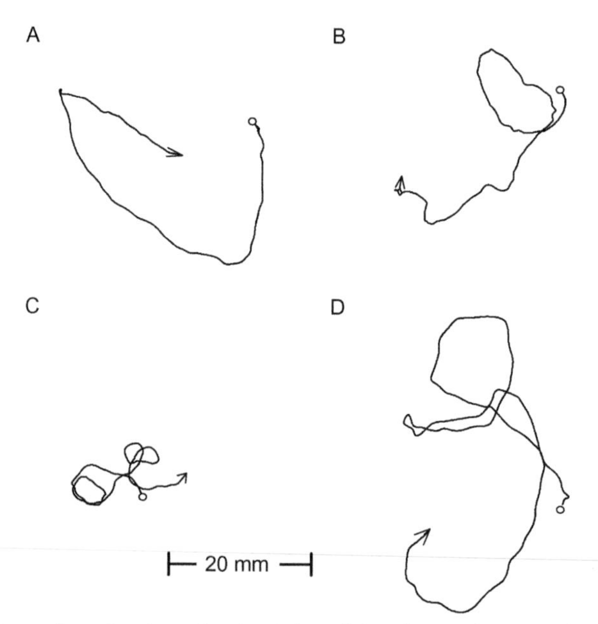

Figure 1.5 Examples of trajectories (and fractal *D* values) of neonatal *Sitona lepidus* larvae moving in CO_2 atmospheres of (A) 0 ppm, $D = 1.247$; (B) 380 ppm, $D = 1.289$; (C) 1000 ppm, $D = 1.677$; and (D) 2500 ppm, $D = 1.299$. Larval starting point indicated by a circle and direction of movement indicated by an arrow. The more tortuous path of *S. lepidus* larvae moving in CO_2 concentrations of 1000 ppm (C) is reflected in the high *D* value. *Reproduced from Johnson et al. (2006b).*

herbivores that locate their hosts in systems where plants are grown in monoculture strips, separated by bare ground (e.g. maize).

CO_2 can also interact with other chemicals in the rhizosphere. For example, the European cockchafer *Melolontha melolontha* orientated towards emissions of CO_2, but this behavioural attraction disappeared in the presence of other root exudates (Reinecke et al., 2008). Here, it was suggested that respiratory emissions of CO_2 from plant roots were used by *M. melolontha* to locate its preferred host plant, dandelion (*Taraxacum* sect. ruderalia), but that roots released other metabolites to mask these emissions.

In addition to CO_2 affecting behaviour, CO_2 in the soil can also affect root herbivore physiology. Because root-feeding arthropods experience high concentrations of CO_2 in the soil, they are less affected by fluctuations in CO_2 concentrations than aboveground herbivores (Staley and Johnson, 2008). Root feeders can, however, also show repellence to CO_2, at concentrations of 1.5% and 6.5%, which has been observed in *C. destructor* (Doane et al., 1975) and *O. sulcatus* (Klingler, 1958), respectively. Escape from pockets of high CO_2 concentrations could depend on the size of the organism, with smaller individuals having to exploit existing soil cracks, whereas larger individuals can physically burrow through the soil to relocate (Cherry and Porter, 1992; Marinissen and Bok, 1988). Mobilisation aside, high concentrations can slow development, induce anaesthesia or be fatally toxic (Bernklau and Bjostad, 1998a). But these responses could be order- or even species-specific. For example, while some Collembola species are paralysed in high CO_2 and low O_2 atmospheres (Zinkler, 1966), researchers found that the scarab *Sphaeridium scarabaeoides*, a soil-dwelling dung beetle, maintained normal respiration in a similar environment, probably because its natural environmental atmosphere has similar anoxic concentrations (Holter, 1994). Additionally, Bernklau and Bjostad (1998a) reported that when soil CO_2 concentrations were artificially increased in the field, the Western corn rootworm (*D. v. virgifera*) became disorientated and less able to locate host plant roots. In aboveground insects, high concentrations of CO_2 cause disorientation by interfering with the bioelectrical responses of the nervous system (Nicolas and Sillans, 1989), and while not explicitly studied in root herbivores, this seems likely to be the case too.

4.2. Plant-mediated effects

While large increases in soil CO_2 concentrations can be directly deleterious to root herbivores, it seems more likely that increases in atmospheric CO_2

will have greater indirect effects on root herbivores via plant-mediated effects (Bardgett, 2005; Staley and Johnson, 2008). The physiological responses of aboveground herbivores to changes in host plant quality under elevated CO_2 are now well understood (Zavala et al., 2013), but the same cannot be said for root herbivores. At present, only thee studies, to our knowledge, report on how elevated atmospheric CO_2 affects root-feeding insects via changes in root quality (Johnson and McNicol, 2010; Johnson et al., 2011; Salt et al., 1996). These studies show differing responses of root herbivores to increases in atmospheric CO_2. Salt et al. (1996) reported that root-feeding aphid physiology was largely unaffected by elevated CO_2. In contrast, elevated CO_2 promoted root nodulation in white clover (*Trifolium repens*), which was associated with increased abundance (Fig. 1.6A) and faster development rates (Fig. 1.6B) of *S. lepidus* larvae (Johnson and McNicol, 2010). First instar *S. lepidus* larval performance is highly dependent on nodule availability (Gerard, 2001), so it seems highly likely that this change in nodulation underpinned the observed changes in *S. lepidus* physiology. In contrast, vine weevil (*O. sulcatus*) abundance and larval body mass decreased by 33% and 23%, respectively, under elevated CO_2, which was correlated with a reduction in root mass in plants grown under elevated CO_2 (Johnson et al., 2011).

Elevated atmospheric CO_2 often results in deterioration in host plant quality for root feeders, which in turn alters feeding behaviour (Newman et al., 2011). In particular, nitrogen concentrations decrease relative to

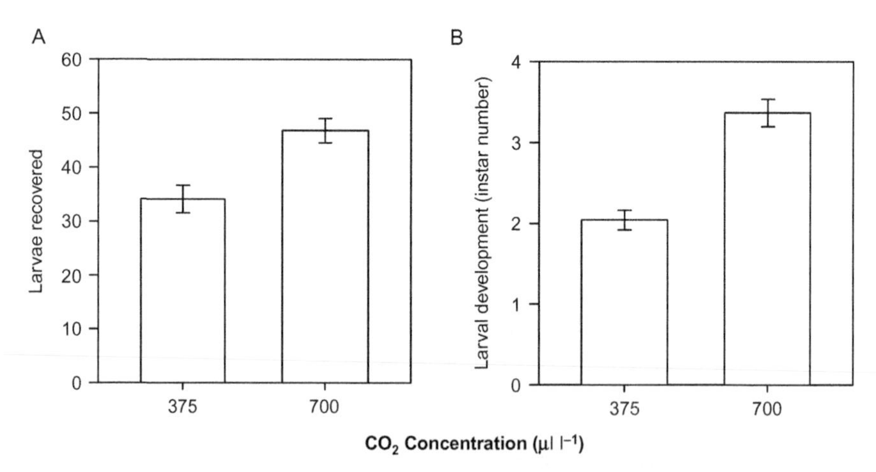

Figure 1.6 Reponses of *Sitona lepidus* larvae (mean ± SE) when caged on plants grown in CO_2 concentrations of either 375 or 700 μl l^{-1}: (A) number of larvae recovered and (B) developmental instar of larvae. *Redrawn from Johnson and McNicol (2010).*

carbon concentrations in the foliage and roots of many plants grown under elevated CO_2 conditions (Luo et al., 2006). Since nitrogen is often the limiting factor in many insect diets (Mattson, 1980; White, 1993), this imposes physiological constraints on insect herbivore development. As a consequence, many aboveground insect herbivores undertake increased (compensatory) rates of feeding (Docherty et al., 1996; Johnson and McNicol, 2010). More recently, this has been observed in a root herbivore, the dusty pasture scarab (*Sericesthis nigrolineata*), which also began to feed in a compensatory manner following increased C:N in grass roots grown under elevated CO_2 (Johnson, unpublished). In this instance, elevated CO_2 had a disproportionately negative effect on a C_3 grass species by reducing nitrogen concentrations much more than in the C_4 grass tested. As a consequence, the performance of the root herbivore was much more adversely affected by elevated CO_2 when feeding on C_3 grass roots (Johnson, unpublished). A similar trend has been observed for shoot herbivores (Barbehenn et al., 2004a,b), which raises the prospect that predicted increases in atmospheric CO_2 might induce compensatory feeding and reduce root herbivore performance on C_3 plants more than C_4 plants.

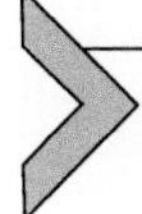

5. SOIL pH

5.1. Effects on larval physiology

Soil pH is affected by a plethora of natural and anthropogenic processes (Rowell, 1988). The impacts of soil pH on root herbivore behaviour and physiology are not well studied however, and there seem to be a few trends (Brown and Gange, 1990). In general, slightly acidic soils seem to have positive effects on root herbivore performance (Johnson et al., 2010; Li et al., 2007; Polivka, 1960a,b; Wessel and Polivka, 1952). For example, both the survival and larval mass of *Diaprepes abbreviatus* decreased (Fig. 1.7) as soil pH increased from 4.8 to 5.1 (Li et al., 2007). The paucity of studies in this area makes it difficult to predict the causes, but one possibility is that slightly acidic soils act as a sterilising agent that helps protect insects from soil pathogens (sensu Neuvonen et al., 1990). A more plausible explanation, however, is that some root exudates reduce the pH of the rhizosphere so herbivores may have simply adapted to these conditions (Bais et al., 2006; Bertin et al., 2003; Brown and Gange, 1990). That said, many roots exude HCO^{3-} and OH^- ions and therefore increase rhizosphere pH (Rowell, 1988), so this remains entirely speculative.

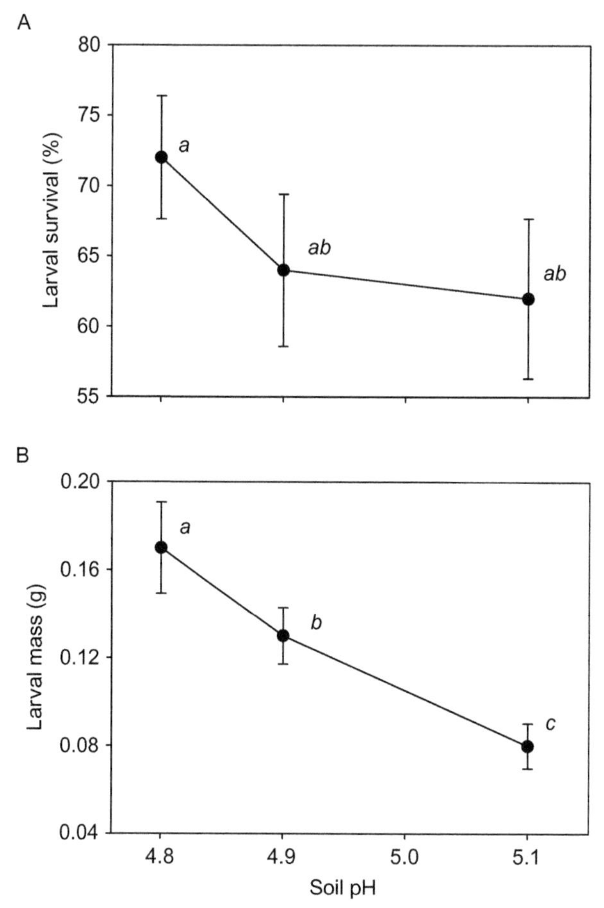

Figure 1.7 Responses of larval *Diaprepes* abbreviates (mean ± SE) to soil pH: (A) survival and (B) body mass. Statistically significant differences between soil pH conditions indicated with lowercase letters. *Created from data reported in Li et al. (2007).*

Moreover, results for the same species conflict with one another. For example, survival of Japanese beetles (*Popillia japonica*) was higher in slightly acidic soils (pH 4.5–5.0) compared to neutral pH (6.8–7.0) soils (Polivka, 1960a,b; Wessel and Polivka, 1952). Given this fact, it might be hypothesised that maternal insects would preferentially lay eggs in such acidic soils (Johnson et al., 2006a). This was not the case; maternal insects oviposited equally in soils ranging from pH 5.0 to 7.9 (Vittum and Morzuch, 1990). Moreover, Vittum (1984) and Vittum and Tashiro (1980) found no response of *P. japonica* larvae to soil pH over a similar range. In addition to survival and oviposition, pH can have additional consequences on the transcription of

DNA. For example, the common soil arthropod *Folsomia candida* showed upregulation of the ATPase gene as soil pH declined, possibly to counteract intracellular pH changes emanating from the gut (de Boer et al., 2010). This could lead to loss of ATP for the maintenance of physiological processes or disrupt moulting cycles. However, some arthropods are actually able to survive a large range of pH (2–9) (van Straalen and Verhoef, 1997). Reproduction, however, can decrease in soils where there is an increase in pH past a certain threshold, which in *F. candida* is between 7.02 and 7.29 (Crommentuijn et al., 1997).

5.2. Effects on egg development

Soil pH can affect the eggs from which root-feeding larvae emerge, although there is even less information about the effect of soil pH on egg development. Eggs of the clover root weevil (*S. lepidus*) hatched slightly earlier in slightly acidic (pH 5) soil, compared with neutral (pH 7) and acidic (pH 9) soil (Johnson et al., 2010). It was unclear why acidic soils would speed up egg hatching time, but it was speculated that it could be linked to the egg case (chorion) dissolving more readily under acidic conditions and easing the emergence of the larvae, which has been observed for foliar insects (e.g. Heliövaara et al., 1992).

6. SOIL TEXTURE AND STRUCTURE

Put simply, the composition of mineral particles in soil is described as soil texture, and the arrangement of these particles is termed soil structure (Brady and Weil, 2007; Brown and Gange, 1990). Both of these properties directly affect root herbivore physiology, largely by shaping other abiotic attributes of the soil. For instance, fine-textured soils (e.g. those with high silt and clay content) retain water much better than coarser soils (e.g. those with high sand content), which have greater porosity (Brady and Weil, 2007). This therefore has major effects on root herbivore physiology because of the important effects of soil moisture on root herbivores (see Section 2 of this chapter). Because finer-textured soils have greater water retention, root herbivores are much less prone to desiccation, and there are multiple examples of root herbivore performance and survival being enhanced in such environments (Cherry and Allsopp, 1991; Lummus et al., 1983; Marrone and Stinner, 1984; Pacchioli and Hower, 2004; Régnière et al., 1981; Turpin and Peters, 1971).

6.1. Oviposition behaviour of maternal insects

Sandier soils are usually associated with negative impacts on root herbivore physiology, and there are several examples of survival being reduced in such soils (e.g. see Brown and Gange, 1990). This may arise through physical abrasion of sand grains against the root herbivore body (Rogers et al., 2000) or more likely reduced incidences of cracks in the soil, which many root herbivores need to move around in the soil matrix (Villani and Wright, 1990). Finer-textured soils, such as clays, frequently form cracks during drying and, as mentioned in a previous section, can be exploited by root herbivores and greatly enhance their ability to move longer distances (Pacchioli and Hower, 2004; Villani and Wright, 1990).

Given that soil structure and texture is important for the performance of root herbivores, it might be expected that maternal insects would oviposit in the most favourable soils, as predicted by the preference–performance hypothesis (Jaenike, 1978), which can be applied to insects with root-feeding larval stages (Johnson et al., 2006a). There is some support for this; larvae of the root-feeding flea beetle (*Longitarsus bethae*) performed better in clay soils, which were amongst the most preferred oviposition sites of the maternal insects. Cracking was a key factor in this behavioural response, with maternal insects laying almost 3× as many eggs in cracked soils as uncracked soils (Simelane, 2007). Cracking appears to be a major means by which root herbivores gain access to roots, especially when they feed on specific parts of the root system such as legume root nodules (Pacchioli and Hower, 2004). Another study with the cabbage root fly (*Delia radicum*) demonstrated a maternal preference for ovipositing in soil comprising particles between 0.6 and 0.8 mm in diameter (Havukkala, 1982; Traynier, 1967). This range of soil particles possibly represents an optimal environment for larvae, big enough to move around in without causing damage from abrasion.

6.2. Effects on larval behaviour

Because root herbivores are constantly in physical contact with the soil environment, texture and structure clearly have the capacity to affect behaviour. The most obvious of these is locomotion and burrowing capacity behaviour (Johnson et al., 2007; Villani and Wright, 1990; Zhang et al., 2006). It is intuitive that more compacted soils will slow burrowing capacity and be energetically more costly than those that are densely aggregated. Villani et al. (1999) provide an excellent review of the physiological and morphological adaptations that soil-dwelling insects deploy to navigate their way through soils with differing textures and structures. In particular, they distinguish between those insects living in the soil surface or litter (epedaphic) and those that live more deeply, particularly in soil pores or voids (euedaphic). Epedaphic insects possess bodies that act like wedges to displace soil to the sides of their bodies, while euedaphic insects (the majority of root herbivores) dig between and within soil pores using mandibles and legs (Villani et al., 1999). Having said that, many root herbivores do not possess appendages that aid excavation, but rather have body shapes (e.g. elongated, flattened or cylindrical) that allow them to move in between fissures and cracks in the soil. More often than not, the antennae, legs and tarsi are shorted or absent altogether (Eisenbeis and Wichard, 1987), which most likely facilitates this type of movement.

Measurements of movement rates of root herbivores are scarce, but expectedly, they have been shown to move more slowly in compacted or fine-textured soils than more porous and coarser soils. Western corn rootworm (*D. v. virgifera*) can move 80–100 cm in field soil (Short and Luedtke, 1970; Suttle et al., 1967), though movement becomes restricted to <5 cm when soil bulk density increases beyond 1.1 Mg m^{-3} (Gustin and Schumacher, 1989). Increasing density from 1.1 to 1.5 Mg m^{-3} reduces the motility of rootworms by up to 90% (Strnad and Bergman, 1987b). In a study that used x-ray microtomography to observe the detailed movement patterns of neonatal *Sitona lepidus* larvae, Johnson et al. (2004b) recorded movement rates of 1.02 mm h^{-1} in a highly compacted and fine-textured soil (<75 μm particles with a dry bulk density of 1.6 Mg m^{-3}). In a parallel study, in which neonatal movement was measured in soil with slightly lower bulk density (particles <2 mm, 1.24 Mg m^{-3} dry bulk density), burrowing speeds increased to 1.5 mm h^{-1} (Johnson et al., 2004a).

6.3. Direct and indirect effects of soil compaction

As discussed earlier, soil compaction reduces the mobility of root herbivores because of higher bulk density and lower air-filled porosity and connectivity. As a consequence, root herbivore survival is often greatly reduced (Ellsbury et al., 1994; Strnad and Bergman, 1987b). Indeed, the use of rollers to compact soil has been advocated as an effective means of controlling turfgrass root herbivores when insecticide treatment is not available or practical (Potter, 1998; Stewart et al., 1988). In addition to placing physical constraints on movement, soil compaction could affect sensory ecology of root herbivores and, therefore, their behaviour. Soil compaction reduces air permeability and, therefore, reduces the diffusion of host plant volatiles, which are used for the location of roots within the soil. For example, a gaseous molecule can diffuse through 1 m of air more rapidly than a 1 mm film of water within a soil pore (Payne and Gregory, 1988). The chemical ecology of root herbivores remains poorly understood compared to their aboveground counterparts, but given that there is a wealth of literature concerning the effects of soil texture and structure on gaseous root exudates (Brady and Weil, 2007; Payne and Gregory, 1988), it seems inevitable that host plant location will be strongly affected by soil texture and function.

7. FERTILISATION AND SOIL NUTRIENTS

Root herbivores feature prominently in agricultural ecosystems (Blackshaw and Kerry, 2008), which are often subject to fertilisation treatments that elevate soil nutrient concentrations. As Erb and Lu (2013) note, soil nutrients are unlikely to directly affect root herbivores, since the insects obtain all of their nutrients from the root tissue they consume. Instead, the effects of fertilisation and improvements in soil nutrients are likely to indirectly affect root herbivores via changes in root quality (but see Radcliffe, 1970 for an exception). Several studies have investigated this with most generally reporting positive effects on herbivore performance and physiology. There are, however, several examples of negative and neutral effects of soil fertilisation on herbivores in the literature.

The majority of studies report that the application of fertilisers positively affects the physiology of root herbivores. This is probably due to the improved nutritional quality, nitrogen concentrations in particular, of host plants for root herbivores. For example, rice weevil (*Lissorhoptrus oryzophilus*) performance and populations increased following the application of nitrogen fertiliser (Way et al., 2006). Likewise, grass grubs (*Costelytra zealandica*) and green June bugs (*Cotinis nitida*) also benefited from application of fertiliser (Potter et al., 1996; Wightman, 1974). Similar patterns were reported for Western corn rootworm larvae (Foster and Harris, 1997; Spike and Tollefson, 1988). In addition to increasing host plant nutritional quality and root biomass, application of fertiliser can have positive effects on root herbivores via changes in plant composition in mixed plant communities. For example, Frew et al. (2013) established that application of nitrogen fertiliser promoted nutritionally superior grass species (e.g. *Microlaena stipoides*) while reducing colonisation of less nutritious species (e.g. *Setaria incrassata*) in the understory of a eucalypt plantation. This change in botanical composition towards more nutritious host plants was linked to the increased abundance of root-feeding scarabs (*Sericesthis nigrolineata*).

In addition to improving the nutritional quality of roots, Erb and Lu (2013) speculated that fertilisation could reduce concentrations of defensive compounds in the roots, which could potentially affect root herbivore physiology and behaviour. There is limited information about root defences generally (van Dam, 2009) and very few studies concerning the effects of fertilisers on root defences; hence, this is difficult to determine. However, in a recent paper, Hol (2011) concluded that fertilisation of *Senecio* spp. with nitrogen, phosphorus and potassium actually reduced concentrations of pyrrolizidine alkaloid in the roots.

The outcome of nutrient application on root herbivore physiology can be predicted more easily in some systems than others. For example, using two legumes—lucerne (*Medicago sativa*) and red clover (*Trifolium pratense*) (Wolfson, 1987)—increased soil nitrogen concentrations, which had the predictable effect of reducing root nodulation because plants were less dependent on biological fixation of atmospheric nitrogen by rhizobial bacteria. It is common for legumes to reduce nodulation when organic forms of nitrogen are plentiful (Frame and Newbould, 1986). Lower number of root nodules negatively affected the performance of *Sitona hispidulus* larvae, whose development is dependent on nodule herbivory (Quinn and Hower, 1986) in common with other *Sitona* spp. larvae (Gerard, 2001). Wireworm (*Agriotes* spp.) development and population size were similarly negatively affected by sustained application of inorganic nitrogen (Edwards, 1977) in a long-term grassland experiment, although the underlying reasons were never determined.

In other cases, root herbivore physiology was largely unchanged by application of fertilisers, though root herbivory itself was much reduced. For example, application of manure to the soil prompted chafer larvae to switch from feeding on roots to the organic matter associated with the mature (Radcliffe, 1970). The performance of chrysomelid larvae feeding on maize was unchanged following application of inorganic nitrogen, mostly due to increased root growth and tolerance of root herbivory (Hill et al., 1948; Lilly and Gunderson, 1952). Similarly, root-feeding scarab larvae have been observed to be unaffected by fertiliser application (Prestidge et al., 1985).

The timing of fertiliser application could also be important, as suggested by Brown and Gange (1990). If fertiliser is applied prior to colonisation by root herbivores, then an abundant and nutritionally superior food supply could be more beneficial than application of fertilisers after colonisation. To our knowledge, this has not been directly tested, though there seems to be indirect evidence from related studies (see references in Brown and Gange, 1990). In conclusion, generalisations about the effects of soil nutrients on root herbivore physiology and behaviour remain difficult to make, though the application of fertilisers seems generally beneficial and an important determinant of root herbivore fitness.

8. CONCLUDING REMARKS

On balance, soil moisture appears to be the most important single abiotic factor influencing the behaviour and physiology of root herbivores, although this is tightly linked with soil temperature and so this distinction is arguable. The strongest effect of soil moisture is through osmotic changes from inside an organism to its environment. However, root herbivores have evolved physiological mechanisms to prevent desiccation and suffocation under drought and flooding, respectively. Additionally, root feeders show behavioural responses to soil moisture with spatial and temporal movement patterns or cell construction or both in some cases. Moisture also interacts with host plant quality and quantity, and the soil atmosphere and pH to affect root herbivore physiology. In some respects, the interactive effects of soil abiotic properties are more important and more relevant than individual effects, since no single property affects root herbivores in isolation. For example, on the face of it, soil pH seems to have few effects on root herbivores, but taken in combination with other factors, it had significant effects on physiology (Johnson et al., 2010). Studies in this chapter have generally focused on the effects of one factor at a time, which probably arises from a variety of pragmatic reasons. While it could be argued that increasing factorial treatments in any ecological study will make them more realistic, we suggest that the case is especially compelling for soil-dwelling herbivores. Firstly, compared to aboveground herbivores, belowground herbivores have comparatively long lifecycles which they spend in a relatively stable environment that is in constant physical contact with their bodies. Secondly, root herbivores have limited abilities to disperse from unfavourable environments (e.g. Murray et al., 2010), unlike many aboveground counterparts which readily do so (Schneider, 1962). So, for instance, the interactive effects of air temperature and air humidity on aboveground herbivores are transient and something that they can possibly disperse from, but the interactive effects of soil temperature and moisture will be sustained and harder for root herbivores to escape from rapidly.

Related to this is the issue of investigating the effects of the soil environment on root herbivores under field conditions to gain a realistic insight into the net effect of any such properties. In particular, natural enemies of root herbivores are also likely to be affected by soil conditions and so may modify the net effect of the soil conditions on root herbivore populations. A cautionary example of this is provided by Frew et al. (2013) in which the interactive effects of irrigation and fertilisation on scarab populations were examined in a 5 ha eucalypt plantation understory. While irrigation was generally favourable for scarab performance, it has also promoted populations of entomopathogenic nematodes, which effectively controlled scarab populations. As a consequence, irrigation had no overall effect on root herbivore populations.

The majority of root herbivores are the juvenile stages of insects that reproduce aboveground, and as such, their location in the soil is highly dependent on the oviposition choices made by the maternal insects (Johnson et al., 2006a). Chemical cues originating from the plant, soil and other organisms have been considered to be the principal mechanism by which maternal insects make such choices (Clark et al., 2011, 2012; Cosme et al., 2011). In this chapter, we report several incidences of the physical texture and structure of the soil shaping this behaviour (see Section 6.1). Soil structure and texture is a major determinant of larval movement and survival in the soil (Villani et al., 1999) and may therefore represent a strong selective pressure for maternal insects. Our understanding of the sensory cues (e.g. visual, olfactory and physical cues) and mechanisms by which maternal insects perceive the physical condition of the soil is poor, but the striking oviposition preferences reported in Section 6 suggest that these are well developed.

Progress in the field of root herbivore physiology and behaviour has undoubtedly been hindered by the difficulties making detailed behavioural observations of insects in their natural habitat—the opaque soil matrix (Hunter, 2001). New technologies for noninvasive study, including acoustic detection (Mankin et al., 2008), x-ray tomography (Johnson et al., 2004b), and root volatile detection with PTR-MS (Danner et al., 2012), are gradually becoming more sophisticated and, crucially, more accessible, which could help redress this. In writing this chapter, we hope to stimulate interest in this expanding area of insect physiology.

ACKNOWLEDGEMENTS

We are grateful to Prof. Ted Turlings, Dr. Phil Murray and Dr. Ivan Hiltpold for useful suggestions in the preparation of this manuscript. We are also thankful to Philip Smith for proofreading this chapter.

REFERENCES

Agrios, G.N., 2005. Plant Pathology. Academic Press, San Diego, CA.

Ahearn, G., 1970. The control of water loss in desert tenebrionid beetles. J. Exp. Bot. 53, 573–595.

Allen, O.N., Allen, E.K., 1981. The Leguminosae: A Source Book of Characteristics, Uses and Nodulation. University of Wisconsin Press, Madison, WI.

Allsopp, P.G., 1992. Sugars, amino acids, and ascorbic acid as phagostimulants for larvae of *Antitrogus parvulus* and *Lepidiota negatoria* (Coleoptera: Scarabaeidae). J. Econ. Entomol. 85, 106–111.

Bais, H.P., Weir, T.L., Perry, L.G., Gilroy, S., Vivanco, J.M., 2006. The role of root exudates in rhizosphere interations with plants and other organisms. Annu. Rev. Plant Biol. 57, 233–266.

Barbehenn, R.V., Karowe, D.N., Chen, Z., 2004a. Performance of a generalist grasshopper on a C_3 and a C_4 grass: compensation for the effects of elevated CO_2 on plant nutritional quality. Oecologia 140, 96–103.

Barbehenn, R.V., Karowe, D.N., Spickard, A., 2004b. Effects of elevated atmospheric CO_2 on the nutritional ecology of C_3 and C_4 grass-feeding caterpillars. Oecologia 140, 86–95.

Bardgett, R.D., 2005. The Biology of Soil: A Community and Ecosystem Approach. Oxford University Press, Oxford, UK.

Barlow, F., Hadaway, A.B., 1956. Effect of changes in humidity on the toxicity and distribution of insecticides sorbed by some dried soils. Nature 178, 1299–1300.

Barlow, F., Hadaway, A.B., 1958. Studies on aqueous suspensions of insecticides. Part VI. Further notes on the sorption of insecticides by soils. Bull. Entomol. Res. 49, 315–331.

BassiriRad, H., Griffin, K.L., Reynolds, J.F., Strain, B.R., 1997. Changes in root NH_4^+ and NO_3^- absorption rates of loblolly and ponderosa pine in response to CO_2 enrichment. Plant Soil 190, 1–9.

Beament, J.W.L., 1961. The water relations of insect cuticle. Biol. Rev. 36, 281–320.

Belgnaoui, S., Barra, J.A., 1988. Anhydrobiosis in the collembola *Folsomides angularis*: cytochemical investigations of glycogen status and glycogen phosphorylase activity. Biol. Struct. Morphog. 1, 147–153.

Benefer, C.M., Knight, M.E., Ellis, J.S., Hicks, H., Blackshaw, R.P., 2012. Understanding the relationship between adult and larval *Agriotes* distributions: the effect of sampling method, species identification and abiotic variables. Appl. Soil Ecol. 53, 39–48.

Bernays, E.A., Chapman, R.F., 1994. Host-Plant Selection by Phytophagous Insects. Chapman and Hall, New York, NY.

Bernklau, E.J., Bjostad, L.B., 1998a. Behavioral responses of first-instar western corn rootworm (Coleoptera: Chrysomelidae) to carbon dioxide in a glass bead bioassay. J. Econ. Entomol. 91, 444–456.

Bernklau, E.J., Bjostad, L.B., 1998b. Reinvestigation of host location by western corn rootworm larvae (Coleoptera: Chrysomelidae): CO_2 is the only volatile attractant. J. Econ. Entomol. 91, 1331–1340.

Bernklau, E.J., Fromm, E.A., Bjostad, L.B., 2004. Disruption of host location of western corn rootworm larvae (Coleoptera: Chrysomelidae) with carbon dioxide. J. Econ. Entomol. 97, 330–339.

Bertin, C., Yang, X.H., Weston, L.A., 2003. The role of root exudates and allelochemicals in the rhizosphere. Plant Soil 256, 67–83.

Blackshaw, R.P., Kerry, B.R., 2008. Root herbivory in agricultural ecosystems. In: Johnson, S.N., Murray, P.J. (Eds.), Root Feeders—An Ecosystem Perspective. CABI, Wallingford, UK, pp. 35–53.

Bosio, C.F., McCrea, K.D., Nitao, J.K., Abrahamson, W.G., 1990. Defense chemistry of *Solidago altissima*: effects on the generalist herbivore *Trichoplusia ni* (Lepidoptera: Noctuidae). Environ. Entomol. 19, 465–468.

Boynton, D., Compton, O.C., 1944. Normal seasonal changes of oxygen and carbon dioxide percentages in gas from the larger pores of three orchard subsoils. Soil Sci. 57, 107–118.

Brady, N.C., Weil, R.R., 2007. The Nature and Properties of Soils. Prentice Hall, Englewood Cliffs, NJ.

Braunberger, P.G., Abbott, L.K., Robson, A.D., 1996. Infectivity of arbuscular mycorrhizal fungi after wetting and drying. New Phytol. 134, 673–684.

Brockhurst, M.A., Buckling, A., Rainey, P.B., 2006. Spatial heterogeneity and the stability of host-parasite coexistence. J. Evol. Biol. 19, 374–379.

Brown, V.K., Gange, A.C., 1990. Insect herbivory below ground. Adv. Ecol. Res. 20, 1–58.

Bruck, D.J., 2005. Ecology of *Metarhizium anisopliae* in soilless potting media and the rhizosphere: implications for pest management. Biol. Control. 32, 155–163.

Brust, G.E., House, G.J., 1990. Effects of soil moisture, no-tillage and predators on southern corn rootworm (*Diabrotica undecimpunctata howardi*) survival in corn agroecosystems. Agric. Ecosyst. Environ. 31, 199–215.

Burrows, W.C., Larson, W.E., 1962. Effect of amount of mulch on soil temperature and early growth of corn. Agron. J. 54, 19–23.

Bursell, E., 1964. Environmental aspects: humidity. In: Rockstein, M. (Ed.), Physiology of Insecta, vol I. Academic Press, New York, NJ, pp. 323–361.

Buxton, P.A., 1930. Evaporation from the meal-worm (Tenebrio: Coleoptera) and atmospheric humidity. Proc. R. Soc. Lond B Biol. 106, 560–577.

Campbell, R.E., 1937. Temperature and moisture preferences of wireworms. Ecology 18, 479–489.

Campbell, J.F., Lewis, E., Yoder, F., Gaugler, R., 1996. Entomopathogenic nematode (Heterorhabditidae and Steinernematidae) spatial distribution in turfgrass. Parasitology 113, 473–482.

Chaves, M.M., Maroco, J.P., Pereira, J.S., 2003. Understanding plant responses to drought—from genes to the whole plant. Funct. Plant Biol. 30, 239–264.

Cherry, R.H., Allsopp, P.G., 1991. Soil texture and distribution of *Antitrogus parvulus* Britton, *Lepidiota crinita* Brenske and *L. negatoria* Blackburn (Coleoptera, Scarabaeidae) in South Queensland sugarcane fields. J. Aust. Entomol. Soc. 30, 89–92.

Cherry, R.H., Porter, P.S., 1992. Respiration and behavior of a sugarcane grub, *Ligyrus subtropicus* (Coleoptera: Scarabaeidae) under flooded conditions. J. Entomol. Sci. 27, 71–77.

Clark, K.E., Hartley, S.E., Johnson, S.N., 2011. Does mother know best? The preference–performance hypothesis and parent–offspring conflict in aboveground–belowground herbivore life cycles. Ecol. Entomol. 36, 117–124.

Clark, K.E., Hartley, S.E., Brennan, R.M., MacKenzie, K., Johnson, S.N., 2012. Investigating preference–performance relationships in aboveground–belowground life cycles: a laboratory and field study with the vine weevil (*Otiorhynchus sulcatus*). Bull. Entomol. Res. 102, 63–70.

Colinet, H., Hance, T., 2009. Male reproductive potential of *Aphidius colemani* (Hymenoptera: Aphidiinae) exposed to constant or fluctuating thermal regimens. Environ. Entomol. 38, 242–249.

Collier, R.H., Finch, S., Phelps, K., Thompson, A.R., 1991. Possible impact of global warming on cabbage root fly (*Delia radicum*) activity in the UK. Ann. Appl. Biol. 118, 261–271.

Cooke, B.M., Jones, D.G., Kaye, B., 2006. The Epidemiology of Plant Diseases. Springer, Dordrecht, The Netherlands.

Cooper, P.D., 1983. Components of evaporative water loss in the desert tenebrionid beetles *Eleodes armata* and *Cryptoglossa verrucosa*. Physiol. Entomol. 56, 47–55.

Cosme, M., Stout, M.J., Wurst, S., 2011. Effect of arbuscular mycorrhizal fungi (*Glomus intraradices*) on the oviposition of rice water weevil (*Lissorhoptrus oryzophilus*). Mycorrhiza 21, 651–658.

Crommentuijn, T., Doornekamp, A., Van Gestel, C.A.M., 1997. Bioavailability and ecological effects of cadmium on *Folsomia candida* (Willem) in an artificial soil substrate as influenced by pH and organic matter. Appl. Soil Ecol. 5, 261–271.

Currie, A.F., Murray, P.J., Gange, A.C., 2006. Root herbivory by *Tipula paludosa* larvae increases colonization of *Agrostis capillaris* by arbuscular mycorrhizal fungi. Soil Biol. Biochem. 38, 1994–1997.

Danner, H., Samudrala, D., Cristescu, S.M., Van Dam, N.M., 2012. Tracing hidden herbivores: time-resolved non-invasive analysis of belowground volatiles by proton-transfer-reaction mass spectrometry (PTR-MS). J. Chem. Ecol. 38, 785–794.

Dautel, H., 1999. Water loss and metabolic water in starving *Argas reflexus* nymphs (Acari: Argasidae). J. Insect Physiol. 45, 55–63.

Davidson, R.L., Wiseman, J.R., Wolfe, V.J., 1972. Environmental stress in the pasture scarab *Sericesthis nigrolineata* Boisd. II. Effects of soil moisture and temperature on survival of first-instar larvae. J. Appl. Ecol. 9, 799–806.

Davis, P.M., Brenes, N., Allee, L.L., 1996. Temperature dependent models to predict regional differences in corn rootworm (Coleoptera: Chrysomelidae) phenology. Environ. Entomol. 25, 767–775.

de Boer, T.E., Holmstrup, M., van Straalen, N.M., Roelofs, D., 2010. The effect of soil pH and temperature on *Folsomia candida* transcriptional regulation. J. Insect Physiol. 56, 350–355.

de Vries, J., 1969. In situ determination of physical properties of the surface layer of field soils. Soil Sci. Soc. Am. J. 33, 349–353.

Dicke, M., Burrough, P.A., 1988. Using fractal dimensions for characterizing tortuosity of animal trails. Physiol. Entomol. 13, 393–398.

Dindal, D.L., 1990. Soil Biology Guide. Wiley, New York, NY.

Dixon, A.F.G., Watson, A., 1970. Quality and availability of food for a sycamore aphid population. Animal populations in relation to their food resources. In: A Symposium of the British Ecological Society, Aberdeen 24–28 March 1969. Blackwell, Oxford and Edinburgh, pp. 271–287.

Doane, J.F., Lee, Y.W., Westcott, N.D., Klingler, J., 1975. The orientation response of *Ctenicera destructor* and other wireworms (Coleoptera: Elateridae) to germinating grain and to carbon dioxide. Can. Entomol. 107, 1233–1251.

Docherty, M., Hurst, D.K., Holopainen, J.K., Whittaker, J.B., Lea, P.J., Watt, A.D., 1996. Carbon dioxide-induced changes in beech foliage cause female beech weevil larvae to feed in a compensatory manner. Glob. Chang. Biol. 2, 335–341.

Dowdy, W., 1944. The influence of temperature on vertical migration of invertebrates inhabiting different soil types. Ecology 25, 449–460.

Doyle, J.J., Luckow, M.A., 2003. The rest of the iceberg. Legume diversity and evolution in a phylogenetic context. Plant Physiol. 131, 900–910.

Dreibelbis, F.R., Post, F.A., 1942. An inventory of soil water relationships on woodland, pasture, and cultivated soils. Soil Sci. Soc. Am. J. 6, 462–473.

Edwards, C.A., 1977. Investigations into the influence of agricultural practice on soil invertebrates. Ann. Appl. Biol. 87, 515–520.

Edwards, A.C., Cresser, M.S., 1992. Freezing and its effect on chemical and biological properties of soil. Adv. Soil Sci. 18, 59–79.

Edwards, E.J., Benham, D.G., Marland, L.A., Fitter, A.H., 2004. Root production is determined by radiation flux in a temperate grassland community. Glob. Chang. Biol. 10, 209–227.

Ehlers, W., Farmer, W.J., Spencer, W.F., Letey, J., 1969a. Lindane diffusion in soils: II. Water content, bulk density, and temperature effects. Soil Sci. Soc. Am. J. 33, 505–508.

Ehlers, W., Letey, J., Spencer, W.F., Farmer, W.J., 1969b. Lindane diffusion in soils: I. Theoretical considerations and mechanism of movement. Soil Sci. Soc. Am. J. 33, 501–504.

Eisenbeis, G., Wichard, W., 1987. Atlas on the Biology of the Soil Arthropods. Springer-Verlag London Ltd, London, UK.

Elkinton, J.S., Cardé, R.T., 1984. Odor dispersion. In: Bill, W.J., Cardé, R.T. (Eds.), Chemical Ecology of Insects. Chapman and Hall, London, UK, pp. 73–91.

Ellsbury, M.M., Schumacher, T.E., Gustin, R.D., Woodson, W.D., 1994. Soil compaction effect on corn rootworm populations in maize artificially infested with eggs of western corn rootworm (Coleoptera: Chrysomelidae). Environ. Entomol. 23, 943–948.

Erb, M., Lu, J., 2013. Soil abiotic factors influence interactions between belowground herbivores and plant roots. J. Exp. Bot. 64, 1295–1303.

Erb, M., Köllner, T.G., Degenhardt, J., Zwahlen, C., Hibbard, B.E., Turlings, T.C.J., 2011. The role of abscisic acid and water stress in root herbivore-induced leaf resistance. New Phytol. 189, 308–320.

Evans, A.C., 1943. Value of the pF scale of soil moisture for expressing the soil moisture relations of wireworms. Nature 152, 21–22.

Falconer, D.S., 1945. On the behaviour of wireworms of the genus *Agriotes* Esch. (Coleoptera, Elateridae) in relation to temperature. J. Exp. Biol. 21, 17–32.

Farmer, W.J., Jensen, C.R., 1970. Diffusion and analysis of carbon-14 labeled dieldrin in soils. Soil Sci. Soc. Am. J. 34, 28–31.

Fierer, N., Jackson, R.B., 2006. The diversity and biogeography of soil bacterial communities. Proc. Natl. Acad. Sci. U. S. A. 103, 626–631.

Fisher, J., Keaster, A.J., Fairchild, M.L., 1975. Seasonal vertical movement of wireworm larvae in Missouri: influence of soil temperature on the genera *Melanotus* Escholtz and *Conoderus* Escholtz. Ann. Entomol. Soc. Am. 68, 1071–1073.

Fitter, A.H., Self, G.K., Brown, T.K., Bogie, D.S., Graves, J.D., Benham, D., Ineson, P., 1999. Root production and turnover in an upland grassland subjected to artificial soil warming respond to radiation flux and nutrients, not temperature. Oecologia 120, 575–581.

Flerchinger, G.N., Pierson, F.B., 1991. Modeling plant canopy effects on variability of soil temperature and water. Agr. Forest Meteorol. 56, 227–246.

Foster, S.P., Harris, M.O., 1997. Behavioral manipulation methods for insect pest-management. Annu. Rev. Entomol. 42, 123–146.

Frame, J., Newbould, P., 1986. Agronomy of white clover. Adv. Agron. 40, 1–88.

Frew, A., Nielsen, U.N., Reigler, M.R., Johnson, S.N., 2013. Do eucalypt plantation management practices create understory reservoirs of scarab beetle pests in the soil? Forest Ecol. Manag. 306, 275–280.

Galbreath, R.A., 1975. Water balance across the cuticle of a soil insect. J. Exp. Biol. 62, 115–120.

Galbreath, R.A., 1988. Orientation of grass grub *Costelytra zealandica* (Coleoptera: Scarabaeidae) to a carbon dioxide source. New Zeal. Entomol. 11, 6–7.

Gange, A.C., Brown, V.K., Sinclair, G.S., 1994. Reduction of black vine weevil larval growth by vesicular arbuscular mycorrhizal infection. Entomol. Exp. Appl. 70, 115–119.

Gerakis, P.A., Guerrero, F.P., Williams, W.A., 1975. Growth, water relations and nutrition of three grassland annuals as affected by drought. J. Appl. Ecol. 12, 125–135.

Gerard, P.J., 2001. Dependence of *Sitona lepidus* (Coleoptera: Curculionidae) larvae on abundance of white clover *Rhizobium* nodules. Bull. Entomol. Res. 91, 149–152.

Gerba, C.P., Bitton, G., 1994. Microbial pollutants: their survival and transport pattern to groundwater. In: Groundwater Pollution Microbiology. Krieger Publishing Company, Malabar, FL, pp. 65–88.
Giojalas, L.C., 1993. Ultrastructural variations in the spermiogenesis of *Triatoma infestans* induced by temperature changes. J. Morphol. 216, 17–27.
Gouinguené, S.P., Turlings, T.C.J., 2002. The effects of abiotic factors on induced volatile emissions in corn plants. Plant Physiol. 129, 1296–1307.
Goverde, M., Bazin, A., Shykoff, J.A., Erhardt, A., 1999. Influence of leaf chemistry of *Lotus corniculatus* (Fabaceae) on larval development of *Polyommatus icarus* (Lepidoptera: Lycaenidae): effects of elevated CO_2 and plant genotype. Funct. Ecol. 13, 801–810.
Grant, J.A., Villani, M.G., 2003. Soil moisture effects on entomopathogenic nematodes. Environ. Entomol. 32, 80–87.
Gregory, P.J., 2006. Roots, rhizosphere and soil: the route to a better understanding of soil science? Eur. J. Soil Sci. 57, 2–12.
Grewal, P.S., Lewis, E.E., Gaugler, R., 1997. Response of infective stage parasites (Nematoda: Steinernematidae) to volatile cues from infected hosts. J. Chem. Ecol. 23, 503–515.
Griffin, D.M., 1963. Soil moisture and the ecology of soil fungi. Biol. Rev. 38, 141–166.
Griffin, D.M., 1981. Water and microbial stress. Adv. Microb. Ecol. 5, 91–136.
Guenzi, W.D., Beard, W.E., 1970. Volatilization of lindane and DDT from soils. Soil Sci. Soc. Am. J. 34, 443–447.
Guimaraes, V.F., Cruz, I.V., Hagler, A.N., Mendonça-Hagler, L.C., van Elsas, J.D., 1997. Transport of a genetically modified *Pseudomonas fluorescens* and its parent strain through undisturbed tropical soil cores. Appl. Soil Ecol. 7, 41–50.
Gustin, R.D., Schumacher, T.E., 1989. Relationship of some soil pore parameters to movement of first instar western corn rootworm (Coleoptera, Chrysomelidae). Environ. Entomol. 18, 343–346.
Hadaway, A.B., Barlow, F., 1951. Sorption of solid insecticides by dried mud. Nature 167, 854.
Hadley, N., 1978. Cuticular permeability of desert tenebrionid beetles: correlations with epicuticular hydrocarbon composition. Insect Biochem. 8, 17–22.
Haile, F.J., 2001. Drought stress, insects, and yield loss. In: Peterson, R.K.D., Higley, L.G. (Eds.), Biotic Stress and Yield Loss. CRC Press, Boca Raton, FL, pp. 117–134.
Hallett, P.D., Bengough, A.G., 2012. Managing the soil physical environment for plants. In: Gregory, P.J., Nortcliff, S. (Eds.), Soil Conditions and Plant Growth. Blackwell Publishing Ltd, Oxford, UK, pp. 238–268.
Hamilton, C.C., 1917. The behaviour of some soil insects in gradients of evaporating powder of air, carbon dioxide and ammonia. Biol. Bull. 32, 159–182.
Hardie, K., Leyton, L., 1981. The influence of vesicular-arbuscular mycorrhiza on growth and water relations of red clover. New Phytol. 89, 599–608.
Harris, C.R., 1966. Influence of soil type on the activity of insecticides in soil. J. Econ. Entomol. 59, 1221–1224.
Harris, C.R., 1972. Factors influencing the effectiveness of soil insecticides. Annu. Rev. Entomol. 17, 177–198.
Harris, C.R., Lichtenstein, E.P., 1961. Factors affecting the volatilization of insecticidal residues from soils. J. Econ. Entomol. 54, 1038–1045.
Harrison-Murray, R.S., Lal, R., 1979. High soil temperature and the response of maize to mulching in the lowland humid tropics. In: Lal, R., Greenland, D.J. (Eds.), Soil Physical Properties and Crop Production in the Tropics. Wiley, Chichester, UK, pp. 285–304.
Harvey, R.W., George, L.H., Smith, R.L., LeBlanc, D.R., 1989. Transport of microspheres and indigenous bacteria through a sandy aquifer: results of natural-and forced-gradient tracer experiments. Environ. Sci. Technol. 23, 51–56.

Hassan, S.T., Hilditch, J.A., 1976. Survival of Larvae of *Anoplognathus porosus* (Dalman) and *Sericesthis nigrolineata* Boisd. (Coleoptera: Scarabaeidae). J. Appl. Ecol. 13, 333–339.
Havukkala, I., 1982. Deterring oviposition of the cabbage root fly, *Delia radicum* (Diptera: Anthomyiidae), by non-chemical methods. Acta Entomol. Fenn. 40, 9–15.
Heckathorn, S.A., DeLucia, E.H., 1994. Drought-induced nitrogen retranslocation in perennial C_4 grasses of tallgrass prairie. Ecology 75, 1877–1886.
Heliövaara, K., Vaisanen, R., Varama, M., 1992. Acidic precipitation increases egg survival in *Neodiprion sertifer*. Entomol. Exp. Appl. 62, 55–60.
Hilborn, R., 1975. The effect of spatial heterogeneity on the persistence of predator-prey interactions. Theor. Popul. Biol. 8, 346–355.
Hill, R.E., Hixson, E., Muma, M.H., 1948. Corn rootworm tests with benzene hexachloride, DDT, nitrogen fertilizers and crop rotations. J. Econ. Entomol. 67, 748–750.
Hillel, D., 1982. Introduction to Soil Physics. Academic Press, New York, NY.
Hiltpold, I., Turlings, T.C.J., 2008. Belowground chemical signaling in maize: when simplicity rhymes with efficiency. J. Chem. Ecol. 34, 628–635.
Hinton, H.E., 1960. Cryptobiosis in the larva of *Polypedilum vanderplanki* Hint. (Chironomidae). J. Insect Physiol. 5, 286–300.
Hol, W.H.G., 2011. The effect of nutrients on pyrrolizidine alkaloids in *Senecio* plants and their interactions with herbivores and pathogens. Phytochem. Rev. 10, 119–126.
Holmstrup, M., Sørensen, L.I., Bindesbøl, A., Hedlund, K., 2007. Cold acclimation and lipid composition in the earthworm *Dendrobaena octaedra*. Comp. Biochem. Physiol. A Mol. Integr. Physiol. 147, 911–919.
Holter, P., 1994. Tolerance of dung insects to low oxygen and high carbon dioxide concentrations. Eur. J. Soil Biol. 30, 187–193.
Howe, R.W., 1967. Temperature effects on embryonic development in insects. Annu. Rev. Entomol. 12, 15–42.
Hu, G., St. Leger, R.J., 2002. Field studies using a recombinant mycoinsecticide (*Metarhizium anisopliae*) reveal that it is rhizosphere competent. Appl. Environ. Microb. 68, 6383–6387.
Huang, X.P., Mack, T.P., 2001. Artificial carbon dioxide source to attract lesser cornstalk borer (Lepidoptera: Pyralidae) larvae. J. Econ. Entomol. 94, 860–867.
Huang, X.P., Mack, T.P., 2002. Collection and determination of lesser cornstalk borer (Lepidoptera: Pyralidae) larval attractant from peanut plants. Environ. Entomol. 31, 15–21.
Huberty, A.F., Denno, R.F., 2004. Plant water stress and its consequences for herbivorous insects: a new synthesis. Ecology 85, 1383–1398.
Humber, R.A., 2008. Evolution of entomopathogenicity in fungi. J. Invert. Pathol. 98, 262–266.
Hunter, M.D., 2001. Out of sight, out of mind: the impacts of root-feeding insects in natural and managed systems. Agr. Forest Entomol. 3, 3–9.
Hunter, M.D., Price, P.W., 1992. Playing chutes and ladders: heterogeneity and the relative roles of bottom-up and top-down forces in natural communities. Ecology 73, 723–732.
Ingham, E.R., Coleman, D.C., 1984. Effects of streptomycin, cycloheximide, Fungizone, captan, carbofuran, cygon, and PCNB on soil microorganisms. Microb. Ecol. 10, 345–358.
Jackson, T.A., Klein, M.G., 2006. Scarabs as pests: a continuing problem. Coleopt. Soc. Monogr. Patricia Vaurie Ser. 5, 102–119.
Jaenike, J., 1978. On optimal oviposition behaviour in phytophagous insects. Theor. Pop. Biol. 14, 350–356.
Jaronski, S.T., 2007. Soil ecology of the entomopathogenic Ascomycetes: a critical examination of what we (think) we know. In: Ekesi, S., Maniania, N.K. (Eds.), Use of

Entomopathogenic Fungi in Biological Pest Management. Research Signpost, Trivandrum, India, pp. 91–144.

Jewett, D.K., Bjostad, L.B., 1996. Dichloromethane attracts diabroticite larvae in a laboratory behavioral bioassay. J. Chem. Ecol. 22, 1331–1344.

Jindra, M., Sehnal, F., 1990. Linkage between diet humidity, metabolic water production and heat dissipation in the larvae of *Galleria mellonella*. Insect Biochem. 20, 389–395.

Johnsen, S., Gutierrez, A.P., Jørgensen, J., 1997. Overwintering in the cabbage root fly *Delia radicum*: a dynamic model of temperature-dependent dormancy and post-dormancy development. J. Appl. Ecol. 34, 21–28.

Johnson, S.N., Gregory, P.J., 2006. Chemically-mediated host-plant location and selection by root-feeding insects. Physiol. Entomol. 31, 1–13.

Johnson, S.N., McNicol, J.W., 2010. Elevated CO_2 and aboveground-belowground herbivory by the clover root weevil. Oecologia 162, 209–216.

Johnson, S.N., Nielsen, U., 2012. Foraging in the dark—chemically mediated host location by root-feeding insects. J. Chem. Ecol. 38, 604–614.

Johnson, S.N., Gregory, P.J., Murray, P.J., Zhang, X., Young, I.M., 2004a. Host plant recognition by the root-feeding clover weevil, *Sitona lepidus* (Coleoptera: Curculionidae). Bull. Entomol. Res. 94, 433–439.

Johnson, S.N., Read, D.B., Gregory, P.J., 2004b. Tracking larval insect movement within soil using high resolution X-ray microtomography. Ecol. Entomol. 29, 117–122.

Johnson, S.N., Birch, A.N.E., Gregory, P.J., Murray, P.J., 2006a. The 'mother knows best' principle: should soil insects be included in the preference–performance debate? Ecol. Entomol. 31, 395–401.

Johnson, S.N., Zhang, X.X., Crawford, J.W., Gregory, P.J., Hix, N.J., Jarvis, S.C., Murray, P.J., Young, I.M., 2006b. Effects of carbon dioxide on the searching behaviour of the root-feeding clover weevil *Sitona lepidus* (Coleoptera: Curculionidae). Bull. Entomol. Res. 96, 361–366.

Johnson, S.N., Crawford, J.W., Gregory, P.J., Grinev, D.V., Mankin, R.W., Masters, G.J., Murray, P.J., Wall, D.H., Zhang, X.X., 2007. Non-invasive techniques for investigating and modelling root-feeding insects in managed and natural systems. Agr. Forest Entomol. 9, 39–46.

Johnson, S.N., Gregory, P.J., McNicol, J.W., Oodally, Y., Zhang, X., Murray, P.J., 2010. Effects of soil conditions and drought on egg hatching and larval survival of the clover root weevil (*Sitona lepidus*). Appl. Soil Ecol. 44, 75–79.

Johnson, S.N., Barton, A.T., Clark, K.E., Gregory, P.J., McMenemy, L.S., Hancock, R.D., 2011. Elevated atmospheric carbon dioxide impairs the performance of root-feeding vine weevils by modifying root growth and secondary metabolites. Glob. Chang. Biol. 17, 688–695.

Jones, E.W., 1951. Laboratory studies on the moisture relations of *Limonius* (Coleoptera: Elaterida). Ecology 32, 284–293.

Jones, O.T., 1979. The responses of carrot fly larvae, *Psila rosae*, to components of their physical environment. Ecol. Entomol. 4, 327–334.

Jones, O.T., Coaker, T.H., 1977. Orientated responses of carrot fly larvae, *Psila rosae*, to plant odours, carbon dioxide and carrot root volatiles. Physiol. Entomol. 2, 189–197.

Jones, O.T., Coaker, T.H., 1978. A basis for host plant finding in phytophagous larvae. Entomol. Exp. Appl. 24, 272–284.

Jones, O.T., Coaker, T.H., 1979. Responses of carrot fly larvae, *Psila rosae*, to the odorous and contact-chemostimulatory metabolites of host and non-host plants. Physiol. Entomol. 4, 353–360.

Jones, E.W., Shirck, F.H., 1942. The seasonal vertical distribution of wireworms in the soil in relation to their control in the Pacific northwest. J. Agric. Res. 65, 125–142.

Jupp, A.P., Newman, E.I., 1987. Morphological and anatomical effects of severe drought on the roots of *Lolium perenne* L. New Phytol. 105, 393–402.
Karban, R., 1989. Community organization of *Erigeron glaucus* folivores: effects of competition, predation, and host plant. Ecology 70, 1028–1039.
Kause, A., Haukioja, E., Hanhimäki, S., 1999. Phenotypic plasticity in foraging behavior of sawfly larvae. Ecology 80, 1230–1241.
Kaya, H.K., Lacey, L.A., 2007. Introduction to microbial control. In: Field Manual of Techniques in Invertebrate Pathology. Springer, Dordrecht, The Netherlands, pp. 3–7.
Keen, B.A., 1931. The Physical Properties of the Soil. Longmans, London, UK.
Kieft, T.L., 1987. Microbial biomass response to a rapid increase in water potential when dry soil is wetted. Soil Biol. Biochem. 19, 119–126.
King, J.S., Thomas, R.B., Strain, B.R., 1996. Growth and carbon accumulation in root systems of *Pinus taeda* and *Pinus ponderosa* seedlings as affected by varying CO_2, temperature and nitrogen. Tree Physiol. 16, 635–642.
Klein, M.G., 1992. Use of *Bacillus popilliae* in Japanese beetle control. In: Glare, T.R., Jackson, T.A. (Eds.), Use of Pathogens in Scarab Pest Management. Intercept, Hampshire, UK, pp. 179–189.
Klein, M.G., Bedding, R., 1993. Biological control of scarabs with entomopathogenic nematodes. In: Bedding, R., Akhurst, R., Kaya, H. (Eds.), Nematodes and the Biological Control of Insect Pests. CSIRO, Melbourne, Australia, pp. 49–51.
Klein, M.G., Grewal, P.S., Jackson, T.A., Koppenhöfer, A.M., 2007. Lawn, turf and grassland pests. In: Field Manual of Techniques in Invertebrate Pathology. Springer, Dordrecht, The Netherlands, pp. 655–675.
Klingler, J., 1957. Über die Bedeutung des Kohlendioxyds fur die Orientierung der Larven von *Otiorrhynchus sulcatus* F., *Melolontha* und *Agriotes* (Col.) im Boden (Vorläufige Mitteilung). Mitt. Schweiz. Entomol. Ges. 30, 317–322.
Klingler, J., 1958. Die Bedeutung der Kohlendioxyd-Ausscheidung der Wurzeln für die Orientierung der Larven von *Otiorrhynchus sulcatus* F. und anderer bodenbewohnender phytophager Insektenarten. Mitt. Schweiz. Entomol. Ges. 31, 205–269.
Klingler, J., 1965. On the orientation of plant nematodes and of some other soil animals. Nematologica 11, 4–18.
Klingler, J., 1966. The location of the CO_2 chemoreceptors in the larva of *Otiorrhynchus sulcatus*. Entomol. Exp. Appl. 9, 271–277.
Koricheva, J., Larsson, S., Haukioja, E., 1998. Insect performance on experimentally stressed woody plants: a meta-analysis. Annu. Rev. Entomol. 43, 195–216.
Kostal, V., Simek, P., 1998. Changes in fatty acid composition of phospholipids and triacylglycerols after cold-acclimation of an aestivating insect prepupa. J. Comp. Physiol. B. 168, 453–460.
Kramer, P.J., 1983. Water Relations of Plants. Academic Press, New York, NY.
Krysan, J.L., 1999. Selected topics in the biology of *Diabrotica*. In: Cox, M.L. (Ed.), Advances in Chrysomelidae Biology. Backhuys, Leiden, The Netherlands, pp. 479–513.
Kukal, O., Dawson, T.E., 1989. Temperature and food quality influences feeding behavior, assimilation efficiency and growth rate of arctic woolly-bear caterpillars. Oecologia 79, 526–532.
Lafrance, J., 1968. The seasonal movements of wireworms (Coleoptera: Elateridae) in relation to soil moisture and temperature in the organic soils of southwestern Quebec. Can. Entomol. 100, 801–807.
Laznik, Ž., Trdan, S., 2013. An investigation on the chemotactic responses of different entomopathogenic nematode strains to mechanically damaged maize root volatile compounds. Exp. Parasitol. 134, 349–355.
Lees, A.D., 1943a. On the behaviour of wireworms of the genus *Agriotes* Esch. (Coleoptera, Elateridae): I. Reactions to humidity. J. Exp. Biol. 20, 43–53.

Lees, A.D., 1943b. On the behaviour of wireworms of the genus *Agriotes* Esch. (Coleoptera, Elateridae): II. Reactions to moisture. J. Exp. Biol. 20, 54–60.
Lewis, L.C., Bing, L.A., 1991. *Bacillus thuringiensis* Berliner and *Beauveria bassiana* (Balsamo) Vuillimen for European corn borer control: program for immediate and season-long suppression. Can. Entomol. 123, 387–393.
Li, D.P., Holdom, D.G., 1993. Effect of soil matric potential on sporulation and conidial survival of *Metarhizium anisopliae* (Deuteromycotina: Hyphomycetes). J. Invertebr. Pathol. 62, 273–277.
Li, H., McCoy, C.W., Syvertsen, J.P., 2007. Controlling factors of environmental flooding, soil pH and *Diaprepes abbreviatus* (L.) root weevil feeding in citrus: larval survival and larval growth. Appl. Soil Ecol. 35, 553–565.
Lilly, J.H., Gunderson, H., 1952. Fighting the corn rootworm. Iowa Farm Sci. 6, 18–19.
Liu, W., Zhang, Z., Wan, S., 2009. Predominant role of water in regulating soil and microbial respiration and their responses to climate change in a semiarid grassland. Glob. Chang. Biol. 15, 184–195.
Ludwig, J.W., Harper, J.L., 1958. The influence of environment on seed and seedling mortality: VIII. The influence of soil colour. J. Ecol. 46, 381–389.
Ludwig, J.W., Bunting, E.S., Harper, J.L., 1957. The influence of environment on seed and seedling mortality: III. The influence of aspect on maize germination. J. Ecol. 45, 205–224.
Lummus, P.F., Smith, J.C., Powell, N.L., 1983. Soil-moisture and texture effects on survival of immature Southern corn rootworms, *Diabrotica undecimpunctata howardi* Barber (Coleoptera, Chrysomelidae). Environ. Entomol. 12, 1529–1531.
Luo, Y.Q., Hui, D.F., Zhang, D.Q., 2006. Elevated CO_2 stimulates net accumulations of carbon and nitrogen in land ecosystems: a meta-analysis. Ecology 87, 53–63.
MacDonald, P.J., Ellis, C.R., 1990. Survival time of unfed, first instar western corn rootworm (Coleoptera, Chrysomelidae) and the effects of soil type, moisture, and compaction on their mobility in soil. Environ. Entomol. 19, 666–671.
Madsen, E.L., Alexander, M., 1982. Transport of rhizobium and *Pseudomonas* through soil. Soil Sci. Soc. Am. J. 46, 557–560.
Maelzer, D.A., 1961. The effect of temperature and moisture on the immature stages of *Aphodius tasmaniae* Hope (Scarabaeidae) in the lower South-east of South Australia. Aust. J. Zool. 9, 173–202.
Mankin, R.W., Johnson, S.N., Grinev, D.V., Gregory, P.J., 2008. New experimental techniques for studying root herbivory. In: Johnson, S.N., Murray, P.J. (Eds.), Root Feeders—An Ecosystem Perspective. CABI, Wallingford, UK, pp. 20–32.
Marinissen, J.C.Y., Bok, J., 1988. Earthworm-amended soil structure: its influence on Collembola populations in grassland. Pedobiologia 32, 243–252.
Marrone, P.G., Stinner, R.E., 1984. Influence of soil physical factors on survival and development of the larvae and pupae of the bean leaf beetle, *Cerotoma trifurcata* (Coleoptera, Chrysomelidae). Can. Entomol. 116, 1015–1023.
Marshall, T.J., Holmes, J.W., 1979. Soil Physics. Cambridge University Press, London, UK.
Matthiessen, J.N., 1999. Late immature mortality is the major influence on reproductive success of African black beetle, *Heteronychus arator* (Fabricius)(Coleoptera: Scarabaeidae), in a Mediterranean-climate region of Australia. Aust. J. Entomol. 38, 348–353.
Mattson, W.J., 1980. Herbivory in relation to plant nitrogen content. Annu. Rev. Ecol. Syst. 11, 119–161.
Mattson, W.J., Haack, R.A., 1987. The role of drought in outbreaks of plant-eating insects. Bioscience 37, 110–118.
Mayhew, P.J., 1997. Adaptive patterns of host-plant selection by phytophagous insects. Oikos 79, 417–428.
McCulloch, J.S.G., 1959. Soil temperatures near Nairobi 1954–1955. Q. J. Roy. Meteor. Soc. 85, 51–56.

McNew, G.L., 1960. The nature, origin, and evolution of parasitism. Plant Pathol. 2, 2–66.
Mellanby, K., 1932. The effect of atmospheric humidity on the metabolism of the fasting mealworm (*Tenebrio molitor* L., Coleoptera). Proc. R. Soc. B Biol. Sci. 111, 376–390.
Mellanby, K., French, R.A., 1958. The importance of drinking water to larval insects. Entomol. Exp. Appl. 1, 116–124.
Mortimer, S.R., van der Putten, W.H., Brown, V.K., 1999. Insect and nematode herbivory under ground: interactions and role in vegetation succession. In: Olff, H., Brown, V.K., Drent, R.H. (Eds.), Herbivores: Between Plants and Predators. Blackwell Science, Oxford, UK, pp. 205–238.
Murdoch, W.W., 1977. Stabilizing effects of spatial heterogeneity in predator-prey systems. Theor. Popul. Biol. 11, 252–273.
Murray, P.J., Gregory, P.J., Granger, S.J., Headon, D.M., Johnson, S.N., 2010. Dispersal of soil-dwelling clover root weevil (*Sitona lepidus* Gyllenhal, Coleoptera: Curculionidae) larvae in mixed plant communities. Appl. Soil Ecol. 46, 422–425.
Neuvonen, S., Saikkonen, K., Haukioja, E., 1990. Simulated acid-rain reduces the susceptibility of the European pine sawfly (*Neodiprion sertifer*) to its nuclear polyhedrosis virus. Oecologia 83, 209–212.
Newman, J.A., Abnand, M., Henry, H.A.L., Hunt, S., Gedalof, Z., 2011. Climate Change Biology. CABI, Oxfordshire, UK.
Nguyen, T.M., Bressac, C., Chevrier, C., 2012. Heat stress affects male reproduction in a parasitoid wasp. J. Insect Physiol. 59, 248–254.
Nicolas, G., Sillans, D., 1989. Immediate and latent effects of carbon dioxide on insects. Annu. Rev. Entomol. 34, 97–116.
Pacchioli, M.A., Hower, A.A., 2004. Soil and moisture effects on the dynamics of early instar clover root curculio (Coleoptera: Curculionidae) and biomass of alfalfa root nodules. Environ. Entomol. 33, 119–127.
Parker, W.E., Howard, J.J., 2001. The biology and management of wireworms (*Agriotes* spp.) on potato with particular reference to the UK. Agric. For. Entomol. 3, 85–98.
Payne, D., Gregory, P.J., 1988. The soil atmosphere. In: Wild, A. (Ed.), Russell's Soil Conditions and Plant Growth. Longman, Harlow, UK, pp. 298–314.
Pessarakli, M., 1999. Handbook of Plant and Crop Stress. Marcel Dekker Inc., New York, NY
Piyaphongkul, J., Pritchard, J., Bale, J., 2012. Heat stress impedes development and lowers fecundity of the brown planthopper *Nilaparvata lugens* (Stål). PLos One 7, e47413.
Polivka, J.B., 1960a. Effect of lime applications to soil on Japanese beetle larval population. J. Econ. Entomol. 53, 476–477.
Polivka, J.B., 1960b. Grub population in turf areas with pH levels in Ohio soils. J. Econ. Entomol. 53, 860–863.
Potter, D.A., 1981. Seasonal emergence and flight of northern and southern masked chafers in relation to air and soil temperature and rainfall patterns. Environ. Entomol. 10, 793–797.
Potter, D.A., 1998. Destructive Turfgrass Insects: Biology, Diagnosis and Control. Ann Arbor Press, Chelsea, MI.
Potter, D.A., Powell, A.J., Spicer, P.G., Williams, D.W., 1996. Cultural practices affect root-feeding white grubs (Coleoptera: Scarabaeidae) in turfgrass. J. Econ. Entomol. 89, 156–164.
Preisser, E.L., Dugaw, C.J., Dennis, B., Strong, D.R., 2006. Plant facilitation of a below-ground predator. Ecology 87, 1116–1123.
Prestidge, R.A., Van Der Zijpp, S., Badan, D., 1985. Effects of plant species and fertilizers on grass grub larvae, *Costelytra zealandica*. New Zeal. J. Agr. Res. 28, 409–417.
Quinn, M.A., Hall, M.H., 1992. Compensatory response of a legume root-nodule system to nodule herbivory by *Sitona hispidulus*. Entomol. Exp. Appl. 64, 167–176.

Quinn, M.A., Hower, A.A., 1986. Effects of root nodules and taproots on survival and abundance of *Sitona hispidulus* (Coleoptera, Curculionidae) on *Medicago sativa*. Ecol. Entomol. 11, 391–400.

Radcliffe, J.E., 1970. Some effects of grass grub (*Costelytra zealandica* (White)) larvae on pasture plants. New Zeal. J. Agr. Res. 13, 87–104.

Rasmann, S., Köllner, T.G., Degenhardt, J., Hiltpold, I., Toepfer, S., Kuhlmann, U., Gershenzen, J., Turlings, T.C.J., 2005. Recruitment of entomopathogenic nematodes by insect-damaged maize roots. Nature 434, 732–737.

Régnière, J., Brooks, W.M., 1978. Entomogenous microorganisms associated with the japanese beetle, *Popillia japonica*, in eastern North Carolina. J. Invertebr. Pathol. 32, 226.

Régnière, J., Rabb, R.L., Stinner, R.E., 1979. *Popillia japonica* (Coleoptera: Scarabaeidae): a mathematical model of oviposition in heterogeneous agroecosystems. Can. Entomol. 111, 1271–1280.

Régnière, J., Rabb, R.L., Stinner, R.E., 1981. *Popillia japonica* (Coleoptera, Scarabaeidae)—effect of soil-moisture and texture on survival and development of eggs and first instar grubs. Environ. Entomol. 10, 654–660.

Rehner, S.A., Buckley, E., 2005. A *Beauveria* phylogeny inferred from nuclear ITS and EF1-α sequences: evidence for cryptic diversification and links to *Cordyceps* teleomorphs. Mycologia 97, 84–98.

Reinecke, A., Mueller, F., Hilker, M., 2008. Attractiveness of CO_2 released by root respiration fades on the background of root exudates. Basic Appl. Ecol. 9, 568–576.

Robert, C.A., Erb, M., Duployer, M., Zwahlen, C., Doyen, G.R., Turlings, T.C., 2012. Herbivore-induced plant volatiles mediate host selection by a root herbivore. New Phytol. 194, 1061–1069.

Roberts, D.W., Humber, R.A., 1981. Entomogenous fungi. In: Cole, G.T., Kenrick, B. (Eds.), The Biology of Conidial Fungi. Academic Press, New York, NY, pp. 201–236.

Robertson, L.N., 1987. Food habits of pasture wireworm, *Conoderus exsul* (Coleoptera: Elateridae). New Zeal. J. Zool. 14, 535–542.

Rogers, C.D., 2011. Ecological and molecular investigation of wheat bulb fly (*Delia coarctata*, Fallén, Diptera: Anthomyiidae) for the advancement of population monitoring and control methodologies. PhD Thesis, University of Edinburgh, UK.

Rogers, S., McCoy, C.W., Graham, J.H., 2000. Larval growth of *Diaprepes abbreviatus* (Coleoptera: Curculionidae) and resulting root injury to three citrus varieties in two soil types. J. Econ. Entomol. 93, 380–387.

Rowell, D.L., 1988. Soil acidity and alkalinity. In: Wild, A. (Ed.), Russell's Soil Conditions and Plant Growth. Longman, Harlow, UK, pp. 844–898.

Russell, E.W., 1973. Soil Conditions and Plant Growth. Longman, London, UK.

Russell, E.W., 1988. Russell's Soil Conditions and Plant Growth. Longman Scientific and Technical, Harlow, UK.

Salisbury, E.J., 1939. Ecological aspects of meteorology. Q. J. Roy. Meteor. Soc. 65, 337–358.

Salt, D.T., Fenwick, P., Whittaker, J.B., 1996. Interspecific herbivore interactions in a high CO_2 environment: root and shoot aphids feeding on *Cardamine*. Oikos 77, 326–330.

Sanderson, M.A., Elwinger, G.F., 2002. Plant density and environment effects on orchardgrass—white clover mixtures. Crop Sci. 42, 2055–2063.

Sardans, J., Peñuelas, J., 2007. Drought changes phosphorus and potassium accumulation patterns in an evergreen Mediterranean forest. Funct. Ecol. 21, 191–201.

Schenk, H.J., Jackson, R.B., 2002. Rooting depths, lateral root spreads and below-ground/above-ground allometries of plants in water-limited ecosystems. J. Ecol. 90, 480–494.

Schmidt, E.L., 1979. Initiation of plant root-microbe interactions. Annu. Rev. Microbiol. 33, 355–376.

Schneider, F., 1962. Dispersal and migration. Annu. Rev. Entomol. 7, 223–242.

Schofield, R.K., 1940. Note on the freezing of soil. Q. J. Roy. Meteor. Soc. 66, 167–170.
Schoonhoven, L.M., Van Loon, J.J.A., Dicke, M., 2005. Insect-Plant Biology. Oxford University Press, Oxford, UK.
Schrag, S.J., Mittler, J.E., 1996. Host-parasite coexistence: the role of spatial refuges in stabilizing bacteria-phage interactions. Am. Nat. 148, 348–377.
Scriber, J.M., 1979. Effects of leaf-water supplementation upon post-ingestive nutritional indices of forb-, shrub-, vine-, and tree-feeding lepidoptera. Entomol. Exp. Appl. 25, 240–252.
Scriber, J.M., 1984. Host-plant suitability. In: Bell, W., Cardé, R.T. (Eds.), Chemical Ecology of Insects. Chapman and Hall, London, UK, pp. 159–202.
Seastedt, T.R., Hayes, D.C., Petersen, N.J. 1986. Effects of vegetation, burning and mowing on soil macroarthropods of tallgrass prairie. In: Proceedings of the Ninth North American Prairie Conference. Tri-College University Center for Environmental Studies, Fargo, ND, USA, pp.99–102.
Serraj, R., Sinclair, T.R., Allen, L.H., 1998. Soybean nodulation and N_2 fixation response to drought under carbon dioxide enrichment. Plant Cell Environ. 21, 491–500.
Short, D.E., Luedtke, R.J., 1970. Larval migration of the western corn rootworm. J. Econ. Entomol. 63, 325–326.
Sijm, D., Kraaij, R., Belfroid, A., 2000. Bioavailability in soil or sediment: exposure of different organisms and approaches to study it. Environ. Pollut. 108, 113–119.
Simelane, D.O., 2007. Influence of soil texture, moisture, and surface cracks on the performance of a root-feeding flea beetle, *Longitarsus bethae* (Coleoptera: Chrysomelidae), a biological control agent for *Lantana camara* (Verbenaceae). Environ. Entomol. 36, 512–517.
Slansky, F., 1993. Nutritional ecology: the fundamental quest for nutrients. In: Stamp, N.E., Casey, T.M. (Eds.), Caterpillars: Ecological and Evolutionary Constraints on Foraging. Chapman and Hall, New York.
Smith, G.E.P., Kinnison, A.F., Carns, A.G., 1931. Irrigation investigation in young grapefruit orchards on the Yuma Mesa. Arizona Agric. Exper. Sta. Tech. Bull. 37, 413–591.
Söndgerath, D., Müller-Pietralla, W., 1996. A model for the development of the cabbage root fly (*Delia radicum* L.) based on the extended Leslie model. Ecol. Model. 91, 67–76.
Spencer, W.F., Cliath, M.M., Farmer, W.J., 1969. Vapor density of soil-applied dieldrin as related to soil-water content, temperature, and dieldrin concentration. Soil Sci. Soc. Am. J. 33, 509–511.
Spike, B.P., Tollefson, J.J., 1988. Western corn rootworm (Coleoptera, Chrysomelidae) larval survival and damage potential to corn subjected to nitrogen and plant-density treatments. J. Econ. Entomol. 81, 1450–1455.
Staley, J.T., Johnson, S.N., 2008. Climate change impacts on root herbivores. In: Johnson, S.N., Murray, P.J. (Eds.), Root Feeders—An Ecosystem Perspective. CABI, Wallingford, UK, pp. 192–213.
Stanhill, G., 1966. Diffuse sky and cloud radiation in Israel. Sol. Energy 10, 96–101.
Stewart, K.M., Vantoor, R.F., Crosbie, S.F., 1988. Control of grass grub (Coleoptera: Scarabaeidae) with rollers of different design. New Zeal. J. Exp. Agric. 16, 141–150.
Streeter, J.G., 2003. Effects of drought on nitrogen fixation in soybean root nodules. Plant Cell Environ. 26, 1199–1204.
Strnad, S.P., Bergman, M.K., 1987a. Distribution and orientation of western corn-rootworm (Coleoptera, Chrysomelidae) larvae in corn roots. Environ. Entomol. 16, 1193–1198.
Strnad, S.P., Bergman, M.K., 1987b. Movement of first-instar western corn rootworms (Coleoptera: Chrysomelidae) in soil. Environ. Entomol. 16, 975–978.
Strnad, S.P., Bergman, M.K., Fulton, W.C., 1986. First instar western corn rootworm (Coleoptera, Chrysomelidae) response to carbon dioxide. Environ. Entomol. 15, 839–842.

Strong, D.R., Maron, J.L., Connors, P.G., Whipple, A., Harrison, S., Jefferies, R.L., 1995. High mortality, fluctuation in numbers, and heavy subterranean insect herbivory in bush lupine, *Lupinus arboreus*. Oecologia 104, 85–92.

Strong, D.R., Kaya, H.K., Whipple, A.V., Child, A.L., Kraig, S., Bondonno, M., Dyer, K., Maron, J.L., 1996. Entomopathogenic nematodes: natural enemies of root-feeding caterpillars on bush lupine. Oecologia 108, 167–173.

Strong, D.R., Whipple, A.V., Child, A.L., Dennis, B., 1999. Model selection for a subterranean trophic cascade: root-feeding caterpillars and entomopathogenic nematodes. Ecology 80, 2750–2761.

Suttle, P.J., Musick, G.J., Fairchild, M.L., 1967. Study of larval migration of western corn rootworm. J. Econ. Entomol. 60, 1226–1228.

Tang, C.S., Cai, W.F., Kohl, K., Nishimoto, R.K., 1995. Allelopathy: organisms, processes, and applications. In: Plant Stress and Allelopathy. ACS Symp. Series. American Chemical Society, Washington DC, pp. 142–157.

Thorpe, W.H., Crombie, A.C., Hill, R., Darrah, J.H., 1946. The behaviour of wireworms in response to chemical stimulation. J. Exp. Biol. 23, 234–266.

Tongma, S., Kobayashi, K., Usui, K., 2001. Allelopathic activity of Mexican sunflower [*Tithonia diversifolia* (Hemsl.) A. Gray] in soil under natural field conditions and different moisture conditions. Weed Biol. Manag. 1, 115–119.

Traynier, R.M.M., 1967. Stimulation of oviposition by the cabbage root fly *Erioischia brassicae*. Entomol. Exp. Appl. 10, 401–412.

Turlings, T.C., Hiltpold, I., Rasmann, S., 2012. The importance of root-produced volatiles as foraging cues for entomopathogenic nematodes. Plant Soil 358, 51–60.

Turpin, F.T., Peters, D.C., 1971. Survival of Southern and Western corn rootworm larvae in relation to soil texture. J. Econ. Entomol. 64, 1448–1451.

van Dam, N.M., 2009. Belowground herbivory and plant defenses. Annu. Rev. Ecol. Syst. 40, 373–391.

van Elsas, J.D., Trevors, J.T., Van Overbeek, L.S., 1991. Influence of soil properties on the vertical movement of genetically-marked *Pseudomonas fluorescens* through large soil microcosms. Biol. Fert. Soils 10, 249–255.

van Herk, W.G., Vernon, R.S., 2013. Wireworm damage to wheat seedlings: effect of temperature and wireworm state. J. Pest. Sci. 86, 63–75.

van Straalen, N.M., Verhoef, H.A., 1997. The development of a bioindicator system for soil acidity based on arthropod pH preferences. J. Appl. Ecol. 34, 217–232.

van Wijk, W.R., Larson, W.E., Burrows, W.C., 1959. Soil temperature and the early growth of corn from mulched and unmulched soil. Soil Sci. Soc. Am. J. 23, 428–434.

Vega, F.E., Goettel, M.S., Blackwell, M., Chandler, D., Jackson, M.A., Keller, S., Koike, M., Maniania, N.K., Monzón, A., Ownley, B.H., 2009. Fungal entomopathogens: new insights on their ecology. Fungal Ecol. 2, 149–159.

Veihmeyer, F.J., Hendrickson, A.H., 1950. Soil moisture in relation to plant growth. Annu. Rev. Plant Physiol. 1, 285–304.

Verhoef, H.A., Witteveen, J., 1980. Water balance in Collembola and its relation to habitat selection; cuticular water loss and water uptake. J. Insect Physiol. 26, 201–208.

Villani, M.G., Gould, F., 1986. Use of radiographs for movement analysis of the corn wireworm, *Melanotus communis* (Coleoptera, Elateridae). Environ. Entomol. 15, 462–464.

Villani, M.G., Wright, R.J., 1988. Use of radiography in behavioral studies of turfgrass-infesting scarab grub species (Coleoptera: Scarabaeidae). Faculty Publications: Paper 110. Department of Entomology, University of Nebraska, Lincoln, USA.

Villani, M.G., Wright, R.J., 1990. Environmental influences on soil macroarthropod behavior in agricultural systems. Annu. Rev. Entomol. 35, 249–269.

Villani, M.G., Allee, L.L., Diaz, A., Robbins, P.S., 1999. Adaptive strategies of edaphic arthropods. Annu. Rev. Entomol. 44, 233–256.

Vittum, P.J., 1984. Effect of lime application on Japanese beetle (Coleoptera, Scarabaeidae) grub populations in Massachusetts soils. J. Econ. Entomol. 77, 687–690.
Vittum, P.J., Morzuch, B.J., 1990. Effect of soil pH on Japanese beetle (Coleoptera, Scarabaeidae) oviposition in potted turfgrass. J. Econ. Entomol. 83, 2036–2039.
Vittum, P.J., Tashiro, H., 1980. Effect of soil pH on survival of Japanese beetle (Coleoptera, Scarabaeidae) and European chafer larvae. J. Econ. Entomol. 73, 577–579.
Vittum, P.J., Villani, M.G., Tashiro, H., 1987. Turfgrass Insects of the United States and Canada. Ann Arbor Press, Chelsea, MA.
von Städler, E., 1971. The orientation and host plant selection of *Psilae rosae* F. (Diptera: Psilidae). Z. Angew. Entomol. 69, 425–438.
Wade, P., 1954. Soil fumigation. I. The sorption of ethylene dibromide by soils. J. Sci. Food Agr. 5, 184–192.
Wade, P., 1955. Soil fumigation. III. The sorption of ethylene dibromide by soils at low moisture contents. J. Sci. Food Agr. 6, 1–3.
Way, M.O., Reay-Jones, F.P.F., Stout, M.J., Tarpley, L., 2006. Effects of nitrogen fertilizer applied before permanent flood on the interaction between rice and rice water weevil (Coleoptera: Curculionidae). J. Econ. Entomol. 99, 2030–2037.
Weibull, J., 1987. Seasonal changes in the free amino acids of oat and barley phloem sap in relation to plant growth stage and growth of *Rhopalosiphum padi*. Ann. Appl. Biol. 111, 729–737.
Wessel, R.D., Polivka, J.B., 1952. Soil pH in relation to Japanese beetle populations. J. Econ. Entomol. 45, 733–735.
Wharton, G.W., 1985. Water balance of insects. In: Kerkut, G.A., Gibert, L.I. (Eds.), Comprehensive Insect Physiology, Biochemistry and Pharmacology. Pergamon Press, Oxford, UK, pp. 565–601.
White, T.C.R., 1993. The Inadequate Environment: Nitrogen and the Abundance of Animals. Springer-Verlag, Heidelberg, Germany.
Whitehead, D.C., 1983. The influence of frequent defoliation and of drought on nitrogen and sulfur in the roots of perennial ryegrass and white clover. Ann. Bot. 52, 931–934.
Wigglesworth, V.B., 1956. The physiology of insect cuticle. Annu. Rev. Entomol. 2, 37–54.
Wightman, J.A., 1974. Rearing *Costelytra zealandica* (Coleoptera: Scarabaeidae) 4. Some effects of different larval densities and food availability on larval survival and weight change. New Zeal. J. Zool. 1, 217–223.
Winston, P.W., Beament, J.W.L., 1969. An active reduction of water level in insect cuticle. J. Exp. Biol. 50, 541–546.
Wolcott, A.R. 1970. Retention of pesticides by organic materials in soils. In: Pesticides in the Soil; Ecology, Degradation and Movement: Papers at an International Symposium Organised by Michigan State University, East Lansing.
Wolfson, J.L., 1987. Impact of rhizobium nodules on *Sitona hispidulus*, the clover root curculio. Entomol. Exp. Appl. 43, 237–243.
Wu, K., Li, M., 1992. Nutritional ecology of the cotton bollworm, *Heliothis armigera* (Hubner): effects of dietary sugar concentrations on development and reproduction. Acta Entomol. Sin. 35, 47–52.
Xu, Q., Huang, B., 2000a. Effects of differential air and soil temperature on carbohydrate metabolism in creeping bentgrass. Crop Sci. 40, 1368–1374.
Xu, Q., Huang, B., 2000b. Growth and physiological responses of creeping bentgrass to changes in air and soil temperatures. Crop Sci. 40, 1363–1368.
Zachariassen, K.E., 1991. Routes of transpiratory water loss in a dry habitat Tenebrionid beetle. J. Exp. Biol. 157, 425–437.
Zacharuk, R.Y., 1962. Seasonal behavior of larvae of *Ctenicera* spp. and other wireworms (Coleoptera: Elateridae), in relation to temperature, moisture, food, and gravity. Can. J. Zool. 40, 697–718.

Zalucki, M.P., Brower, L.P., 1992. Survival of first instar larvae of *Danaus plexippus* (Lepidoptera: Danainae) in relation to cardiac glycoside and latex content of *Asclepias humistrata* (Asclepiadaceae). Chemoecology 3, 81–93.

Zangerl, A., Bazzaz, F., 1992. Theory and pattern in plant defense allocation. In: Fritz, R.S., Simms, E.L. (Eds.), Plant Resistance to Herbivores and Pathogens. University of Chicago Press, Chicago, IL, pp. 363–391.

Zavala, J.A., Nabity, P.D., DeLucia, E.H., 2013. An emerging understanding of mechanisms governing insect herbivory under elevated CO_2. Annu. Rev. Entomol. 58, 79–97.

Zhang, M., Crocker, R.L., Mankin, R.W., Flanders, K.L., Brandhorst-Hubbard, J.L., 2003. Acoustic identification and measurement of activity patterns of white grubs in soil. J. Econ. Entomol. 96, 1704–1710.

Zhang, X., Johnson, S.N., Gregory, P.J., Crawford, J.W., Young, I.M., Murray, P.J., Jarvis, S.C., 2006. Modelling the movement and survival of the root-feeding clover weevil, *Sitona lepidus*, in the root-zone of white clover. Ecol. Model. 190, 133–146.

Zinkler, D., 1966. Vergleichende untersuchungen zur Atmungsphysiologie von Collembolen (Apterygota) und anderen Bodenkleinarthropoden. Z. Vergl. Physiol. 52, 99–144.

CHAPTER TWO

The Role of Plant Primary and Secondary Metabolites in Root-Herbivore Behaviour, Nutrition and Physiology

Matthias Erb[*,†], Meret Huber[*,‡], Christelle A.M. Robert[*,‡], Abigail P. Ferrieri[*,†], Ricardo A.R. Machado[*,†], Carla C.M. Arce[*,†,§]

[*]Root-Herbivore Interactions Group, Max Planck Institute for Chemical Ecology, Jena, Germany
[†]Department of Molecular Ecology, Max Planck Institute for Chemical Ecology, Jena, Germany
[‡]Department of Biochemistry, Max Planck Institute for Chemical Ecology, Jena, Germany
[§]Departamento de Entomologia, Universidade Federal de Viçosa, Viçosa, Minas Gerais, Brazil

Contents

Abstract

Many insect herbivores feed on belowground plant tissues. In this chapter, we discuss how they have adapted to deal with root primary and secondary metabolites. It is becoming evident that root herbivores can use root volatiles and exudates for host location and foraging. Their complex sensory apparatus suggests a sophisticated recognition and signal transduction system. Furthermore, endogenous metabolites trigger attractive or repellent responses in root feeders, indicating that they may specifically fine-tune food uptake to meet their dietary needs. Little evidence for direct toxic effects of root secondary metabolites has accumulated so far, indicating high prevalence of tolerance mechanisms. Root herbivores furthermore facilitate the entry of soil microbes into the roots, which may influence root nutritional quality. Investigating the role of plant metabolites in an ecologically and physiologically relevant context will be crucial to refine our current models on root-herbivore physiology and behaviour in the future.

Advances in Insect Physiology, Volume 45
ISSN 0065-2806
http://dx.doi.org/10.1016/B978-0-12-417165-7.00002-7

1. INTRODUCTION

The ability to take up nutrients and at the same time cope with harmful chemicals is essential for all life on Earth, including insect herbivores. Over evolutionary time, they have evolved strategies to exploit specific feeding niches of their host plants. One of the most remarkable environmental adaptations in this context is their ability to feed on belowground plant organs during the larval stage and then switch to the consumption of leaf material as adults. Members of at least seven insect orders, including many important agricultural pests, spend a part of their life cycle as belowground feeders (Blossey and Hunt-Joshi, 2003; Hunter, 2001).

Although feeding on roots comes with advantages, including protection from harsh environmental conditions, predators and competitors, it also poses a number of major challenges. Host finding and foraging, for example, is complicated by the absence of visual cues and the increased energy required for moving through the soil matrix (Fig. 2.1). Growth is constrained by the low nutritional value and unique blend of bioactive secondary compounds of roots (Kaplan et al., 2008a; Vaughan et al., 2013). Finally, there are many microorganisms, including beneficial, opportunistic and pathogenic fungi and bacteria (Burrows et al., 2009; Van Der Heijden et al., 2008), that need to be dealt with within the soil.

For all these reasons, the capacity of insects to develop on roots is expected to depend on a number of specific metabolic and behavioural adaptations. Studying these processes has the potential to facilitate our understanding of insect physiological processes and may also unravel novel ways to control some of the most devastating agricultural pests. Significant progress in understanding belowground plant–herbivore interactions has been made, and patterns are emerging that indicate general mechanisms of root-herbivore adaptation. For instance, while CO_2 has long been seen as the single most important plant volatile that root feeders use to locate host plants, it is now becoming clear that the perception of root volatiles is much more sophisticated and enables root-feeding larvae to make fine-tuned foraging decisions. Furthermore, it is becoming evident that secondary metabolites not only serve defensive purposes but also can be used by root feeders for host recognition and foraging. Specific host manipulation and detoxification strategies are being unravelled, and we are beginning to understand how closely root herbivores interact with the rhizosphere microbiome.

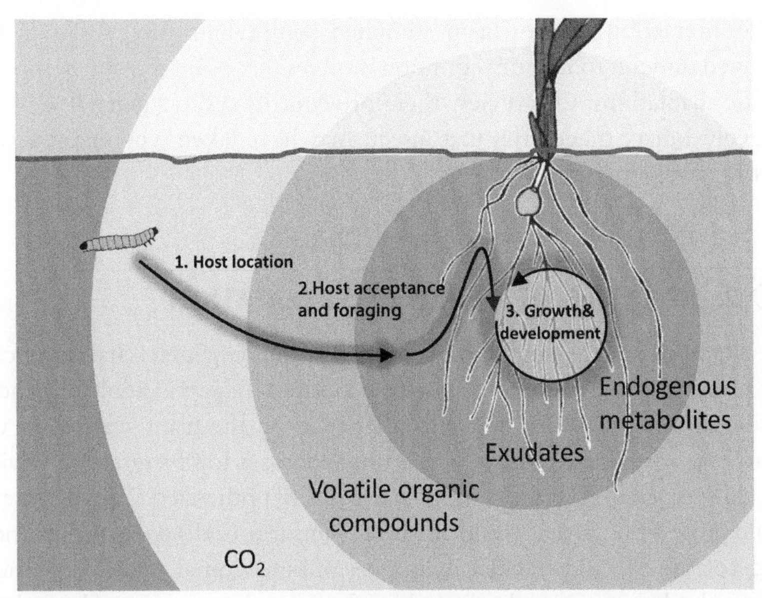

Figure 2.1 *Host location, foraging and development of root herbivores in the soil.* The spatial scale of different plant metabolites is depicted. CO_2 and other small-molecular-weight compound volatiles can diffuse several centimetres in the soil and can be used by root herbivores to locate host plants. Soluble root exudates may also be used for host location and foraging and may be required for host acceptance in direct proximity to the emitting roots. Endogenous primary and secondary metabolites determine the growth and development of the attacking herbivore but may also be used for foraging decisions. (For colour version of this figure, the reader is referred to the online version of this chapter.)

In this chapter, we bring together recent and traditional findings on the role of plant metabolites in root-herbivore interactions. First, we review how larvae can use volatile and nonvolatile cues to find the roots of their host plants. Second, we discuss how primary and secondary metabolites influence host acceptance and foraging. Third, we discuss how the metabolic make-up of the roots determines herbivore growth, development and fitness. Fourth, we review the emerging evidence for an important role of soil microorganisms in determining root-herbivore interactions.

By compiling the literature on these subjects, we hope to complement other recent reviews on belowground herbivory and plant defences (Rasmann and Agrawal, 2008; van Dam, 2009), chemically mediated host plant location by root feeders (Johnson and Gregory, 2006) and root chemistry (Kaplan et al., 2008a). As the current state of the art regarding specific host plant adaptations that shape root-herbivore behaviour at different stages of the interaction has not been compiled comprehensively, so far, it has remained difficult to identify common motives that point to root-herbivore specific adaptations. Our review therefore contributes to a general view on the evolutionary trajectories that insects may have taken when conquering the soil environment.

2. VOLATILE-MEDIATED HOST LOCATION

In contrast to leaf herbivores, many root-feeding insects have to locate host plants upon hatching, as eggs are deposited by gravid adults into adequate oviposition sites in the soil rather than on the plant surface. In the absence of any visual stimuli, larvae must locate a food source by relying on tactile or chemical cues, and it has been hypothesized that the insects should possess the capacity to locate host plants from a distance, for instance, using volatile or exudate cues. A number of behavioural studies document that root herbivores are indeed capable of sensing the presence of host plants long before physical contact. Using X-ray tomography, Johnson et al. (2004) demonstrated that neonates of the clover root weevil (*Sitona lepidus*) can recognize the presence of plant roots in the soil at a distance of up to 6 cm (Johnson et al., 2004). Similar observations have been made for the western corn rootworm (*Diabrotica virgifera virgifera*) (Robert et al., 2012a) and many other root-feeding insects (Johnson and Gregory, 2006).

One signal that elicits a strong behavioural response in the majority of root-feeding insects is carbon dioxide (CO_2). Recent reviews on the role of root volatiles as host location cues compiled over 20 studies that document a major role of CO_2 in long-distance host location (Johnson and Gregory, 2006; Johnson and Nielsen, 2012). The carrot root fly (*Psila rosae*), for instance, was found to be attracted to CO_2, but not to essential oils released from carrot seeds (Städler, 1971). Similarly, prairie grain wireworms (*Ctenicera destructor*) moved towards a variety of CO_2 sources, an effect that was eliminated by passing the airstream over a KOH solution that eliminated CO_2 from the system (Doane et al., 1975; Vaughan et al., 2013). Bernklau and Bjostad (2008) found that the attractiveness of maize extracts to the western corn rootworm can be matched and even overcome by increasing doses of CO_2.

Despite the clear potential of CO_2 to attract a wide variety of root-feeding insects, doubts have been raised about its reliability as a chemotactic signal. First, it has been argued that the soil environment contains a wide variety of CO_2 sources, including degrading organic matter and patches of high microbial turnover (Agus et al., 2010). Second, all plant roots release CO_2, including not only host plants but also toxic nonhost species, rendering the signal unreliable to a foraging insect. Third, CO_2 emission from plant roots shows strong diurnal variation (Hansen, 1977) and mostly vertical concentration gradients (Pline and Dusenbery, 1987), which diminishes its importance as a signal for horizontal root location behaviour (Johnson and Gregory, 2006). Fourth, the integration of nonvolatile signals in an ecologically relevant concentration has been found to abolish the relative attractiveness of CO_2 for some herbivores (Reinecke et al., 2008).

An increasing number of studies document that root feeders can use signals other than CO_2 for host plant location from a distance (Fig. 2.2; Chapter 3). The large pine weevil (*Hylobius abietis*), for instance, was found to be attracted to α-pinene (Nordenhem and Nordlander, 1994), a

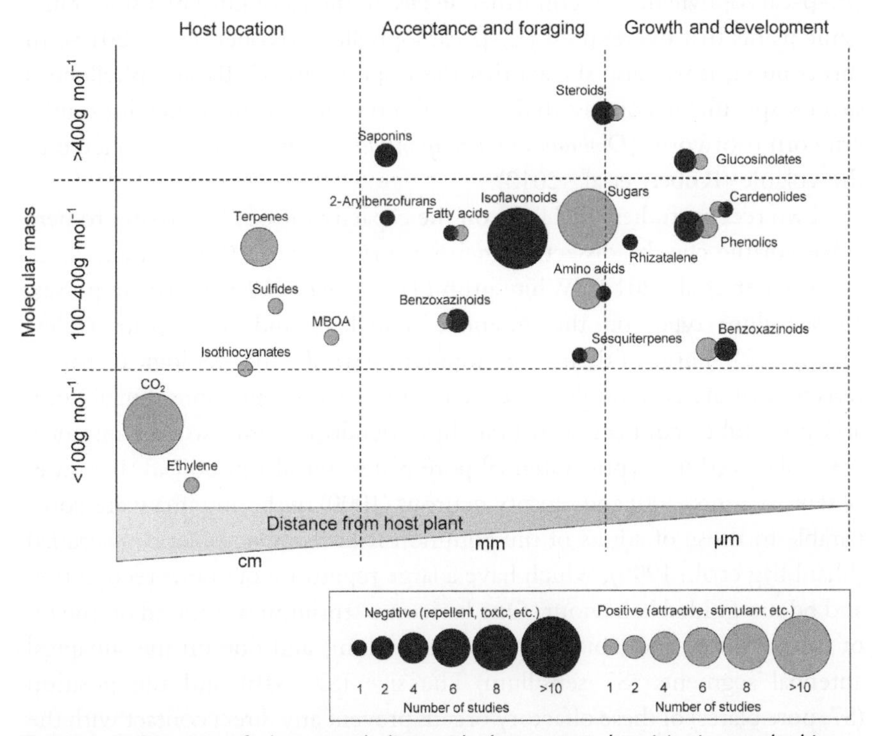

Figure 2.2 *Overview of plant metabolites with demonstrated activity in root-herbivore interactions.* Purple bubbles show negative effects on herbivores and green bubbles stand for positive effects. The size of the bubbles is proportional to the number of studies documenting the effect. (For interpretation of the references to colour in this figure legend, the reader is referred to the online version of this chapter.)

monoterpene emitted by its host plant. The cabbage root fly (*Delia radicum*) is attracted to isothiocyanates, a class of breakdown products of glucosinolates produced by cabbage plants and other crucifers (Koštál, 1992). Equally, the onion root fly (*Delia antiqua*) was found to orient towards sulphides emitted from onion bulbs (Soni and Finch, 1979), and the damage-induced oak volatiles eucalyptol and anisole were found to attract forest cockchafer (*Melolontha hippocastani*) larvae at concentrations below 5 ppb. Also, the western corn rootworm, despite its strong responsiveness to CO_2, has been shown to use other volatile cues. The benzoxazinoid breakdown product 6-methoxy-2-benzoxazolinone (MBOA), for instance, was found to attract western corn rootworm larvae (Bjostad and Hibbard, 1992). The sesquiterpene (*E*)-β-caryophyllene and the hormone ethylene were found to be used by this root feeder as well to distinguish leaf- and root-infested maize plants from uninfested individuals (Robert et al., 2012a). The attractive effect of (*E*)-β-caryophyllene was confirmed *in vivo* in the laboratory by using transgenic plants that overexpress (*E*)-β-caryophyllene (Robert et al., 2013). In this context, it was also shown that the response to (*E*)-β-caryophyllene is species-specific: in contrast to the specialist root feeder, the generalist southern corn rootworm (*Diabrotica undecimpunctata howardi*) was not attracted to the volatile (Robert et al., 2013).

Two recent studies shed light on the capacity of root herbivores to perceive volatile cues by dissecting their sensory apparatus (Eilers et al., 2012; Weissteiner et al., 2012). White grub (*Melolontha melolontha*) larvae possess 17 sensillum types on the antennae, maxillary and labial palps (Eilers et al., 2012). Three of them were found to have the morphological characteristics that are commonly associated with olfactory functions. Small pores and pore tubes connected to branching dendritic segments, for instance, were observed in cryptic antennal pore plates (sensilla placodea). Densities of sensory units (300) and sensory neurons (1000) in the antenna were comparable to those of adults of the common fruit fly (*Drosophila melanogaster*) (Shanbhag et al., 1999), which have a large repertoire of odour recognition and odour-guided behaviour. One basiconic sensillum was found on the tip of the maxillary and labial palps (S10 sensillum) and one on the subapical antennal segments (S7 sensillum).The size (S7, S10) and the position (S7, pore plates) of these olfactory organs prevent any direct contact with the substrate, and they can therefore only be stimulated by volatile compounds. By combining behavioural and electrophysiological assays, the authors could provide evidence that the sensilla placodea of the antennae are involved in the perception of different volatiles, including CO_2. Furthermore, based on the facts that (i) some volatiles elicited responses in the palps and that (ii) palpographic signals are unlikely to be picked up from a single neuron, the authors suggested that gustatory sensilla may have a dual function in olfaction and taste. In the European forest cockchafer (*M. hippocastani*), the antennal lobe, the first brain centre to process olfactory input, was found to contain about 70 glomeruli (Weissteiner et al., 2012). The high number of glomeruli and the large antennal surface covered by antennal pore plates points to a sophisticated olfactory system in this root herbivore as well. Taken together, these two physiological studies clearly support the notion that root-herbivore olfaction is complex and is likely to go way beyond the simple detection of and response to CO_2. Evolutionary aspects of insect olfaction are discussed in detail by Hiltpold in Chapter 3.

From an ecological perspective, several explanations may help to resolve the dispute on the relative importance of CO_2 and volatile organic compounds as root host location cues. First, it was suggested that specialist herbivores may use host-specific volatile organic compounds, while generalists may be more likely to follow CO_2 as a universal cue. However, while this may be true for certain specialists (Nordenhem and Nordlander, 1994; Soni and Finch, 1979), the current literature contains many examples that go against this general assumption. Cockchafer larvae, for example, are polyphagous, yet able to respond to many cues other than CO_2 (Eilers et al., 2012; Weissteiner et al., 2012). The highly monophagous western corn rootworm on the other hand is strongly attracted to CO_2 (Bernklau and Bjostad, 1998), a finding that is also true for other root feeders with a restricted host range. Therefore, little support exists to support a generalist specialist dichotomy. As an alternative hypothesis, based on the observation that most root feeders are attracted to both CO_2 and volatile organic compounds, an integrated model was proposed by Johnson and Gregory (2006). This model proposes that CO_2 is used primarily as a long-distance cue that elicits searching activity, while VOCs are employed at a shorter range to distinguish host plant species and their different quality. Testing this model will require tight spatial control of volatile concentrations in the soil coupled with detailed behavioural observations.

Here, we propose a third, not mutually exclusive hypothesis that may reconcile CO_2 and VOC-mediated attraction. It is based on the fact that respiration, and hence CO_2 release, is an unavoidable consequence of root metabolic activity and that any emission of root VOCs will invariably be accompanied by an elevated CO_2 background. We propose that CO_2 can be used as a background odour by root feeders, which they integrate with other volatile cues to optimize their searching behaviour. In the absence of any other cues, which is an unlikely situation in the rhizosphere, root herbivores will use CO_2 as an attractant, while in the absence of CO_2, VOCs would be less likely to trigger searching behaviour. This hypothesis is supported by research on aboveground perception and attraction of insects to plant odours (Schröder and Hilker, 2008), where CO_2 was found to enhance the attraction of the common fruit fly (*D. melanogaster*) to apple cider vinegar (Faucher et al., 2013). Also, many of the earlier-mentioned root-herbivore studies are compatible with the notion that CO_2 acts synergistically with other volatiles to attract belowground insects. The western corn rootworm, for instance, is attracted to the sesquiterpene (*E*)-β-caryophyllene only when an appropriate plant background is provided (Robert et al., 2012a but see Chapter 3). The finding that removing CO_2 from a volatile blend reduces wireworm (*Ctenicera destructor*) attraction also speaks in favour of CO_2 as a necessary background odour source (Doane et al., 1975). Apart from the studies conducted with the western corn rootworm, little information is available that allows testing of the CO_2 background hypothesis (but see Turlings et al., 2012), and more detailed experiments are clearly required to explore volatile-mediated host location belowground in more detail.

3. HOST SELECTION VIA SOLUBLE ROOT EXUDATES

Upon successful localization of a host plant, insects have to make a decision to start feeding or to move on to search for another root system to feed on. As roots are constantly exuding large amounts of organic carbon and other soluble metabolites into the rhizosphere, it can be expected that 'tasting exudates' provides a means for root herbivores to assess host plant susceptibility without having to chew through the root epidermis (Box 2.1).

Although direct evidence for the utilization of soluble exudates in host selection processes is scarce, several studies provide indirect support for this concept. Wireworms (*Agriotes* spp.), for instance, are attracted to malic acid (Thorpe et al., 1947), which is exuded by roots to mobilize phosphate (Hoffland et al., 1992). Western corn rootworm feeding is stimulated by MBOA (Bjostad and Hibbard, 1992), a partially soluble breakdown product of benzoxazinoids that is released by maize roots (Robert et al., 2012c). The western corn rootworm has also been shown to be repelled by the exuded precursor molecule 2,4-dihydroxy-7-methoxy-1,4-benzoxazin-3-one

BOX 2.1 Volatile vs. soluble root exudates: A blurred dichotomy

Contrary to the phyllosphere, the distinction between volatile and soluble exudates is much less clear in the rhizosphere. Many typical volatile compounds, for instance, are likely to diffuse through the soil at the interface between the soluble and vapour phase. The sesquiterpene (*E*)-β-caryophyllene, for instance, diffuses better at 5% soil humidity than lower or higher levels (Hiltpold and Turlings, 2008). On the other hand, soluble compounds may behave like volatiles, as illustrated by the soluble benzoxazinoid breakdown product 6-methoxy-2-benzoxazolinone (MBOA), which can be perceived as a volatile by the western corn rootworm (Bjostad and Hibbard, 1992). The situation with root herbivores is further complicated by the finding that the gustatory sensilla of white grubs (*M. melolontha*) may be able to perceive both gaseous and soluble molecules (Eilers et al., 2012). From a chemical point of view, volatile organic compounds are roughly defined as organic chemicals with high vapour pressure and low boiling points. As, for most investigated chemicals, no information is available on their state of aggregation in the soil, we rely on boiling points and the type of bioassay (i.e. olfactometer vs. contact experiments) to distinguish volatile vs. soluble exudates in this chapter.

(DIMBOA) (Xie et al., 1992a), even though this finding was later contested by the fact that the herbivore actually prefers to feed on maize mutants that produce higher levels of benzoxazinoids (Robert et al., 2012c). Also, formononetin, an isoflavonoid exuded from root nodules (Mathesius, 2001), was found to be attractive to the clover root weevil (*S. lepidus*) at concentrations of 0.01 M (Johnson et al., 2005). A noteworthy interaction between CO_2 and soluble cues was observed in a study involving the European cockchafer (*M. melolontha*) and the common dandelion (*Taraxacum officinale*) (Reinecke et al., 2008): while synthetic CO_2 attracted the root herbivore, adding *T. officinale* or an aqueous rhizosphere extract caused this effect to disappear, suggesting that exudate cues are integrated by the herbivore to yield a specific behavioural response.

As most experiments on the behavioural activity of root exudates have been conducted using root extracts or purified compounds, it remains unclear whether soluble signals are used for short-distance host location or whether they are actively involved in host-acceptance processes and small-scale foraging decisions. Formononetin exudation, for example, may be a reliable indicator of active nodulation and may actively guide foraging decisions by the clover root weevil (*S. lepidus*) (Johnson et al., 2005). Similarly, high benzoxazinoid exudation is associated with crown roots of maize plants (*Zea mays*), which are more nutritious than other root types (Robert et al., 2012c) and may therefore be used as reliable foraging signals. To conclusively demonstrate the role of phytochemicals in exudate-mediated host selection and foraging, classical pharmacological approaches may be difficult to implement as the simulation of exudation patterns is challenging. A promising alternative research strategy may be the use of mutants or transgenic plants with altered levels of exudate metabolites. This approach has been used in thale cress (*Arabidopsis thaliana*) and maize to demonstrate the role of malic acid and benzoxazinoids in the recruitment of beneficial microbes (Neal et al., 2012; Rudrappa et al., 2008). However, one drawback of this approach is that many exudate metabolites also accumulate in the roots themselves, and changing their biosynthesis may alter other aspects of the root metabolic make-up. A more targeted approach would therefore be to alter the expression of the transport proteins that are responsible for the release of the chemicals from the cells into the apoplast and rhizosphere. The first transporters involved in the exudation of specific plant metabolites, including an ABC transporter that is responsible for the release of strigolactones, have been identified in petunia (*Petunia hybrida*) (Kretzschmar et al., 2012) and thale cress (Badri et al., 2009, 2012) and may provide a starting point for such an approach in root-herbivore model systems.

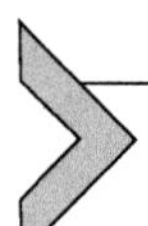

4. THE ROLE OF ENDOGENOUS PRIMARY METABOLITES IN HOST ACCEPTANCE AND FORAGING

Apart from sensing root exudates, root herbivores can assess the quality of their host plant by biting into the roots and directly tasting their food (Johnson and Gregory, 2006; Johnson et al., 2011; van Dam, 2009; Watts et al., 2011). This ability is essential to any heterotrophic organism that needs to achieve a balance of nutrients for successful development (Dethier, 1973; Schoonhoven, 1974) and may be particularly important for root feeders that feed from low-quality food matrices. Screening compounds for their potential phagostimulatory or deterrent effects on belowground herbivores generally involves measuring biting responses towards or the ingestion of artificial substrate treated with the chemical of interest. This approach was pioneered by Thorpe et al. (1947) and Crombie and Darrah (1947) who pretreated filter paper discs with chemical solutions that stimulated biting responses of the wireworms *Agriotes lineatus*, *A. sputator* and *A. obscurus*. Despite being developed over 50 years ago, similar bioassays are still used today to understand behavioural aspects of important agricultural pests, including the Pacific coast wireworms (*Limonius canus*) (Horton et al., 2012), turnip root flies (*D. floralis*) (Hopkins et al., 1993) and the western corn rootworm (Bernklau et al., 2011).

Approximately 80% of the primary metabolites analysed with the above bioassay have been classified as 'phagostimulants' due to their ability to increase consumption rates by herbivores when applied to artificial diet. In general, low-molecular-weight compounds, such as alcohols, esters and aldehydes, appear to have attractive properties, while larger hydrocarbons are considered to be repellent. Despite this overarching pattern, other studies suggest that biochemical relatedness may be more important than functional groups in determining attractiveness or repellency (Finch and Skinner, 1974; Mochizuki et al., 1989; Soni and Finch, 1979; Weissteiner and Schütz, 2006). Here, we discuss the effects of three major classes of essential elements, endogenous sugars, amino acids and fatty acids, to illustrate the specific behavioural patterns that primary metabolites induce in root-feeding insects.

Sugars account for nearly half of all phagostimulatory chemicals identified to date (Johnson and Nielsen, 2012) and have been shown to stimulate feeding by a number of root-feeding insects (Johnson and Gregory, 2006). An early survey by Sutherland (1971) found that 10 out of 13 sugars could enhance biting and feeding responses of grass grub larvae (*Costelytra zealandica*) feeding on treated agar-cellulose discs, with sucrose eliciting the strongest response. Similar observations have been made for pruinose scarabs (*Sericesthis geminata*) (Wensler and Dudzinski, 1972), the cane beetle (*Antitrogus parvulus*), the negatoria canegrub (*Lepidiota negatoria*) (Allsopp, 1992) and Japanese beetles (*Popillia japonica*) (Ladd, 1988). Grass grub larvae (*C. zealandica*) feed on a wide variety of plant species (Radcliffe, 1970), which may explain their broad responsiveness to sugars. In general, studies with oligophagous insects find a more restricted range of carbohydrates to be effective phagostimulants (Augustine et al., 1964; Beck, 1957; Gothilf and Beck, 1967; Heron, 1965; Hsiao and Fraenkel, 1968; Ito, 1960; Tsuneyama et al., 2005). An investigation of feeding responses by Japanese beetles (*P. japonica*) found that three disaccharides, sucrose, maltose and trehalose, and two hexoses, fructose and glucose, were strong feeding stimulants for third-instar larvae feeding at concentrations of 0.1 M (Ladd, 1988). All of these sugars except trehalose also elicited feeding responses in adult beetles (Ladd, 1986), suggesting that chemoreception is retained during metamorphosis. Artificial diet experiments by Sutherland and Hillier (1976) on the other hand found that the root-feeding African black beetle larvae (*Heteronychus arator*) respond most strongly to maltose while adult beetles, which also feed at the base of plants, exhibit positive feeding responses to many different sugars, including maltose, glucose and fructose (Sutherland and Hillier, 1976), indicating that sugar perception may change over the course of insect development as the relation to the part of the plant or host may also vary. These results, and later studies by Sutherland (1983) who found that 0.1 M maltose induces twice as much feeding by black beetle larvae compared to responses elicited by several amino acids and inorganic salt mixtures, strongly suggest that soluble sugars play a prominent role in stimulating feeding by root herbivores.

Sugars have been shown to act synergistically with other metabolites in eliciting feeding behaviour by above (Bartlet et al., 1994; Hsiao and Fraenkel, 1968; Shanks and Doss, 1987)- and belowground phytophagous insects (Crombie and Darrah, 1947). An early study by Crombie and Darrah (1947) found that while individual solutions of 0.5% glucose and 0.126% sucrose failed to elicit significant changes to orientation or biting response of wireworms (*Agriotes* spp.) compared to a control solution, combining these two sugars increased the number of bites taken in 24 h by approximately 70% when compared to the number of bites elicited by each sugar alone. Similar responses have been observed for wireworms (*Agriotes* spp.) presented with mixtures of glucose and the peptone, triolein or tannins (Crombie and Darrah, 1947). Sucrose was also found to act synergistically with L-ascorbic acid to induce feeding responses in grass grubs (Sutherland and Hillier, 1973), more than doubling the number of faecal pellets produced by third-instar larvae during 24 h exposure compared to larvae feeding on sucrose or L-ascorbic acid alone. In contrast, bioassays by Horton et al. (2012) found that presenting larvae of Pacific coast wireworms (*L. canus*) with sucrose in combination with plant essential oils including winter savoury, tea tree and cedar wood may lead to additive, synergistic and antagonistic effects on feeding behaviour, depending on the additive. The addition of a stimulant blend consisting glucose, fructose, sucrose and linoleic acid (Bernklau and Bjostad, 2008) overcame the normally repellent effects of thiamethoxam insecticide against western corn rootworm larvae (Bernklau et al., 2011). In both 30 min and 4 h bioassays, approximately 60% of larvae stayed on discs treated with thiamethoxam and the feeding stimulant blend for every concentration of the insecticide tested, compared to 5–35% of larvae that stayed when the feeding stimulant was absent (Bernklau et al., 2011). Interestingly, the presence of the stimulant blend was also found to improve the efficacy of thiamethoxam against rootworm larvae; in 4 h bioassays, the concentration of thiamethoxam required for 50% mortality was 2800 pg/mL for the insecticide alone but was reduced by more than 100,000-fold to 0.0075 pg/mL when the feeding stimulant blend was added. The ability of sugars to mask deterrent effects of noxious compounds has also been demonstrated for above ground herbivores (Cocco and Glendinning, 2012; Shields and Mitchell, 1995), highlighting the potential similarities that may exist between chemically mediated host plant foraging above- and belowground.

Proteins and free amino acids constitute major nutrients for insects, as they are needed to supply nitrogen in a digestible form. Furthermore, up to 10 amino acids (arginine, isoleucine, leucine, lysine, histidine, methionine, phenylalanine, threonine, tryptophan and valine) are essential for insects (Dadd, 1977). Therefore, an insect's performance directly depends on its ability to detect and feed on plants that contain these necessary amino acids. Most amino acids in plants occur in the form of proteins. Until today, little evidence has accumulated that insects can detect and taste proteins, with the exception of those that contain terminal sequences that resemble polypeptide neurotransmitters (Chapman, 2003). To our knowledge, no study to date has investigated protein sensing and its effect on root feeders' host acceptance and foraging, but several studies highlight the role of amino acids in this context. Sutherland and Hillier (1973) tested the individual effect of 20 amino acids, at 0.01 and 0.1 M, on the feeding behaviour of the grass grub larvae. After 24 h of starvation, the larvae were fed an agar/cellulose medium containing one of the 20 ubiquitous amino acids. The number of faecal pellets after 24 h of feeding was recorded as a parameter of feeding. Six amino acids induced higher number of faecal pellets at a concentration of 0.1 M: L-alanine, L-serine, L-threonine, L-aspartic acid and L-glutamine (Sutherland and Hillier, 1973). Only L-aspartic acid and L-glutamic acid enhanced the feeding at 0.01 M. Using the same methodology, Sutherland (1983) was also able to demonstrate a stimulating effect for 0.1 M L-alanine, 0.01 M L-aspartic acid and 0.01 M L-glutamic acid on black beetle larvae (*H. arator*) (Sutherland, 1983). Allsopp (1992) tested the effect of 18 amino acids at 0.01 M concentrations in cellulose acetate discs on the foraging of the negatoria canegrub (*L. negatoria*). By assessing the percentage of the disc eaten by the larvae, the author found that L-cysteine, L-glutamic acid and L-histidine monohydrochloride stimulated the feeding of the larvae (Allsopp, 1992). In the same study, the author demonstrated that only L-tyrosine (0.01 M) stimulated feeding by larvae of the cane beetle (Allsopp, 1992). L-alanine as well as two of the essential amino acids,

L-isoleucine and L-leucine, on filter papers stimulated the feeding behaviour of the pruinose scarab at low concentrations, while they inhibited food intake at high concentrations (Wensler and Dudzinski, 1972). Clearly, many root-feeding insects have the ability to perceive and respond to the presence of amino acids. Yet, it remains unclear why only a specific subset of them elicits behavioural activity. We speculate here that in general, the perception of free amino acids may enable root feeders to assess host plant quality or root identity. The western corn rootworm, for instance, develops better on crown than primary roots of maize. Crown roots contain both higher protein and free amino acid levels, and amino acids could therefore be used as a proxy for the nutritional quality of the plants (Robert et al., 2012c).

Fatty acids such as linoleic acid and oleic acid were reported to act as phagostimulants for root herbivores as well. Their role was discovered following the finding that the liquid pressed from germinating maize roots contains phagostimulants for the larvae of the western corn rootworm (Bernklau and Bjostad, 2005). An ethyl acetate and an aqueous fraction of the extract elicited feeding together but not individually, suggesting that both polar and nonpolar compounds are required for stimulation (Bernklau and Bjostad, 2008). Subsequent bioassays with sugars, amino acids, diacids and free fatty acids demonstrated that the stimulatory activity of a sugar mix (glucose/fructose/sucrose; 30:4:4 mg/mL) was strongly enhanced by applying linoleic (3 mg/mL) or oleic acid (0.3 mg/mL) (Bernklau and Bjostad, 2008). It is interesting to note that the same ecologically relevant concentrations of fatty acids lead to 70% of mortality when tested alone (Bernklau and Bjostad, 2008). This is consistent with other studies reporting insecticidal effects of both fatty acids against aboveground insects (Guang et al., 1991; Harada et al., 2000; Ramsewak et al., 2001; Yang et al., 1992). The mechanism behind this *in vitro* phenomenon remains to be determined.

Taken together, the earlier evidence demonstrates that root-feeding herbivores can sense primary metabolites, including sugars, amino acids and fatty acids, as a means of assessing host plant quality. Carbohydrates are shown to act primarily as phagostimulants (Chapter 3). To our knowledge, no sugar has been shown to elicit deterrent effects on root-feeding herbivores. The mechanisms by which sugars are perceived by root feeders and how or why this perception changes over the course of an insect's development warrant further study. It appears from the literature that the affinity for amino acid classes is species-specific. For example, L-alanine, L-isoleucine and L-leucine, which are all amino acids with hydrophobic side chains, were shown to stimulate feeding by pruinose scarabs (Wensler and Dudzinski, 1972), but not of the negatoria canegrub (Allsopp, 1992). Given that roots of different species differ in their primary metabolome, one might expect that root herbivores use them to distinguish suitable from unsuitable host plants and to differentiate between parts of the root system (e.g. root nodules and root meristems) as they take exploratory bites. While the current bioassays enable measurement of the influence of individual compounds on root-herbivore behaviour, they are not sufficient to simulate natural situations. The fact that physiological levels of fatty acids lead to excessive mortality of the highly specialized western corn rootworm *in vitro*, for instance, illustrates this point (Bernklau and Bjostad, 2008). An alternative approach may be the use of mutants and transgenic plants. The fatty acid dehydrogenase (fad) mutants of thale cress, for instance, have significantly altered linoleic acid levels (Okuley et al., 1994). However, in many cases, this approach is not without problems due to pleiotropic effects associated with perturbing the biosynthesis of primary metabolites (Yang et al., 2006).

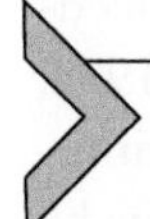

5. THE ROLE OF ENDOGENOUS SECONDARY METABOLITES IN HOST ACCEPTANCE AND FORAGING

Plant roots produce a diverse array of endogenous secondary compounds that trigger behavioural responses in root-feeding insects. Sutherland et al. (1980), for instance, found that grass grubs were deterred by isoflavonoids and other phytoalexins that are commonly found in legumes. Artificial diets containing pisatin, phaseolin, medicarpin, maackiain, vestitol, phaseollinisoflavan and 2′-methoxyphaseollinisoflavan were specifically avoided by these larvae. Also, African black beetle (*H. arator*) larvae were deterred by the flavonoids phaseolin, medicarpin, maackiain, vestitol, coumestrol, genistein and biochanin (Sutherland et al., 1980). Phaseolin, phaseollinisoflavan, medicarpin, vestitol, maackiain and 2′-methoxyphaseollinisoflavan have been found in legume roots at concentrations ranging from 1 to 6 μg/g FW (Gaynor et al., 1986; Liu et al., 1995; Russell et al., 1978; Stevenson et al., 1997; Sutherland et al., 1980) and pisatin at concentrations between 40 and 500 μg/g (Cannesan et al., 2011; Christenson and Hadwiger, 1973). Bioassays with artificial diets containing these flavonoids suggest that concentrations commonly found in roots (between 30 ng/g and 55 μg/g) are sufficient to reduce grass grub feeding by 50% (FD50) (Lane et al., 1985, 1987; Russell et al., 1978).

Lane et al. (1987) investigated the biological activity of root-derived isoflavonoids of lupin legumes (*Lupinus angustifolius*). Of the nine isoflavones that were isolated and tested, four showed significant deterrent properties against grass grubs at concentrations of 2 μg/g: licoisoflavone A, licoisoflavone B, 2-hydroxygenistein and luteone.

Apart from flavonoids, a number of other secondary metabolites have been demonstrated to deter root herbivores. Two-arylbenzofurans contained in the roots of the common sainfoin (*Onobrychis viciifolia*), for instance, have been shown to deter grass grubs at a physiological concentration of 2 μg/g (Russell et al., 1984). Similarly, saponins (Sutherland et al., 1982) 20-hydroxyecdysone (20E) (Schmelz et al., 2002) and the benzoxazinoid DIMBOA (Xie et al., 1990) were shown to exhibit a strong antifeedant activity against root herbivores *in vitro*.

In contrast to the classical phytochemical studies that focused on repellent compounds, a number of studies demonstrate that endogenous secondary metabolites may in fact stimulate feeding by belowground herbivores. As discussed in the previous sections, benzoxazinoids, which are highly concentrated in maize roots (up to several mg/g FW), are an attractant and feeding stimulant for the western corn rootworm (Bjostad and Hibbard, 1992; Robert et al., 2012c), and isoflavonoids found in white clover roots stimulate the clover root weevil (*S. lepidus*) (Johnson et al., 2005). Also, feeding by the spotted cucumber beetle was found to be stimulated by bitter and toxic cucurbitacins (De Heer and Tallamy, 1991). Furthermore, as breakdown products of glucosinolates are attractive for the cabbage root fly (Koštál, 1992), it is possible that they are also acting as feeding stimulants, similarly to what has been observed for crucifer specialists aboveground (Nielsen et al., 1979). The currently available studies suggest that feeding stimulation by secondary metabolites occurs in root herbivores with a restricted host range, and it is conceivable that host plant specialization determines the degree of behavioural adaptations to secondary compounds as host recognition signals.

Just as with the investigated primary metabolites, two major issues remain to be resolved in order to understand the role of endogenous secondary metabolites in determining root-herbivore foraging and behaviour. First, the ecological relevance of the behavioural assay needs to be improved. A majority of tests were conducted in artificial substrates, and only few studies have attempted to observe root-herbivore behaviour in a natural soil context. Given the structural and chemical composition of the soil, it is well possible that the full behavioural range that root herbivores display will only become apparent once they are exposed to the behaviourally active compounds in such an environment. Pioneering work in that respect comes from Johnson et al. (2007). The authors used X-ray tomography to track insect movement in a soil matrix. Although no detailed activity tests with plant metabolites were carried out in this study, it would in theory be possible to adapt this system to test individual root chemicals for behavioural activity. Second, the role of plant metabolites must be assessed *in planta*. Most behavioural assays have been conducted with individual compounds, neglecting the fact that endogenous metabolites occur within a complex plant matrix consisting not only of many different metabolites but also of enzymes that modify the chemical structure of compounds upon tissue disruption. Plant–insect interactions have benefitted tremendously from the availability of molecular and genetic plant resources, including mutants and transgenic plants with an altered metabolism that enable investigation of the importance of phytochemicals in their natural context (Jander and Howe, 2008). Unfortunately, research on root-herbivore interactions has not taken full advantage of these tools yet, possibly because the major molecular plant models are not typically attacked by root-feeding insects. Exceptions include the use of transgenic and mutant maize to investigate the role of root phytochemicals (Degenhardt et al., 2009) as well as studies in the thale cress with the opportunistic root-feeding fungus gnat (*Bradysia* spp.) (Vaughan et al., 2013).

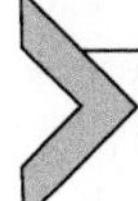

6. FOOD-QUALITY ASPECTS OF ROOT-HERBIVORE INTERACTIONS I: PRIMARY METABOLITES

Apart from its effects on herbivore foraging and host acceptance, the metabolic make-up of the roots directly determines how well the attacking insect will grow and develop. The major dietary factors for insects in general include nitrogen in the form of proteins or amino acids, carbohydrates, lipids, sterols, minerals and vitamins (Awmack and Leather, 2002). As discussed in a previous section, root herbivores respond strongly to many dietary constituents, illustrating how important nutrient foraging is for belowground plant feeders. In this section, we review some of the general nutritional challenges that insects face when feeding on the roots (nitrogen, carbohydrates and fatty acids) and discuss possible adaptations that herbivores may have developed during their transition to root-feeding behaviour.

Traditionally, nitrogen (N) has been regarded as the major growth-limiting factor for insect herbivores. This notion is based on the inherent imbalance in N concentrations between insects and their host plants (Slansky and Feeny, 1977). Depending on species, developmental and ontogenetic stage, N content per dry mass ranges between 0.03% and 7% for plants and between 7% and 14% for phytophagous insects (Mattson, 1980). Although roots of some plants contain lower N levels than leaves (Hol et al., 2003; Murray et al., 1996), a phenomenon that has been proposed as a general pattern (van Dam, 2009), this does not seem to hold true across plant species. Red oaks (*Quercus rubra*), for instance, have similar N levels in the leaves and roots (2.3%) (Frost and Hunter, 2008), and the spotted knapweed (*Centaurea maculosa*) contains about 4% N in both roots and shoots (Newingham et al., 2007). Interestingly, however, attack of the spotted knapweed (*Centaurea maculosa*) by the root-feeding larvae of the sulphur knapweed moth (*Agapeta zoegana*) leads to a dramatic reduction of root N levels, which again may enforce N limitation for the herbivore (Newingham et al., 2007). That root-feeding insects need to forage actively for N is indicated by behavioural studies (see sections earlier). The western corn rootworm was found to feed preferentially on maize crown roots, which contain higher protein and free amino acid levels (Robert et al., 2012c). And the first-instar clover weevil larva feeds preferentially on nodules (Bigger, 1930; Hackell and Gerard, 2004), which have pronouncedly higher N levels than the rest of the root system (Murray et al., 1996). Western corn rootworm attack was also found to increase free amino acid levels in maize roots (Robert et al., 2012b). Whether root herbivores manipulate the metabolism of their host plant to acquire additional N remains to be determined. Even though N in the form of proteins and free amino acids may be just as limiting for root-herbivore growth as for leaf herbivores, few studies have specifically investigated the role of this element belowground. Way et al. (2006) treated maize plants with different levels of N fertilizer ranging from 34 to 202 kg N/ha and recorded larval and pupal densities in different plots (Way et al., 2006). Overall, higher doses of N increased larval densities of the rice water weevil (*Lissorhoptrus oryzophilus*). However, as N fertilization also increases adult feeding and oviposition (Jiang and Cheng, 2003), it is unclear whether this effect is due to improved larval nutrition or an increase in oviposition by the leaf-feeding adults. Overall, we hypothesize that adaptations of root feeders to N as a limiting factor should mirror the strategies that leaf-feeding insects typically use, including foraging for N-rich tissues (Kimmerer and Potter, 1987), compensatory feeding (Awmack and Leather, 2002) and the acquisition of nutritional symbioses (Douglas, 2003).

As the primary source of energy, carbohydrates are an indispensable part in insect nutrition. In particular, the ratio between proteins and carbohydrates determines insect growth (Lee et al., 2002). Starch, mono- and disaccharides can be expected to play a particularly important role in this context, as they are much easier to break down and digest than more complex molecules like cellulose. Strikingly, the concentrations of accessible carbohydrates are often lower in the roots than the leaves of plants. In the wild tobacco (*Nicotiana attenuata*), for instance, the total concentration of sugars and starch is 50% lower in the roots (Machado et al., 2013). Similar patterns have been observed in bean (*Phaseolus vulgaris*) (Cakmak et al., 1994), maize (Braun et al., 2006; Robert et al., 2012c), thale cress (Zeeman and Rees, 1999) and strawberry (*Fragaria* × *ananassa*) (Eshghi et al., 2007). In contrast, five kiwi species (*Actinidia* spp.) exhibited no consistent allocation pattern, but substantial seasonal variation with a depletion of root carbohydrates towards the middle of the growing season and replenishment at the end (Boldingh et al., 2000). A simple physiological explanation for this pattern may be that carbon is assimilated and primarily stored in the leaves and only transported to the roots as a source of energy for growth, development and storage. In some perennial plants, carbohydrates can be stored in belowground organs for overwintering: Starch concentrations in potato tubers, for instance, can reach 15% fresh weight. While belowground feeders that specialize on these types of plants (Rondon, 2010) are unlikely to be carbohydrate-limited, it is possible that root feeders that live on plants that do not store carbon suffer from low sugar and starch levels of their host tissues. Again, the strong stimulatory effect of many sugars on root feeders (as discussed earlier) may serve as an argument that supports this hypothesis. Unfortunately, while the multidimensional nutritional spaces, including optimal carbon/nitrogen ratios, are well understood for leaf feeders (Joern and Behmer, 1997), very little is known for root herbivores in this respect.

We hypothesize here that root herbivores may have developed several strategies to cope with low starch and sugar levels of their host plants. Firstly, they could efficiently metabolize cellulose either by microbial symbiosis or by the production of endogenous proteins. Secondly, root-feeding insects may actively mix diets and satisfy their carbohydrate needs by feeding on carbon-storage plants temporarily. Thirdly, they may manipulate the plant to increase root carbon allocation. Cellulose digestion is rare in insects having been detected in about 20 families in eight orders (Martin, 1983, 1991). Most cellulose digesters are xylophagous insects, notably the termites with a digestion efficiency reaching 99% (Esenther and Kirk, 1974), the larvae of siricid wood wasps and anobiid, buprestid and cerambycid beetles (Martin, 1991) and the omnivorous silverfish (*Ctenolepisma lineata*) (Lasker and Giese, 1956), firebrat (*Thermobia domestica*) (Treves and Martin, 1994; Zinkler et al., 1986) and several roaches (Wharton and Wharton, 1965). Interestingly, the root-feeding larvae of the European rhinoceros beetle (*Oryctes nasicornis*) have a cellulose-digesting efficiency reaching 68% (Rössler, 1961) and the Australian pruinose scarab larvae were reported to exhibit cellulase activity (Soo Hoo and Dudzinski, 1967). Notably, evidence for substantial cellulose digestion by leaf-feeding arthropods is lacking (Martin, 1991; Prins and Kreulen, 1991) although weak cellulase activity was detected in the desert locust (*Schistocerca gregaria*) (Evans and Payne, 1964), the migratory locust (*Locusta migratoria migratorioides*) (Morgan, 1975) and the Mexican leaf beetle (*Epilachna varivestis*) (Taylor, 1985). These findings support the hypothesis that insects have evolved digestive adaptations to reduce carbohydrate limitations imposed by their food source and it is possible that these are common among root-feeding insects. A second possibility is that root herbivores may feed on carbon-storage plants temporarily. Diet mixing is well documented for aboveground herbivores. Unfortunately, because observing root herbivores in natural plant communities is difficult, very little is known out of host-switching behaviour and nutrient uptake optimization belowground. A third possible adaptation is that root herbivores may benefit from herbivore-induced carbon allocation to the roots. Two studies have found higher sucrose concentrations in root-herbivore-attacked belowground plant organs (Pierre et al., 2012; Robert et al., 2012b). Furthermore, some root feeders induce water stress in their host plant (Erb et al., 2009; Masters et al., 1993), a condition that is known to increase carbon allocation belowground (Ibrahim et al., 1997). It is tempting to speculate that the induction of mild water stress may help root herbivores gain access to leaf carbohydrates by hijacking a plant reallocation response that is geared towards stabilizing water supply via an increase in root proliferation.

Apart from sugars, fatty acids can also be used to satisfy the energy requirements of an insect's metabolism. Certain fatty acids, including sterols (Friend, 1958; House, 1961) and polyunsaturated fatty acids (PUFAs) (Dadd, 1973), are essential for insect development. Linolenic, linoleic and oleic acid are often the most common PUFAs in plants. There has been a controversy in the past as to whether linolenic and linoleic acid can be used interchangeably in insect nutrition. Hoppers of the Acrididae family (*Locusta* spp.) had a similar growth and survival rate when fed on either linoleic or linolenic acid at 5 mg/g of diet (Dadd, 1961). However, it became increasingly clear that in some Lepidoptera and Hymenoptera, proper metamorphosis only took place when linolenic acid was added to artificial diets (Dadd, 1973): in the wax moth (*Galleria mellonella*), all emerging adults looked normal when reared with 5 mg linolenic acid per 6 g artificial diet, while proper adult formation dropped to one third when linolenic acid was substituted with linoleic acid. The major sources of PUFAs in the leaves are cell and chloroplast membranes. It is therefore not surprising that PUFA patterns in the roots differ substantially from the aboveground tissues. For instance, in eight out of the nine tested species of Fabaceae (*Astragalus* spp.), the concentration of linoleic acid was 30–80% higher in roots than shoots, while the concentration of linolenic acid was more than twofold lower in below- than aboveground tissue in six of these species (Keskin and Kacar, 2012). This shift from linolenic to linoleic acid was also observed in Solanaceae (*Capsicum* spp. and *Solanum* spp.) (Lyons and Lippert, 1966; Ouariti et al., 1997) and Poaceae (*Agrostis* cultivars) (Larkindale and Huang, 2004). Given that linolenic acid is essential for the development of many insects (Dadd, 1973), it is possible that root herbivores, at least in extreme cases, may be limited by the available root pools. Possible counter adaptations might include (i) a reduced demand of PUFAs as shown for the common fruit fly (Rapport et al., 1984), (ii) an improved exploitation of the available PUFAs, for example, by breaking down cell walls, (iii) endogenous production of PUFAs as shown for linoleic acid in the pea aphid (*Acyrthosiphon pisum*) (De Renobales et al., 1986) or (iv) association with beneficial microbes providing PUFAs. We are unaware of experimental evidence for any of these hypotheses for belowground feeding insects.

From the three examples earlier, it becomes clear that the nutritional composition of roots differs substantially from the leaves. The evolution of root-feeding behaviour was therefore likely accompanied by physiological adaptations that enabled insects to optimize growth and development on their new substrate (see Chapter 3). We propose that an experimental approach that combines evolutionary and physiological aspects may reveal interesting adaptive patterns that may improve our general understanding of insect nutrition and plasticity. In this context, it is noteworthy that many root herbivores only spend their larval stage belowground, while adults feed on foliage and flowers. It can, therefore, be expected that there is a major developmental shift in dietary physiology as the root feeders mature.

Comparing the dietary requirements of larval and adult stages within the same species may prove particularly useful to unravel adaptations associated with a root-feeding habit.

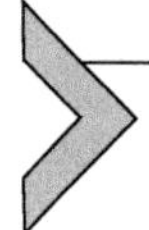

7. FOOD-QUALITY ASPECTS OF ROOT-HERBIVORE INTERACTIONS II: SECONDARY METABOLITES

The diversity of secondary metabolites in roots rivals the diversity in aboveground plant parts (Rasmann and Agrawal, 2008; van Dam, 2009), including many classes of secondary metabolites that reduce plant quality for herbivores such as alkaloids (Dawson, 1941), glucosinolates (Kaplan et al., 2008b), terpenoids (Malcolm, 1991), phenolics and benzoxazinoids (Niemeyer, 2009). It is therefore likely that, apart from influencing herbivore behaviour, these metabolites will also determine the nutritional value of the roots for root feeders, for instance, by reducing the efficacy of digestive enzymes (Houseman et al., 1992) or by interfering with essential functions in the insect body (Dobler et al., 2011). Although in many cases behavioural and nutritional effects are difficult to separate experimentally (Stout, 2013), it remains important to distinguish these two biological dimensions in order to understand plant–insect interactions from a mechanistic and evolutionary point of view. We therefore address behavioural and food-quality effects separately in this chapter. Compared to the situation aboveground, several rhizosphere-specific aspects need to be taken into account when studying root secondary metabolites in a nutritional context:

- Although roots produce the same classes of secondary metabolites as the leaves, their secondary metabolite blends differ both quantitatively and qualitatively. Some secondary compounds, for instance, are exclusively produced in the roots. Given that the activity of a metabolite is directly determined by its specific structure rather than by chemical classification (Mazoir et al., 2008), extrapolating data from leaf-herbivore interaction studies to the roots is problematic.
- The toxicity of secondary metabolites depends in many cases directly on their nutritional context. Nitrogen limitation, for instance, increases the toxicity of tannins (Raubenheimer, 1992). As the primary metabolome of the roots is different from the leaves, secondary metabolites may therefore have a different effect belowground than aboveground.
- Roots are colonized by a multitude of microorganisms that live at the root surface and in the extracellular spaces. Many microorganisms process plant secondary compounds (Neal et al., 2012), and root herbivores

may therefore encounter a cocktail of plant compounds that is partially digested by microorganisms.

- Roots are constantly wounded: Lateral roots and root hairs break through the epidermis of their mother roots and thereby directly destroy cells. Given that many secondary metabolites are only activated upon tissue rupturing, for instance, by deglycosylation (Hopkins et al., 2009; Rask et al., 2000), it can be expected that root feeders will encounter higher concentrations of preactivated secondary compounds.

Few examples from the literature clearly demonstrate negative effects of root secondary metabolites on the food quality of roots for belowground herbivores. Schmelz et al. (1998, 2002) showed that in spinach (*Spinacia oleracea*), the steroid insect moulting hormone homologue 20-hydroxyecdysone (20E) was induced after feeding by the larvae of the dark-winged fungus gnat (*Bradysia impatiens*), wounding and methyl jasmonate (MeJA) application. Apart from a repellent effect of 20E, the authors also found that larval survival dropped and prepupae formation increased dramatically when 20E was added in physiological concentrations to potato-based diets (Schmelz et al., 2002). Given the homology of 20E to insect moulting hormones, it is likely that this compound interferes with insect development directly and reduces the food quality of its host plant beyond its repellent effect. Cucurbitacins, another class of defensive steroids produced by the *Cucurbitaceae,* on the other hand, had no negative effects on the southern corn rootworm (Halaweish et al., 1999). On the contrary, the southern corn rootworm larvae even grew slightly better when feeding on a cucurbitacin-rich squash cultivar (Halaweish et al., 1999). It remains to be determined whether the stimulation in growth is due to feeding stimulation or a nutritional effect.

In a phylogenetically corrected correlation, total root cardenolide content of 18 milkweed species (*Asclepias* spp.) was negatively associated with milkweed beetle (*Tetraopes tetraophthalmus*) larval performance in the greenhouse (Rasmann and Agrawal, 2011). In line with these findings, the red milkweed beetle larvae survival rate was almost 3× higher on *A. syriaca* plants with low cardenolide contents compared to conspecifics with more than double the cardenolide concentration (Rasmann et al., 2011). As cardenolides inhibit the animal sodium–potassium pump that is vital for membrane potential maintenance, these metabolites likely decrease growth by directly interacting with the insect's physiology. In contrast to these results, the number of larvae of the generalist fungus gnats (*Bradysia* spp.), which spontaneously infested plants in the greenhouse, was not correlated to cardenolide content across 14 milkweed species (Vannette and

Rasmann, 2012). It remains to be determined whether the fungus gnats have evolved physiological mechanisms to cope with cardenolides as shown for aboveground herbivores (Dobler et al., 2012; Zhen et al., 2012) or whether specific cardenolides, rather than the total concentration, determines the performance of this opportunistic root herbivore.

In kale (*Brassica oleracea*), rape (*B. napus*) and swede (*B. napus*), Birch et al. (1992) found that the percentage of pupae formation of the turnip root fly differed between both species and genotypes. However, neither total nor the abundance of any specific glucosinolate was correlated with root fly performance. In both broccoli (*Brassica oleracea* subsp. *italica*) and turnip (*Brassica rapa* subsp. *rapa*), aliphatic and aromatic glucosinolates were not changed in roots upon cabbage fly infestation, while indolyl glucosinolate was upregulated two- to fourfold (Pierre et al., 2012). However, these changes in indolyl glucosinolate were not correlated with larval performance: in broccoli, pupation rate on infested plants was similar to control plants, while in turnip, pupae formation increased by approximately 10%. These results indicate that indolic glucosinolates do not affect the herbivore. Recently, van Leur et al. (2008) investigated cabbage fly larval performance on two natural chemotypes of bittercress (*Barbarea vulgaris*), which differed in the ratio of two major glucosinolate species glucobarbarin ((S)-2-hydroxy-2-phenylethyl-glucosinolate) and gluconasturtiin (2-phenylethyl-glucosinolate). In this greenhouse experiment, pupae mass of the cabbage fly larvae was higher on the cultivar with the low gluconasturtiin ratio. Interestingly, the low gluconasturtiin cultivar contained approximately 15% less total free amino acids and total sugars, indicating that the glucosinolate profile rather than the nutritional quality might account for the differences in larval performance. These results are in line with bioassays showing toxic effects of gluconasturtiin against nematodes (Potter et al., 1998, 2000). Apart from affecting insect behaviour, the breakdown products of glucosinolates are thought to react with amino and sulfhydryl groups of proteins (Kawakish et al., 1987), suggesting that some of the negative effects of glucosinolates are due to antibiosis. Taken together, antibiosis of glucosinolates towards root herbivores remains ambiguous and clearly deserves more attention.

The few studies that explored the role of phenolics on root-herbivore performance showed little toxic effects of these metabolites on belowground feeders. In a greenhouse experiment with 15 sugar cane clones (*Saccharum* spp.) differing in their soluble and cell-wall-bound phenolic composition, there was no correlation between the growth of the sugar cane white grub and the abundance of these compounds (Nutt et al., 2004). Unexpectedly, Johnson et al. (2011) found a positive correlation between growth and survival of the black vine weevil (*Otiorhynchus sulcatus*) and total phenolics in the roots of blackcurrant (*Ribes nigrum*). A similar pattern was observed with black vine weevil larvae (*O. sulcatus*) feeding on raspberry (*Rubus idaeus* cv. Glen Ample) (Clark et al., 2011). It is unclear whether this positive relationship is due to covarying beneficial metabolites or whether phenolics had a positive effect on the insects. Beneficial effects of phenolics were demonstrated in leaf-feeding insects (Bernays and Woodhead, 1982), assumingly by reducing the demand of amino acids in the sclerotized exoskeleton. However, specific phenolic compounds may still be active against root herbivores: for instance, in a comparison of two varieties of sweet potato (*Ipomoea batatas*), fewer larvae of the sweet potato weevil (*Cylas puncticollis*) emerged from a variety with high levels of octadecyl and hexadecyl esters of hydroxycinnamic acids than from a variety with low levels (Stevenson et al., 2009). Incorporation of hydroxycinnamic acids that occur in the latex, hexadecylcaffeic acid and hexadecyl-p-coumaric acid into artificial diet increased mortality of the weevil larvae in a dose-dependent manner, thereby providing indirect evidence for a defensive function of hydroxycinnamic acids against the sweet potato weevil. Cole (1987) reported a positive correlation between plant resistance to the lettuce root aphid (*Pemphigus bursarius*) and the concentration of isochlorogenic acid in lettuce (*Lactuca sativa*). Monitoring of feeding patterns revealed differences in the behaviour of the lettuce root aphid (*Pemphigus bursarius*) on resistant and susceptible lettuce varieties (Cole et al., 1993). In contrast, carrot (*Daucus carota* subsp. *sativus*) seedlings resistant to the carrot root fly (*Psila rosae*) contained lower concentrations of chlorogenic acid than susceptible plants (Cole, 1987). Taken together, the role of phenolics in belowground defence remains ambiguous. Investigations on specific phenolics rather than on total content could help to clarify the situation.

Early field studies in the 1990s suggested a positive correlation between the concentration of benzoxazinoids and resistance against the western corn rootworm in nine maize inbred lines, assumingly mediated by antibiosis (Assabgui et al., 1995). Similarly, there was a negative correlation between root DIMBOA content and the survival and performance of the western corn rootworm in seven inbred maize lines (Xie et al., 1992b). At the beginning of the twenty-first century, however, evidence accumulated that hydroxamic acids were not correlated with western corn rootworm resistance: after infestation of 19 maize lines with the western corn rootworm in the field, Abel et al. (2000) did not find differences in root damage between lines differing in benzoxazinoid concentration. Direct evidence for tolerance of the western corn rootworm against benzoxazinoids recently came from Robert et al. (2012c) by using a 1,4-benzoxazin-3-one-deficient maize mutant and its parental wild-type line. The authors showed that the western corn rootworm larvae grew equally well over 24 h on the two plant genotypes. In addition, the presence of benzoxazinoids was needed by the larvae to localize the most nutritious roots, which also contained the highest amount of benzoxazinoids. All in all, these studies show no negative effect of a prominent secondary chemical on its specialized root herbivore. It remains an open question whether tolerance of the western corn rootworm to benzoxazinoids is a recently evolved counteradaptation against high benzoxazinoid levels in maize. Benzoxazinoids are known to reduce the nutritional value of plants by interfering with insect digestive enzymes (Houseman et al., 1992), and it is, for instance, possible that the western corn rootworm is able to compensate for these inhibitory effects, either by overproducing digestive enzymes or by detoxifying the secondary compounds.

Terpenoids show contrasting effects on root feeders. Genetically modified maize lines that constitutively emit (*E*)-β-caryophyllene and α-humulene showed similar root damage in the field compared to its parental line (Robert et al., 2013). Accordingly, the western corn rootworm emergence rate in the field was the same in transformed and untransformed plants. Moreover, the beetle larvae grew equally well on these plants under controlled conditions (Robert et al., 2013) and emergence was not affected by addition of (*E*)-β-caryophyllene in the lab (Rasmann et al., 2005). In contrast, the southern corn rootworm had a lower emergence rate from the constitutively emitting maize plants in the field, possibly because of a higher abundance of natural enemies (Robert et al., 2013). Taken together, these results provide no evidence for a direct negative effect of sesquiterpenes on the western corn rootworm. Evidence for a direct defensive function of terpenoids in roots recently came from thale cress (Vaughan et al., 2013): silencing the production of the semivolatile diterpene rhizathalene rendered the plants more susceptible to opportunistic fungus gnats (*Bradysia* spp.). Plant biomass was lower in the silenced plants compared to the parental wild-type and T-DNA insertion control when infested with fungus gnats. The addition of rhizathalene to semiartificial diet reduced diet consumption in a dose-dependent manner, indicating that plant resistance was at least partly mediated by antixenosis. As no information about the prevalence of root herbivores in thale cress in nature is available, the ecological and evolutionary context of rhizathalene-mediated resistance remains to be determined.

Specialized leaf-feeding insects can sequester plant secondary metabolites to protect themselves against natural enemies (Opitz and Müller, 2009), and there is some evidence that root feeders are able to pursue similar strategies. The southern corn rootworm, for instance, can sequester cucurbitacins (Tallamy et al., 1998). The secondary metabolites protect larvae and eggs against the entomopathogenic fungus *Metarhizium anisopliae* (Tallamy et al., 1998). As discussed later, protection against soil microbes may be one of the most important advantages of toxin sequestration for root-feeding herbivores. How widespread this strategy is belowground remains to be determined.

In conclusion, roots produce a variety of secondary metabolites. Whereas the impact of such chemicals on food quality has been studied in detail in aboveground systems, there is surprisingly little evidence that they reduce the nutritional value of belowground tissues. This might partly be due to the small number of studies that explicitly tested this hypothesis. We found in total four metabolite classes that exhibit negative effects on belowground feeders, namely, phytoecdysteroids (Schmelz et al., 2002), cardenolides (Rasmann and Agrawal, 2011; Rasmann et al., 2011), hydroxycinnamic acids (Stevenson et al., 2009) and the diterpenoid rhizathalene (Vaughan et al., 2013). In contrast, we found no toxic or even beneficial effects of benzoxazinoids (Robert et al., 2012c) and total phenolics (Clark et al., 2011; Johnson et al., 2011). The role of glucosinolates remains ambiguous. Therefore, the long-standing assumption that secondary metabolites reduce food quality belowground remains largely untested and clearly deserves more attention. *In vitro* bioassays with pure compounds are only a first step to investigate the toxic function of a metabolite. As the toxicity of metabolites often depends on the surrounding matrix, an *in vivo* approach is necessary as well. In aboveground studies, two approaches have been widely used: on one hand, the abundance of a secondary metabolite between clones or species can be correlated to herbivore performance. This approach has also been employed belowground. However, many studies suffer from a low number of species or clones, thereby increasing the probability that covarying factors rather than the metabolites of interest are responsible for the observed pattern. On the other hand, genetic manipulation has successfully been employed in aboveground studies to verify the effect of secondary metabolites on herbivore performance. Notably, there have been only three studies using a mutant approach, providing much insight into the presence (Vaughan et al., 2013) and absence (Robert et al., 2012c, 2013) of toxic effects of secondary metabolites on root feeders. A combination of these two approaches could greatly help to elucidate the role of chemicals for herbivore defence. Furthermore, experimental or observational evidence that belowground herbivores drive the evolution of toxic metabolites in nature, as was recently provided in aboveground systems (Agrawal et al., 2012; Prasad et al., 2012; Züst et al., 2012), is lacking. The lack of such data is also attributable to the fact that most of the studies investigated belowground pest on agricultural plants. Much insight into the importance of secondary metabolites in root-herbivore interactions could be gained by investigating the interaction of natural root-herbivore systems.

8. MICROBIAL INTERACTIONS

One important factor that sets the rhizosphere apart from the phyllosphere is the diversity and abundance of microorganisms: One gram of soil can contain up to 10^{11} individual bacteria, 50,000 bacterial species and 200 m of fungal hyphae (Van Der Heijden et al., 2008). Compared to these high concentrations, atmospheric microbe counts are modest and at least several orders of magnitude lower (10^4 individual bacteria per m^3) (Burrows et al., 2009). In contrast to leaf feeders, root feeders are therefore constantly exposed to a variety of microbes that live around, on and in the roots and that may influence the interaction. In the context of plant metabolites, soil microbes may play an important role in the following ways (Fig. 2.3):

- Microbial colonization of the roots may change the plant primary and secondary metabolism and thereby alter the root nutritional value for insects.
- Soil microbes may be transferred to the plant during feeding and influence the nutritional value of the plants.
- Soil microbes ingested by the root feeders may influence the gut microflora and affect the digestion and uptake of plant metabolites.
- Soil microbes at the root surface may metabolize soluble and volatile root exudates and alter root-herbivore recognition and foraging patterns.

Note that nematodes are sometimes also considered as microbes and can have a strong influence on root-herbivore interactions. However, as a number of recent reviews cover their effects in detail, they will not be discussed here (Hiltpold and Turlings, 2012; Turlings et al., 2012).

Several studies demonstrate that mycorrhizal fungi influence the performance of root-feeding herbivores. Infection of dandelion and strawberry

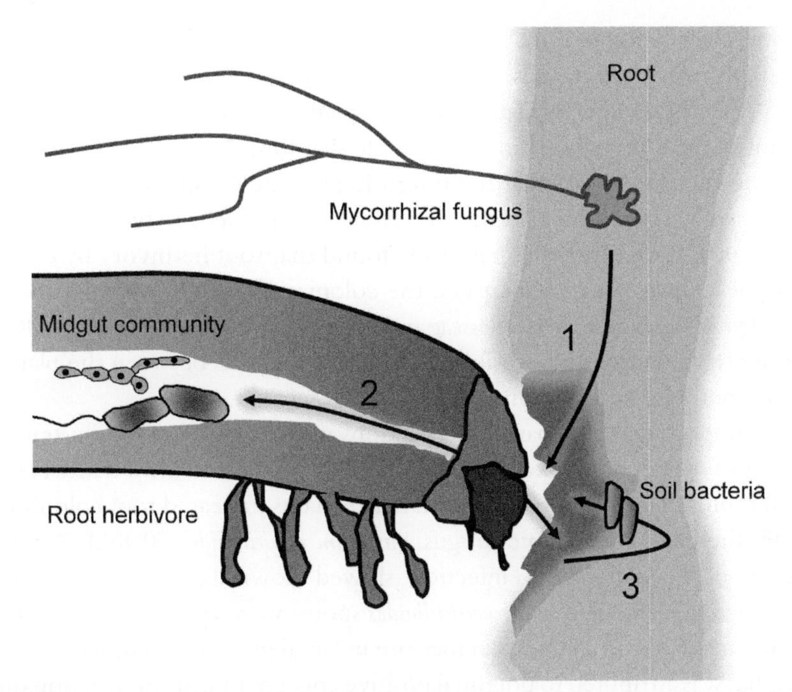

Figure 2.3 *Conceptual overview of the potential role of soil microbes in root-herbivore interactions.* 1. Soil microbes can colonize the plant and influence root herbivores by changing the plants metabolism or via direct interference. 2. Soil microbes in the gut of root herbivores may change the absorption and conversion of plant material. 3. Microbes may invade the wound sites of root-herbivore-attacked plants and thereby influence the interaction. (For colour version of this figure, the reader is referred to the online version of this chapter.)

with the mycorrhizal fungus *Glomus mosseae* reduced the survival of black vine weevil larvae (Gange et al., 1994). Arbuscular mycorrhizal colonization of milkweeds reduced the colonization of the plants by opportunistic fungus gnats (*Bradysia* spp.) (Vannette and Rasmann, 2012). A protective effect of ectomycorrhizae against the black vine weevil larvae was also observed for the Russian larch (*Larix sibirica*) in a 3-year field study (Halldórsson et al., 2000). It remains to be determined whether the protective effect of the fungi is due to changes in the plants metabolism. *Glomus intraradices* colonization, for instance, increased concentrations of hydroxycinnamic acid amides and cyclohexenone derivatives in barley (*Hordeum vulgare*) plants (Peipp et al., 1997), and *G. intraradices* enhanced concentrations of amino and fatty acids as well as isoflavonoids in the roots of the barrel clover (*Medicago truncatula*) (Schliemann et al., 2008). It is, therefore, possible that mycorrhizal fungi increase plant resistance by inducing the production of secondary compounds. However, in the absence of studies that go beyond the mere observation of herbivore performance, other explanations have to be taken into account as well. It is, for example, possible that fungal mycotoxins directly affect the root feeders or that the fungal hyphae alter root structure, morphology and exudation patterns in a way that influences their foraging behaviour. Interestingly, a recent study found that root herbivory by crane fly (*Tipula paludosa*) larvae increased the colonization of the common bent grass (*Agrostis capillaris*) by *G. mosseae* and *G. intraradices* (Currie et al., 2006), pointing to the intriguing possibility of feedback effects between the plant, the root herbivore and the mycorrhizal fungi.

The importance of tripartite interactions in the soil is further illustrated by series of studies conducted with the western corn rootworm. Kurtz et al. (2010) found that rootworm-attacked maize plants increased root colonization by the plant-pathogenic fungus *Fusarium verticillioides* 50-fold. At the same time, *F. verticillioides* infection slowed down larval development (Kurtz et al., 2010). When *F. verticillioides* spores were applied in low densities before rootworm attack, an increase in larval growth was observed, an effect that was attributed to potential positive effect of fungal colonization on root N levels (Kurtz et al., 2010). Interestingly, a follow-up study identified *Fusarium* spp. as the dominant fungal population in the gut of the western corn rootworm, irrespective of the soil type that was used (Dematheis et al., 2012a). This finding opens up the possibility that the western corn rootworm acts as a vector for the disease and suggests that the presence of *Fusarium* spp. in the system may determine both plant and insect performance. Just as with the mycorrhizal fungi, role of plant primary and secondary metabolites in the interaction remains to be determined. However, given how strongly maize roots respond to colonization by pathogens (Balmer et al., 2013), it is likely that *Fusarium*-induced changes in the primary and secondary plant metabolome will influence the behaviour and performance of the western corn rootworm. That root-herbivore behaviour is altered by the interactions between plants and fungal pathogens is demonstrated by a study on the black vine weevil and the entomopathogenic fungus *M. anisopliae* (Kepler and Bruck, 2006). In a two-choice setup, the vine weevil larvae did not show any preference for pots with or without the fungus. However, when host plants (*Picea abies*) were added to the system, the larvae preferred to orient towards host plants with the fungus (Kepler and Bruck, 2006), because of either an interaction of host orientation cues from the plant and the fungus or a fungus-mediated influence on plant metabolism.

Recent research on soil feedbacks demonstrates yet another route by which soil microorganisms influence root-herbivore interactions. Kostenko et al. (2012) found that wireworm (*A. lineatus*) attack of ragwort (*Jacobaea vulgaris*) led to significant changes in the soil fungal community. A second generation of plants cultivated in the conditioned soils had significantly lower concentrations of pyrrolizidine alkaloids (PAs) in the leaves. As PAs are partially synthesized in the roots, it is likely that the metabolic make-up of belowground tissues was also significantly altered by the soil feedback (Kostenko et al., 2012). Alterations of the fungal and bacterial soil microbial community were also found in the rhizosphere of western corn rootworm-attacked maize plants (Dematheis et al., 2012b). As many root herbivores develop over several years, microbe-mediated soil-legacy effects that influence plant secondary metabolites may be an important determinant for their performance and development, as well as final plant productivity (Sonnemann et al., 2013).

Given (i) their high abundance and (ii) their strong influence on herbivore fitness via plant-mediated effects, we propose here that root herbivores have evolved strategies to manage the soil microbial community to their own benefit. Soil-dwelling leaf-cutter ants, for instance, are known to cultivate fungi as a food source (Quinlan and Cherrett, 1979), and beewolf digger wasps (*Philanthus* spp.) protect their brood in the soil by transmitting streptomycetes that excrete protective antibiotic compounds (Kroiss et al., 2010). Root herbivores may use similar strategies, including the maintenance of a gut microbial flora that favours the digestion of root material and the protection of eggs and larvae with antimicrobial compounds. Until today, very little information is available about these possible 'management tools', and further research is clearly warranted.

9. CONCLUSIONS AND OUTLOOK

The current state of research on the role of plant primary and secondary metabolites in root-herbivore interactions highlights a number of adaptations that may enable root herbivores to survive in the soil environment.

First, an increasing number of studies clearly demonstrate that root herbivores possess an exquisite capacity to sense volatile and nonvolatile plant metabolites and use them to adjust their foraging decisions and feeding patterns. Foraging decisions are made both at a distance using volatile and exudate cues and during the feeding process itself. A major open question in this context is how the different cues are integrated spatially, temporally and physiologically. Experiments using molecular approaches in relevant soil environments are likely to yield novel and interesting responses on root-herbivore orientation and foraging.

A second conclusion from our review is that root herbivores are likely to have adapted their digestive physiology to the poor quality of their food source. These adaptations include (i) tolerance to otherwise toxic plant secondary metabolites and (ii) digestive strategies that allow growth and development from diets that are poor in soluble carbohydrates and essential PUFAs. A set of relatively straightforward experiments, for example, using geometric analysis, could shed light on possible shifts in nutritional optima that accompanied the evolution of root-feeding behaviour. The knowledge gained in the process could then serve as a starting point for more detailed studies to pinpoint the mechanisms behind root-herbivore physiology.

A third point that is becoming evident from recent work is that soil microbes may influence the interaction between root-feeding insects and their host plants. The systems described so far are highly dynamic, with the herbivores vectoring or facilitating microbial colonization of the plants followed by feedbacks on their own performance. Although the earlier examples illustrate the potential of soil microbes to influence root-herbivore interactions by changing the metabolism of their host plants, current research efforts are barely scratching the surface of this topic. Understanding tripartite interactions should be a priority of future research, especially because they may be a defining feature of root-herbivore interactions *per se*. The results obtained from such studies could clearly point the way to a better grasp on evolutionary and ecological processes in the rhizosphere.

ACKNOWLEDGEMENTS

We thank Scott Johnson, Ivan Hiltpold and Ted Turlings for the invitation to contribute to this AIIP volume. We are grateful to Philip Smith for proofreading the article. M. E. and C. A. M. R. are supported by a Marie Curie Intra European Fellowship (grant no. 273107 to M. E.) and a Swiss National Science Foundation Fellowship (grant no. 140196 to C. A. M. R.). C.C.M.A. is supported by the Brazilian National Council for Research (CNPq; grant no. 237929/2012-0).

REFERENCES

Abel, C.A., Berhow, M.A., Wilson, R.L., Binder, B.F., Hibbard, B.E., 2000. Evaluation of conventional resistance to European corn borer (Lepidoptera: Crambidae) and western corn rootworm (Coleoptera: Chrysomelidae) in experimental maize lines developed from a backcross breeding program. J. Econ. Entomol. 93, 1814–1821.

Agrawal, A.A., Hastings, A.P., Johnson, M.T.J., Maron, J.L., Salminen, J.P., 2012. Insect herbivores drive real-time ecological and evolutionary change in plant populations. Science 338, 113–116.

Agus, F., Handayani, E., Van Noordwijk, M., Idris, K., Sabiham, S., 2010. Root respiration interferes with peat CO_2 emission measurement. In: 19th World Congress of Soil Science, Soil Solutions for a Changing World, 2010.

Allsopp, P.G., 1992. Sugars, amino acids, and ascorbic acid as phagostimulants for larvae of *Antitrogus parvulus* and *Lepidiota negatoria* (Coleoptera: Scarabaeidae). J. Econ. Entomol. 85, 106–111.

Assabgui, R.A., Arnason, J.T., Hamilton, R.I., 1995. Field evaluations of hydroxamic acids as antibiosis factors in elite maize inbreds to the western corn-rootworm (Coleoptera, Chrysomelidae). J. Econ. Entomol. 88, 1482–1493.

Augustine, M.G., Fisk, F.W., Davidson, R.H., Lapidus, J.B., Cleary, R.W., 1964. Host-plant selection by the Mexican bean beetle, *Epilachna varivestis*. Ann. Entomol. Soc. Am. 57, 127–134.

Awmack, C.S., Leather, S.R., 2002. Host plant quality and fecundity in herbivorous insects. Annu. Rev. Entomol. 47, 817–844.

Badri, D.V., Quintana, N., El Kassis, E.G., Kim, H.K., Choi, Y.H., Sugiyama, A., Verpoorte, R., Martinoia, E., Manter, D.K., Vivanco, J.M., 2009. An ABC transporter mutation alters root exudation of phytochemicals that provoke an overhaul of natural soil microbiota. Plant Physiol. 151, 2006–2017.

Badri, D.V., Chaparro, J.M., Manter, D.K., Martinoia, E., Vivanco, J.M., 2012. Influence of ATP-binding cassette transporters in root exudation of phytoalexins, signals and in disease resistance. Front. Plant Sci. 3, 149. http://dx.doi.org/10.3389/fpls.2012.00149.

Balmer, D., De Papajewski, D.V., Planchamp, C., Glauser, G., Mauch-Mani, B., 2013. Induced resistance in maize is based on organ-specific defence responses. Plant J. 74, 213–225.

Bartlet, E., Parsons, D., Williams, I.H., Clark, S.J., 1994. The influence of glucosinolates and sugars on feeding by the cabbage stem flea beetle, *Psylliodes chrysocephala*. Entomol. Exp. Appl. 73, 77–83.

Beck, S.D., 1957. The European corn borer, *Pyrausta nubilalis* (Hubn.) and its principal host plant. IV. Larval saccharotrophism and host plant resistance. Ann. Entomol. Soc. Am. 50, 247–250.

Bernays, E.A., Woodhead, S., 1982. Plant phenols utilized as nutrients by a phytophagous insect. Science 216, 201–203.

Bernklau, E.J., Bjostad, L.B., 1998. Reinvestigation of host location by western corn rootworm larvae (Coleoptera: Chrysomelidae): CO_2 is the only volatile attractant. J. Econ. Entomol. 91, 1331–1340.

Bernklau, E.J., Bjostad, L.B., 2005. Insecticide enhancement with feeding stimulants in corn for western corn rootworm larvae (Coleoptera: Chrysomelidae). J. Econ. Entomol. 98, 1150–1156.

Bernklau, E.J., Bjostad, L.B., 2008. Identification of feeding stimulants in corn roots for western corn rootworm (Coleoptera: Chrysomelidae) larvae. J. Econ. Entomol. 101, 341–351.

Bernklau, E.J., Bjostad, L.B., Hibbard, B.E., 2011. Synthetic feeding stimulants enhance insecticide activity against western corn rootworm larvae, *Diabrotica virgifera virgifera* (Coleoptera: Chrysomelidae). J. Appl. Entomol. 135, 47–54.

Bigger, J.H., 1930. Notes on the life history of the clover root curculio, *Sitona hispidula* Fab., in Central Illinois. J. Econ. Entomol. 23, 334–342.

Birch, A.N.E., Griffiths, D.W., Hopkins, R.J., Macfarlane Smith, W.H.M., McKinlay, R.G., 1992. Glucosinolate responses of swede, kale, forage and oilseed rape to root damage by turnip root fly (*Delia floralis*) larvae. J. Sci. Food Agr. 60, 1–9.

Bjostad, L.B., Hibbard, B.E., 1992. 6-Methoxy-2-benzoxazolinone—a semiochemical for host location by western corn-rootworm larvae. J. Chem. Ecol. 18, 931–944.

Blossey, B., Hunt-Joshi, T.R., 2003. Belowground herbivory by insects: influence on plants and aboveground herbivores. Annu. Rev. Entomol. 48, 521–547.

Boldingh, H., Smith, G.S., Klages, K., 2000. Seasonal concentrations of non-structural carbohydrates of five *Actinidia* species in fruit, leaf and fine root tissue. Ann. Bot. 85, 469–476.

Braun, D.M., Ma, Y., Inada, N., Muszynski, M.G., Baker, R.F., 2006. Tie-dyed1 regulates carbohydrate accumulation in maize leaves. Plant Physiol. 142, 1511–1522.

Burrows, S., Elbert, W., Lawrence, M., Pöschl, U., 2009. Bacteria in the global atmosphere—part 1: review and synthesis of literature data for different ecosystems. Atmos. Chem. Phys. 9, 9263–9280.

Cakmak, I., Hengeler, C., Marschner, H., 1994. Partitioning of shoot and root dry matter and carbohydrates in bean plants suffering from phosphorus, potassium and magnesium deficiency. J. Exp. Bot. 45, 1245–1250.

Cannesan, M.A., Gangneux, C., Lanoue, A., Giron, D., Laval, K., Hawes, M., Driouich, A., Vicré-Gibouin, M., 2011. Association between border cell responses and localized root infection by pathogenic *Aphanomyces euteiches*. Ann. Bot. 108, 459–469.

Chapman, R.F., 2003. Contact chemoreception in feeding by phytophagous insects. Annu. Rev. Entomol. 48, 455–484.

Christenson, J.A., Hadwiger, L.A., 1973. Induction of pisatin formation in the pea foot region by pathogenic and nonpathogenic clones of *Fusarium solani*. Phytopathology 63, 784–790.

Clark, K.E., Hartley, S.E., Johnson, S.N., 2011. Does mother know best? The preference–performance hypothesis and parent–offspring conflict in aboveground–belowground herbivore life cycles. Ecol. Entomol. 36, 117–124.

Cocco, N., Glendinning, J.I., 2012. Not all sugars are created equal: some mask aversive tastes better than others in an herbivorous insect. J. Econ. Entomol. 215, 1412–1421.

Cole, R.A., 1987. Intensity of radicle fluorescence as related to the resistance of seedlings of lettuce to the lettuce root aphid and carrot to the carrot fly. Ann. Appl. Biol. 111, 629–639.

Cole, R.A., Riggall, W., Morgan, A., 1993. Electronically monitored feeding-behaviour of the lettuce root aphid (*Pemphigus bursarius*) on resistant and susceptible lettuce varieties. Entomol. Exp. Appl. 68, 179–185.

Crombie, A.C., Darrah, J.H., 1947. The chemoreceptors of the wireworm (*Agriotes* spp.) and the relation of activity to chemical constitution. J. Econ. Entomol. 24, 95–109.

Currie, A.F., Murray, P.J., Gange, A.C., 2006. Root herbivory by *Tipula paludosa* larvae increases colonization of *Agrostis capillaris* by arbuscular mycorrhizal fungi. Soil Biol. Biochem. 38, 1994–1997.

Dadd, R.H., 1961. The nutritional requirements of locusts. 5. Observations on essential fatty acids, chlorophyll, nutritional salt mixtures, and the protein or amino acid components of synthetic diets. J. Insect Physiol. 6, 126–145.

Dadd, R.H., 1973. Insect nutrition—current developments and metabolic implications. Annu. Rev. Entomol. 18, 381–420.

Dadd, R.H., 1977. Qualitative requirements and utilization of nutrients: insects. In: Rechcigl, M. (Ed.), Handbook Series in Nutrition and Food. CRC Press, Cleveland, OH, pp. 305–346.

Dawson, R.F., 1941. The localization of the nicotine synthetic mechanism in the tobacco plant. Science 94, 396–397.

Degenhardt, J., Hiltpold, I., Köllner, T.G., Frey, M., Gierl, A., Gershenzon, J., Hibbard, B.E., Ellersieck, M.R., Turlings, T.C.J., 2009. Restoring a maize root signal that attracts insect-killing nematodes to control a major pest. Proc. Natl. Acad. Sci. U. S. A. 106, 13213–13218.

De Heer, C.J., Tallamy, D.W., 1991. Affinity of spotted cucumber beetle (Coleoptera: Chrysomelidae) larvae to cucurbitacins. Environ. Entomol. 20, 1173–1175.

Dematheis, F., Kurtz, B., Vidal, S., Smalla, K., 2012a. Microbial communities associated with the larval gut and eggs of the western corn rootworm. PLoS ONE 7, e44685.

Dematheis, F., Zimmerling, U., Flocco, C., Kurtz, B., Vidal, S., Kropf, S., Smalla, K., 2012b. Multitrophic interaction in the rhizosphere of maize: root feeding of western corn rootworm larvae alters the microbial community composition. PLoS ONE 7, e37288.

De Renobales, M., Ryan, R.O., Heisler, C.R., McLean, D.L., Blomquist, G.J., 1986. Linoleic acid biosynthesis in the pea aphid, *Acyrthosiphon pisum* (Harris). Arch. Insect Biochem. 3, 193–203.

Dethier, V.G., 1973. Electrophysiological studies of gustation in lepidopterous larvae. 2. Taste spectra in relation to food-plant discrimination. J. Comp. Physiol. 82, 103–134.

Doane, J.F., Lee, Y.W., Klingler, J., Westcott, N.D., 1975. The orientation response of *Ctenicera destructor* and other wire worms (Coleoptera: Elateridae) to germinating grain and to carbon dioxide. Can. Entomol. 107, 1233–1252.

Dobler, S., Petschenka, G., Pankoke, H., 2011. Coping with toxic plant compounds—the insect's perspective on iridoid glycosides and cardenolides. Phytochemistry 72, 1593–1604.

Dobler, S., Dalla, S., Wagschal, V., Agrawal, A.A., 2012. Community-wide convergent evolution in insect adaptation to toxic cardenolides by substitutions in the Na, K-ATPase. Proc. Natl. Acad. Sci. U. S. A. 109, 13040–13045.

Douglas, A.E., 2003. The nutritional physiology of aphids. Adv. Insect Physiol. 31, 73–140.

Eilers, E.J., Talarico, G., Hansson, B.S., Hilker, M., Reinecke, A., 2012. Sensing the underground—ultrastructure and function of sensory organs in root-feeding *Melolontha melolontha* (Coleoptera: Scarabaeinae) larvae. PLoS ONE 7, e41357.

Erb, M., Flors, V., Karlen, D., De Lange, E., Planchamp, C., D'Alessandro, M., Turlings, T.C.J., Ton, J., 2009. Signal signature of aboveground-induced resistance upon belowground herbivory in maize. Plant J. 59, 292–302.

Esenther, G.R., Kirk, T.K., 1974. Catabolism of aspen sapwood in *Reticulitermes flavipes* (Isoptera-Rhinotermitidae). Ann. Entomol. Soc. Am. 67, 989–991.

Eshghi, S., Tafazoli, E., Dokhani, S., Rahemi, M., Emam, Y., 2007. Changes in carbohydrate contents in shoot tips, leaves and roots of strawberry (*Fragaria* × *ananassa* Duch.) during flower-bud differentiation. Sci. Horti. 113, 255–260.

Evans, W.A.L., Payne, D.W., 1964. Carbohydrases of the alimentary tract of the desert locust, *Schistocerca gregaria Forsk*. J. Insect Physiol. 10, 657–674.

Faucher, C.P., Hilker, M., De Bruyne, M., 2013. Interactions of carbon dioxide and food odours in Drosophila: olfactory hedonics and sensory neuron properties. PLoS ONE 8, e56361.

Finch, S., Skinner, G., 1974. Studies of the cabbage root fly. Report of the National Vegetable Research Station. National Vegetable Research Station, Wellesbourne.

Friend, W.G., 1958. Nutritional requirements of phytophagous insects. Annu. Rev. Entomol. 3, 57–74.

Frost, C.J., Hunter, M.D., 2008. Herbivore-induced shifts in carbon and nitrogen allocation in red oak seedlings. New Phytol. 178, 835–845.

Gange, A.C., Brown, V.K., Sinclair, G.S., 1994. Reduction of black vine weevil larval growth by vesicular-arbuscular mycorrhizal infection. Entomol. Exp. Appl. 70, 115–119.

Gaynor, D.L., Lane, G.A., Biggs, D.R., Sutherland, O.R.W., 1986. Measurement of grass grub resistance of bean in a controlled environment. New Zeal. J. Exp. Agr. 14, 77–82.

Gothilf, S., Beck, S.D., 1967. Larval feeding behaviour of cabbage looper *Trichoplusia ni*. J. Insect Physiol. 13, 1039–1053.

Guang, Y., Isenhour, D.J., Espelie, K.E., 1991. Activity of maize leaf cuticular lipids in resistance to leaf-feeding by the fall armyworm. Fla. Entomol. 74, 229–236.

Hackell, D.L., Gerard, P.J., 2004. Nodule preference by first instar clover root weevil. New Zeal. Plant Prot. 57, 319–322.

Halaweish, F.T., Tallamy, D.W., Santana, E., 1999. Cucurbitacins: a role in cucumber beetle steroid nutrition? J. Chem. Ecol. 25, 2373–2383.

Halldórsson, G., Sverrisson, H., Eyjólfsdóttir, G.G., Oddsdóttir, E.S., 2000. Ectomycorrhizae reduce damage to Russian Larch by *Otiorhyncus* larvae. Scand. J. Forest Res. 15, 354–358.

Hansen, G.K., 1977. Adaption to photosynthesis and diurnal oscillation of root respiration rates for *Lolium multiflorum*. Physiol. Plant. 39, 275–279.

Harada, K., Suomalainen, M., Uchida, H., Masui, H., Ohmura, K., Kiviranta, J., Niku-Paavola, M.L., Ikemoto, T., 2000. Insecticidal compounds against mosquito larvae from *Oscillatoria agardhii* strain 27. Environ. Toxicol. 15, 114–119.

Heron, R.J., 1965. Role of chemotactic stimuli in feeding behaviour of spruce budworm larvae on white spruce. Can. J. Zool. 43, 247–269.

Hiltpold, I., Turlings, T.C.J., 2008. Belowground chemical signaling in maize: when simplicity rhymes with efficiency. J. Chem. Ecol. 34, 628–635.

Hiltpold, I., Turlings, T.C.J., 2012. Manipulation of chemically mediated interactions in agricultural soils to enhance the control of crop pests and to improve crop yield. J. Chem. Ecol. 38, 641–650.

Hoffland, E., Boogaard, R.V.D., Nelemans, J., Findenegg, G., 1992. Biosynthesis and root exudation of citric and malic acids in phosphate-starved rape plants. New Phytol. 122, 675–680.

Hol, W.H.G., Vrieling, K., Veen, J.A.V., 2003. Nutrients decrease pyrrolizidine alkaloid concentrations in *Senecio jacobaea*. New Phytol. 158, 175–181.

Hopkins, R.J., Griffiths, D.W., Birch, A.N.E., McKinlay, R.G., Hall, J.E., 1993. Relationships between turnip root fly (*Delia floralis*) larval development and the sugar content of swede (*Brassica napus* ssp. *rapifera*) roots. Ann. Appl. Biol. 122, 405–415.

Hopkins, R.J., van Dam, N.M., van Loon, J.J.A., 2009. Role of glucosinolates in insect–plant relationships and multitrophic interactions. Annu. Rev. Entomol. 54, 57–83.

Horton, D., Guedot, C., Landolt, P., 2012. Identification of feeding stimulants for Pacific Coast wireworm by use of a filter paper assay (Coleoptera: Elateridae). J. Entomol. Soc. BC 109, 38–47.

House, H.L., 1961. Insect nutrition. Annu. Rev. Entomol. 6, 13–26.

Houseman, J.G., Campos, F., Thie, N.M.R., Philogene, B.J.R., Atkinson, J., Morand, P., Arnason, J.T., 1992. Effect of the maize-derived compounds DIMBOA and MBOA on growth and digestive processes of European corn-borer (Lepidoptera, Pyralidae). J. Econ. Entomol. 85, 669–674.

Hsiao, T.H., Fraenkel, G., 1968. Influence of nutrient chemicals on feeding behaviour of Colorado potato beetle *Leptinotarsa decemlineata* (Coleoptera: Chrysomelidae). Ann. Entomol. Soc. Am. 61, 44–54.

Hunter, M.D., 2001. Out of sight, out of mind: the impacts of root-feeding insects in natural and managed systems. Agr. Forest Entomol. 3, 3–9.

Ibrahim, L., Proe, M., Cameron, A., 1997. Main effects of nitrogen supply and drought stress upon whole-plant carbon allocation in poplar. Can. J. Forest Res. 27, 1413–1419.

Ito, T., 1960. Effect of sugars on feeding of larvae of the silkworm, *Bombyx mori*. J. Insect Physiol. 5, 95–107.

Jander, G., Howe, G., 2008. Plant interactions with arthropod herbivores: state of the field. Plant Physiol. 146, 801–803.

Jiang, M., Cheng, J., 2003. Feeding, oviposition and survival of overwintered rice water weevil (Coleoptera: Curculionidae) adults in response to nitrogen fertilization of rice at seedling stage. Appl. Entomol. Zool. 38, 543–549.

Joern, A., Behmer, S.T., 1997. Importance of dietary nitrogen and carbohydrates to survival, growth, and reproduction in adults of the grasshopper *Ageneotettix deorum* (Orthoptera: Acrididae). Oecologia 112, 201–208.

Johnson, S.N., Gregory, P.J., 2006. Chemically-mediated host-plant location and selection by root-feeding insects. Physiol. Entomol. 31, 1–13.

Johnson, S.N., Nielsen, U.N., 2012. Foraging in the dark—chemically mediated host plant location by belowground insect herbivores. J. Chem. Ecol. 38, 604–614.

Johnson, S.N., Gregory, P.J., Murray, P.J., Zhang, X., Young, I.M., 2004. Host plant recognition by the root feeding clover weevil, *Sitona lepidus* (Coleoptera: Curculionidae). Bull. Entomol. Res. 94, 433–439.

Johnson, S.N., Gregory, P.J., Greenham, J.R., Zhang, X., Murray, P.J., 2005. Attractive properties of an isoflavonoid found in white clover root nodules on the clover root weevil. J. Chem. Ecol. 31, 2223–2229.

Johnson, S.N., Crawford, J.W., Gregory, P.J., Grinev, D.V., Mankin, R.W., Masters, G.J., Murray, P.J., Wall, D.H., Zhang, X., 2007. Non-invasive techniques for investigating and modelling root-feeding insects in managed and natural systems. Agr. Forest Entomol. 9, 39–46.

Johnson, S.N., Barton, A.T., Clark, K.E., Gregory, P.J., McMenemy, L.S., Hancock, R.D., 2011. Elevated atmospheric carbon dioxide impairs the performance of root-feeding vine weevils by modifying root growth and secondary metabolites. Glob. Change Biol. 17, 688–695.

Kaplan, I., Halitschke, R., Kessler, A., Rehill, B.J., Sardanelli, S., Denno, R.F., 2008a. Physiological integration of roots and shoots in plant defense strategies links above- and belowground herbivory. Ecol. Lett. 11, 841–851.

Kaplan, I., Halitschke, R., Kessler, A., Sardanelli, S., Denno, R.F., 2008b. Constitutive and induced defenses to herbivory in above- and belowground plant tissues. Ecology 89, 392–406.

Kawakishi, S., Kaneko, T., 1987. Interaction of proteins with allyl isothiocyanate. J. Agric. Food Chem. 35, 85–88.

Kepler, R.M., Bruck, D.J., 2006. Examination of the interaction between the black vine weevil (Coleoptera: Curculionidae) and an entomopathogenic fungus reveals a new tritrophic interaction. Environ. Entomol. 35, 1021–1029.

Keskin, C., Kacar, S., 2012. Fatty acid composition of root and shoot samples of some *Astragalus* L. (Fabaceae) taxa growing in the east and southeast of Turkey. Turk. J. Biol. 37, 122–128.

Kimmerer, T.W., Potter, D.A., 1987. Nutritional quality of specific leaf tissues and selective feeding by a specialist leafminer. Oecologia 71, 548–551.

Koštál, V., 1992. Orientation behaviour of newly hatched larvae of the cabbage maggot, *Delia radicum* (L.) (Diptera: Anthomyiidae), to volatile plant metabolites. J. Insect Behav. 5, 61–70.

Kostenko, O., Van De Voorde, T.F.J., Mulder, P.P.J., Van Der Putten, W.H., Bezemer, T.M., 2012. Legacy effects of aboveground–belowground interactions. Ecol. Lett. 15, 813–821.

Kretzschmar, T., Kohlen, W., Sasse, J., Borghi, L., Schlegel, M., Bachelier, J.B., Reinhardt, D., Bours, R., Bouwmeester, H.J., Martinoia, E., 2012. A petunia ABC protein controls strigolactone-dependent symbiotic signalling and branching. Nature 483, 341–344.

Kroiss, J., Kaltenpoth, M., Schneider, B., Schwinger, M.G., Hertweck, C., Maddula, R.K., Strohm, E., Svatoš, A., 2010. Symbiotic streptomycetes provide antibiotic combination prophylaxis for wasp offspring. Nat. Chem. Biol. 6, 261–263.

Kurtz, B., Karlovsky, P., Vidal, S., 2010. Interaction between western corn rootworm (Coleoptera: Chrysomelidae) larvae and root-infecting Fusarium verticillioides. Environ. Entomol. 39, 1532–1538.
Ladd, T.L., 1986. Influence of sugars on the feeding response of Japanese beetles (Coleoptera: Scarabaeidae). J. Econ. Entomol. 79, 668–671.
Ladd, T.L., 1988. Japanese beetle (Coleoptera: Scarabaeidae)—influence of sugars on feeding response of larvae. J. Econ. Entomol. 81, 1390–1393.
Lane, G.A., Biggs, D.R., Russell, G.B., Sutherland, O.R.W., Williams, E.M., Maindonald, J.H., Donnell, D.J., 1985. Isoflavonoid feeding deterrents for *Costelytra zealandica*: structure–activity relationships. J. Chem. Ecol. 11, 1713–1735.
Lane, G.A., Sutherland, O.R.W., Skipp, R.A., 1987. Isoflavonoids as insect feeding deterrents and antifungal components from root of *Lupinus angustifolius*. J. Chem. Ecol. 13, 771–783.
Larkindale, J., Huang, B.R., 2004. Changes of lipid composition and saturation level in leaves and roots for heat-stressed and heat-acclimated creeping bentgrass (*Agrostis stolonifera*). Environ. Exp. Bot. 51, 57–67.
Lasker, R., Giese, A.C., 1956. Cellulose digestion by the silverfish *Ctenolepisma lineata*. J. Econ. Entomol. 33, 542–553.
Lee, K.P., Behmer, S.T., Simpson, S.J., Raubenheimer, D., 2002. A geometric analysis of nutrient regulation in the generalist caterpillar *Spodoptera littoralis* (Boisduval). J. Insect Physiol. 48, 655–665.
Liu, L., Punja, Z.K., Rahe, J.E., 1995. Effect of *Pythium* spp. and glyphosate on phytoalexin production and exudation by bean (*Phaseolus vulgaris* L.) roots grown in different media. Physiol. Mol. Plant Pathol. 47, 391–405.
Lyons, J.M., Lippert, L.F., 1966. Characterization of fatty acids from root and shoot lipids of *Capsicum* species. Lipids 1, 136–140.
Machado, R.A.R., Ferrieri, A.P., Robert, C.A.M., Glauser, G., Kallenbach, M., Baldwin, I.T., Erb, M., 2013. Leaf-herbivore attack reduces carbon reserves and regrowth from the roots via jasmonate and auxin signaling. New Phytol., http://dx.doi.org/10.1111/nph.12438.
Malcolm, S.B., 1991. Cardenolide-mediated interactions between plants and herbivores. In: Rosenthal, G.A., Berenbaum, M.R. (Eds.), Herbivores: Their Interactions with Secondary Metabolites. Academic Press, San Diego, CA, pp. 251–296.
Martin, M.M., 1983. Cellulose digestion in insects. Comp. Biochem. Physiol. 75, 313–324.
Martin, M.M., 1991. The evolution of cellulose digestion in insects. Philos. Trans. R. Soc. B 333, 281–288.
Masters, G.J., Brown, V.K., Gange, A.C., 1993. Plant mediated interactions between aboveground and belowground insect herbivores. Oikos 66, 148–151.
Mathesius, U., 2001. Flavonoids induced in cells undergoing nodule organogenesis in white clover are regulators of auxin breakdown by peroxidase. J. Exp. Bot. 52, 419–426.
Mattson, W.J., 1980. Herbivory in relation to plant nitrogen content. Annu. Rev. Ecol. Syst. 11, 119–161.
Mazoir, N., Benharref, A., Bailén, M., Reina, M., González-Coloma, A., 2008. Bioactive triterpene derivatives from latex of two Euphorbia species. Phytochemistry 69, 1328–1338.
Mochizuki, A., Ishikawa, Y., Matsumoto, Y., 1989. Olfactory response of the larvae of the onion fly, *Hylemya antiqua* Meigen (Diptera, Anthomyiidae) to volatile compounds. Appl. Entomol. Zool. 24, 29–35.
Morgan, M.R.J., 1975. Qualitative survey of carbohydrases of alimentary-tract of migratory locust, *Locusta migratoria migratorioides*. J. Insect Physiol. 21, 1045–1053.
Murray, P.J., Hatch, D.J., Cliquet, J.B., 1996. Impact of insect root herbivory on the growth and nitrogen and carbon contents of white clover (*Trifolium repens*) seedlings. Can. J. Bot. 74, 1591–1595.

Neal, A.L., Ahmad, S., Gordon-Weeks, R., Ton, J., 2012. Benzoxazinoids in root exudates of maize attract *Pseudomonas putida* to the rhizosphere. PLoS ONE 7, e35498.

Newingham, B., Callaway, R., Bassirirad, H., 2007. Allocating nitrogen away from a herbivore: a novel compensatory response to root herbivory. Oecologia 153, 913–920.

Nielsen, J.K., Dalgaard, L., Larsen, L.M., Sørensen, H., 1979. Host plant selection of the horse-radish flea beetle *Phyllotreta armoraciae* (Coleoptera: Chrysomelidae): feeding responses to glucosinolates from several crucifers. Entomol. Exp. Appl. 25, 227–239.

Niemeyer, H.M., 2009. Hydroxamic acids derived from 2-hydroxy-2H-1,4-benzoxazin-3 (4H)-one: key defense chemicals of cereals. J. Agr. Food Chem. 57, 1677–1696.

Nordenhem, H., Nordlander, G., 1994. Olfactory oriented migration through soil by root-living *Hylobius abietis* (L.) larvae (Col., Curculionidae). J. Appl. Entomol. 117, 457–462.

Nutt, K.A., O'Shea, M.G., Allsopp, P.G., 2004. Feeding by sugarcane whitegrubs induces changes in the types and amounts of phenolics in the roots of sugarcane. Environ. Exp. Bot. 51, 155–165.

Okuley, J., Lightner, J., Feldmann, K., Yadav, N., Lark, E., Browse, J., 1994. Arabidopsis FAD2 gene encodes the enzyme that is essential for polyunsaturated lipid synthesis. Plant Cell 6, 147–158.

Opitz, S.W., Müller, C., 2009. Plant chemistry and insect sequestration. Chemoecology 19, 117–154.

Ouariti, O., Boussama, N., Zarrouk, M., Cherif, A., Ghorbal, M.H., 1997. Cadmium- and copper-induced changes in tomato membrane lipids. Phytochemistry 45, 1343–1350.

Peipp, H., Maier, W., Schmidt, J., Wray, V., Strack, D., 1997. Arbuscular mycorrhizal fungus-induced changes in the accumulation of secondary compounds in barley roots. Phytochemistry 44, 581–587.

Pierre, P.S., Dugravot, S., Cortesero, A.M., Poinsot, D., Raaijmakers, C.E., Hassan, H.M., Van Dam, N.M., 2012. Broccoli and turnip plants display contrasting responses to belowground induction by *Delia radicum* infestation and phytohormone applications. Phytochemistry 73, 42–50.

Pline, M., Dusenbery, D.B., 1987. Responses of plant-parasitic nematode *Meloidogyne incognita* to carbon dioxide determined by video camera-computer tracking. J. Chem. Ecol. 13, 873–888.

Potter, M.J., Davies, K., Rathjen, A.J., 1998. Suppressive impact of glucosinolates in *Brassica* vegetative tissues on root lesion nematode *Pratylenchus neglectus*. J. Chem. Ecol. 24, 67–80.

Potter, M.J., Vanstone, V.A., Davies, K.A., Rathjen, A.J., 2000. Breeding to increase the concentration of 2-phenylethyl glucosinolate in the roots of *Brassica napus*. J. Chem. Ecol. 26, 1811–1820.

Prasad, K.V.S.K., Song, B.H., Olson-Manning, C., Anderson, J.T., Lee, C.R., Schranz, M.E., Windsor, A.J., Clauss, M.J., Manzaneda, A.J., Naqvi, I., Reichelt, M., Gershenzon, J., Rupasinghe, S.G., Schuler, M.A., Mitchell-Olds, T., 2012. A gain-of-function polymorphism controlling complex traits and fitness in nature. Science 337, 1081–1084.

Prins, R.A., Kreulen, D.A., 1991. Comparative aspects of plant-cell wall digestion in insects. Anim. Feed Sci. Technol. 32, 101–118.

Quinlan, R.J., Cherrett, J.M., 1979. The role of fungus in the diet of the leaf-cutting ant *Atta cephalotes* (L.). Ecol. Entomol. 4, 151–160.

Radcliffe, J.E., 1970. Some effects of grass grub (*Costelytra zealandica* (White)) larvae on pasture plants. New Zeal. J. Agr. Res. 13, 87–104.

Ramsewak, R.S., Nair, M.G., Murugesan, S., Mattson, W.J., Zasada, J., 2001. Insecticidal fatty acids and triglycerides from *Dirca palustris*. J. Agr. Food Chem. 49, 5852–5856.

Rapport, E.W., Stanley-Samuelson, D., Dadd, R.H., 1984. Ten generations of *Drosophila melanogaster* reared axenically on a fatty-acid-free holidic diet. Arch. Insect Biochem. 1, 243–250.

Rask, L., Andreasson, E., Ekbom, B., Eriksson, S., Pontoppidan, B., Meijer, J., 2000. Myrosinase: gene family evolution and herbivore defense in Brassicaceae. Plant Mol. Biol. 42, 93–113.

Rasmann, S., Agrawal, A.A., 2008. In defense of roots: a research agenda for studying plant resistance to belowground herbivory. Plant Physiol. 146, 875–880.

Rasmann, S., Agrawal, A.A., 2011. Evolution of specialization: a phylogenetic study of host range in the red milkweed beetle (*Tetraopes tetraophthalmus*). Am. Nat. 177, 728–737.

Rasmann, S., Köllner, T.G., Degenhardt, J., Hiltpold, I., Toepfer, S., Kuhlmann, U., Gershenzon, J., Turlings, T.C.J., 2005. Recruitment of entomopathogenic nematodes by insect-damaged maize roots. Nature 434, 732–737.

Rasmann, S., Erwin, A.C., Halitschke, R., Agrawal, A.A., 2011. Direct and indirect root defences of milkweed (*Asclepias syriaca*): trophic cascades, trade-offs and novel methods for studying subterranean herbivory. J. Ecol. 99, 16–25.

Raubenheimer, D., 1992. Tannic acid, protein, and digestible carbohydrate—dietary imbalance and nutritional compensation in locusts. Ecology 73, 1012–1027.

Reinecke, A., Müller, F., Hilker, M., 2008. Attractiveness of CO_2 released by root respiration fades on the background of root exudates. Basic App. Ecol. 9, 568–576.

Robert, C.A.M., Erb, M., Duployer, M., Zwahlen, C., Doyen, G.R., Turlings, T.C.J., 2012a. Herbivore-induced plant volatiles mediate host selection by a root herbivore. New Phytol. 194, 1061–1069.

Robert, C.A.M., Erb, M., Hibbard, B.E., Wade French, B., Zwahlen, C., Turlings, T.C.J., 2012b. A specialist root herbivore reduces plant resistance and uses an induced plant volatile to aggregate in a density-dependent manner. Funct. Ecol. 26, 1429–1440.

Robert, C.A.M., Veyrat, N., Glauser, G., Marti, G., Doyen, G.R., Villard, N., Gaillard, M.D.P., Köllner, T.G., Giron, D., Body, M., Babst, B.A., Ferrieri, R.A., Turlings, T.C.J., Erb, M., 2012c. A specialist root herbivore exploits defensive metabolites to locate nutritious tissues. Ecol. Lett. 15, 55–64.

Robert, C.A.M., Erb, M., Hiltpold, I., Hibbard, B.E., Gaillard, M.D.P., Bilat, J., Degenhardt, J., Cambet-Petit-Jean, X., Turlings, T.C.J., Zwahlen, C., 2013. Genetically engineered maize plants reveal distinct costs and benefits of constitutive volatile emissions in the field. Plant Biotechnol. J. 11, 628–639.

Rondon, S.I., 2010. The potato tuberworm: a literature review of its biology, ecology, and control. Am. J. Potato Res. 87, 149–166.

Rössler, M.E., 1961. Ernährungsphysiologische Untersuchungen an Scarabaeidenlarven (*Oryctes nasicornis* L., *Melolontha melolontha* L.). J. Insect Physiol. 6, 62–74.

Rudrappa, T., Czymmek, K.J., Paré, P.W., Bais, H.P., 2008. Root-secreted malic acid recruits beneficial soil bacteria. Plant Physiol. 148, 1547–1556.

Russell, G.B., Sutherland, O.R.W., Hutchins, R.F.N., Christmas, P.E., 1978. Vestitol: a phytoalexin with insect feeding-deterrent activity. J. Chem. Ecol. 4, 571–579.

Russell, G.B., Shaw, G.J., Christmas, P.E., Yates, M.B., Sutherland, O.R.W., 1984. Two 2-arylbenzofurans as insect feeding deterrents from sainfoin (*Onobrychis viciifolia*). Phytochemistry 23, 1417–1420.

Schliemann, W., Ammer, C., Strack, D., 2008. Metabolite profiling of mycorrhizal roots of *Medicago truncatula*. Phytochemistry 69, 112–146.

Schmelz, E.A., Grebenok, R.J., Galbraith, D.W., Bowers, W.S., 1998. Damage-induced accumulation of phytoecdysteroids in spinach: a rapid root response involving the octadecanoic acid pathway. J. Chem. Ecol. 24, 339–360.

Schmelz, E.A., Grebenok, R.J., Ohnmeiss, T.E., Bowers, W.S., 2002. Interactions between *Spinacia oleracea* and *Bradysia impatiens*: a role for phytoecdysteroids. Arch. Insect Biochem. 51, 204–221.

Schoonhoven, L.M., 1974. What makes a caterpillar eat? The sensory codes underlying feeding behaviour. In: Chapman, R.F., Bernays, E.A., Stoffolano, J.G. (Eds.), Advances in Chemoreception and Behaviour. Springer, New York, pp. 69–97.

Schröder, R., Hilker, M., 2008. The relevance of background odor in resource location by insects: a behavioural approach. Bioscience 58, 308–316.

Shanbhag, S.R., Müller, B., Steinbrecht, R.A., 1999. Atlas of olfactory organs of *Drosophila melanogaster*. 1. Types, external organization, innervation and distribution of olfactory sensilla. Int. J. Insect Morphol. 28, 377–397.

Shanks, C.H., Doss, R.P., 1987. Feeding responses by adults of five species of weevils (Coleoptera: Curculionidae) to sucrose and sterols. Ann. Entomol. Soc. Am. 80, 41–46.

Shields, V.D.C., Mitchell, B.K., 1995. The effects of phagostimulant mixtures on deterrent receptor(s) in two crucifer-feeding lepidopterous species. Philos. Trans. R. Soc. B 347, 459–464.

Slansky, F., Feeny, P., 1977. Stabilization of rate of nitrogen accumulation by larvae of cabbage butterfly on wild and cultivated food plants. Ecol. Monogr. 47, 209–228.

Soni, S., Finch, S., 1979. Laboratory evaluation of sulphur-bearing chemicals as attractants for larvae of the onion fly, *Delia antiqua* (Meigen)(Diptera: Anthomyiidae). Bull. Entomol. Res. 69, 291–298.

Sonnemann, I., Hempel, S., Beutel, M., Hanauer, N., Reidinger, S., Wurst, S., 2013. The root herbivore history of the soil affects the productivity of a grassland plant community and determines plant response to new root herbivore attack. PLoS ONE 8, e56524.

Soo Hoo, C.F., Dudzinski, A., 1967. Digestion by the larva of the pruinose scarab, *Sericesthis geminata*. Entomol. Exp. Appl. 10, 7–15.

Städler, E., 1971. Über die Orientierung und das Wirtswahlverhalten der Möhrenfliege, *Psila rosae* F. (Diptera: Psilidae) I. Larven. Z. Angew. Entomol. 69, 425–438.

Stevenson, P.C., Turner, H.C., Haware, M.P., 1997. Phytoalexin accumulation in the roots of chickpea (*Cicer arietinum* L.) seedlings associated with resistance to fusarium wilt (*Fusarium oxysporum* f.sp. *ciceri*). Physiol. Mol. Plant P 50, 167–178.

Stevenson, P.C., Muyinza, H., Hall, D.R., Porter, E.A., Farman, D.I., Talwana, H., Mwanga, R.O.M., 2009. Chemical basis for resistance in sweetpotato *Ipomoea batatas* to the sweetpotato weevil *Cylas puncticollis*. Pure Appl. Chem. 81, 141–151.

Stout, M.J., 2013. Reevaluating the conceptual framework for applied research on host-plant resistance. Insect Sci. 20, 263–272.

Sutherland, O.R.W., 1971. Feeding behaviour of grass grub *Costelytra zealandica* (White) (Coleoptera: Melolonthinae)—1. Influence of carbohydrates. New Zeal. J. Sci. 14, 18–24.

Sutherland, O.R.W., 1983. The influence of amino-acids and inorganic salts on feeding by larvae of *Heteronychus arator* (Coleoptera: Scarabaeidae). New Zeal. J. Zool. 10, 117–119.

Sutherland, O.R.W., Hillier, J.R., 1973. Feeding behaviour of the grass hub *Costelytra zealandica* (White) (Coleoptera: Melolonthinae). 3: the influence of amino acids, ascorbic acid, and inorganic salts. New Zeal. J. Zool. 1, 211–216.

Sutherland, O.R.W., Hillier, J.R., 1976. The influence of maltose and other carbohydrates on the feeding behaviour of *Heteronychus arator* (Scarabaeidae: Coleoptera). Experientia 32, 701–702.

Sutherland, O.R.W., Russell, G.B., Biggs, D.R., Lane, G.A., 1980. Insect feeding deterrent activity of phytoalexin isoflavonoids. Biochem. Syst. Ecol. 8, 73–75.

Sutherland, O.R.W., Hutchins, R., Greenfield, W., 1982. Effect of lucerne saponins and Lotus condensed tannins on survival of grass grub, *Costelytra zealandica*. New Zeal. J. Zool. 9, 511–514.

Tallamy, D.W., Whittington, D.P., Defurio, F., Fontaine, D.A., Gorski, P.M., Gothro, P.W., 1998. Sequestered cucurbitacins and pathogenicity of *Metarhizium*

anisopliae (Moniliales: Moniliaceae) on spotted cucumber beetle eggs and larvae (Coleoptera: Chrysomelidae). Environ. Entomol. 27, 366–372.
Taylor, E.C., 1985. Cellulose digestion in a leaf eating insect, the Mexican bean beetle, *Epilachna varivestis*. Insect Biochem. 15, 315–320.
Thorpe, W.H., Crombie, A.C., Hill, R., Darrah, J.H., 1947. The behaviour of wireworms in response to chemical stimulation. J. Econ. Entomol. 23, 234–266.
Treves, D.S., Martin, M.M., 1994. Cellulose digestion in primitive hexapods—effect of ingested antibiotics on gut-microbial populations and gut cellulase levels in the firebrat, *Thermobia domestica* (Zygentoma, Lepismatidae). J. Chem. Ecol. 20, 2003–2020.
Tsuneyama, I., Ikejima, S., Iida, N., Tsurui, Y., Tanaka, Y., Ninagi, O., Oshiki, T., 2005. Combined effect of calcium carbonate and sugar on the feeding behaviour of the newly hatched larvae of the silkworm, *Bombyx mori*. J. Seric. Sci. Jpn. 74, 9–14.
Turlings, T.C.J., Hiltpold, I., Rasmann, S., 2012. The importance of root-produced volatiles as foraging cues for entomopathogenic nematodes. Plant Soil 358, 51–60.
Van Dam, N.M., 2009. Belowground herbivory and plant defenses. Annu. Rev. Ecol. Evol. Syst. 40, 373–391.
Van Der Heijden, M.G.A., Bardgett, R.D., Van Straalen, N.M., 2008. The unseen majority: soil microbes as drivers of plant diversity and productivity in terrestrial ecosystems. Ecol. Lett. 11, 296–310.
Van Leur, H., Raaijmakers, C.E., Van Dam, N.M., 2008. Reciprocal interactions between the cabbage root fly (*Delia radicum*) and two glucosinolate phenotypes of *Barbarea vulgaris*. Entomol. Exp. Appl. 128, 312–322.
Vannette, R.L., Rasmann, S., 2012. Arbuscular mycorrhizal fungi mediate below-ground plant-herbivore interactions: a phylogenetic study. Funct. Ecol. 26, 1033–1042.
Vaughan, M.M., Wang, Q., Webster, F.X., Kiemle, D., Hong, Y.J., Tantillo, D.J., Coates, R.M., Wray, A.T., Askew, W., O'Donnell, C., Tokuhisa, J.G., Tholl, D., 2013. Formation of the unusual semivolatile diterpene rhizathalene by the *Arabidopsis* class I terpene synthase TPS08 in the root stele is involved in defense against below-ground herbivory. Plant Cell 25, 1108–1125.
Watts, S.M., Dodson, C.D., Reichman, O.J., 2011. The roots of defense: plant resistance and tolerance to belowground herbivory. PLoS ONE 6, e18463.
Way, M.O., Reay-Jones, F.P.F., Stout, M.J., Tarpley, L., 2006. Effects of nitrogen fertilizer applied before permanent flood on the interaction between rice and rice water weevil (Coleoptera: Curculionidae). J. Econ. Entomol. 99, 2030–2037.
Weissteiner, S., Schütz, S., 2006. Are different volatile pattern influencing host plant choice of belowground living insects? M. D. Gesell. Allg. Ange. Entomol. 15, 51–55.
Weissteiner, S., Huetteroth, W., Kollmann, M., Weißbecker, B., Romani, R., Schachtner, J., Schütz, S., 2012. Cockchafer larvae smell host root scents in soil. PLoS ONE 7, e45827.
Wensler, R.J., Dudzinski, A.E., 1972. Gustation of sugars, amino acids and lipids by larvae of the scarabaeid, *Sericesthis geminata* (Coleoptera). Entomol. Exp. Appl. 15, 155–165.
Wharton, D.R.A., Wharton, M.L., 1965. The cellulase content of various species of cockroaches. J. Insect Physiol. 11, 1401–1405.
Xie, Y.S., Arnason, J.T., Philogene, B.J.R., Lambert, J.D.H., Atkinson, J., Morand, P., 1990. Role of 2,4-dihydroxy-7-methoxy-1,4-benzoxazin-3-one (DIMBOA) in the resistance of maize to western corn-rootworm, *Diabrotica virgifera virgifera* (LeConte) (Coleoptera, Chrysomelidae). Can. Entomol. 122, 1177–1186.
Xie, Y.S., Arnason, J.T., Philogene, B.J.R., Atkinson, J., Morand, P., 1992a. Behavioural responses of western corn rootworm larvae to naturally occurring and synthetic hydroxamic acids. J. Chem. Ecol. 18, 945–957.
Xie, Y.S., Arnason, J.T., Philogene, B.J.R., Olechowski, H.T., Hamilton, R.I., 1992b. Variation of hydroxamic acid content in maize roots in relation to geographic origin of maize

germ plasm and resistance to western corn-rootworm (Coleoptera, Chrysomelidae). J. Econ. Entomol. 85, 2478–2485.

Yang, G., Wiseman, B.R., Espelie, K.E., 1992. Cuticular lipids from silks of 7 corn genotypes and their effect on development of corn-earworm larvae *Helicoverpa zea* (Boddie). J. Agr. Food Chem. 40, 1058–1061.

Yang, M., Zheng, G., Zhang, F., Xu, Y., 2006. FAD2-silencing has pleiotropic effect on polar lipids of leaves and varied effect in different organs of transgenic tobacco. Plant Sci. 170, 170–177.

Zeeman, S.C., Rees, T.A., 1999. Changes in carbohydrate metabolism and assimilate export in starch-excess mutants of *Arabidopsis*. Plant Cell Environ. 22, 1445–1453.

Zhen, Y., Aardema, M.L., Medina, E.M., Schumer, M., Andolfatto, P., 2012. Parallel molecular evolution in an herbivore community. Science 337, 1634–1637.

Zinkler, D., Goetze, M., Fabian, K., 1986. Cellulose digestion in primitive insects Apterygota and oribatid mites. Zool. Beitr. 30, 17–28.

Züst, T., Heichinger, C., Grossniklaus, U., Harrington, R., Kliebenstein, D.J., Turnbull, L.A., 2012. Natural enemies drive geographic variation in plant defenses. Science 338, 116–119.

CHAPTER THREE

Nature, Evolution and Characterisation of Rhizospheric Chemical Exudates Affecting Root Herbivores

Ivan Hiltpold*, Elisa Bernklau†, Louis B. Bjostad†, Nadir Alvarez‡, Nicole E. Miller-Struttmann§, Jonathan G. Lundgren¶, Bruce E. Hibbard||

*Division of Plant Sciences, University of Missouri, Columbia, Missouri, USA
†Department of Bioagricultural Sciences and Pest Management, Colorado State University, Fort Collins, Colorado, USA
‡Department of Ecology and Evolution, University of Lausanne, Lausanne, Switzerland
§Division of Biological Sciences, University of Missouri, Columbia, Missouri, USA
¶USDA-ARS, North Central Agricultural Research Laboratory, Brookings, South Dakota, USA
||USDA-ARS, Plant Genetic Research, University of Missouri, Columbia, Missouri, USA

Contents

Advances in Insect Physiology, Volume 45
ISSN 0065-2806
http://dx.doi.org/10.1016/B978-0-12-417165-7.00003-9

Abstract

Similar to aboveground herbivores, root-feeding insects must locate and identify suitable resources. In the darkness of soil, they mainly rely on root chemical exudations and, therefore, have evolved specific behaviours. Because of their impact on crop yield, most of our knowledge in belowground chemical ecology is biased towards soil-dwelling insect pests. Yet, the increasing literature on volatile-mediated interactions in the ground underpins the great importance of chemical signalling in this ecosystem and its potential in pest control. Here, we explore the ecology and physiology of these chemically based interactions. An evolutionary approach reveals interesting patterns in the response of insects to particular classes of volatile or water-soluble organic compounds commonly emitted by roots. Food web analyses reasonably support that volatiles are used as long-range cues whereas water-soluble molecules serve in host acceptance/rejection by the insect; however, data are still scarce. As a case study, the chemical ecology of *Diabrotica virgifera virgifera* is discussed and applications of belowground signalling in pest management are examined. Soil chemical ecology is an expanding field of research and will certainly be a hub of our understanding of soil communities and subsequently of the management of belowground ecosystem services.

1. INTRODUCTION

Research on plant–insect interactions has historically focused on aboveground niches where numerous interactions among various functional guilds (e.g. herbivores, pollinators, predators and parasitoids) have been described. Decades of research exploring almost all terrestrial ecosystems, ranging from complex natural meadows or native forests to monoculture cropping systems, have shown that arthropods rely on plant-derived cues that are either physical (visual or tactile) or chemical (olfactory or gustatory) to identify, locate and finally accept or reject plants (e.g. Schoonhoven et al., 2005 and references therein). From the perspective of mankind, such interactions can be positive (i.e. pollination by insects) or negative (i.e. yield reductions) and attempts to respectively favour or control certain plant–insect interactions have been conducted in order to optimise (agro-) ecosystem functions and services.

In comparison, over a similar period of time, belowground plant–insect interactions have been largely understudied (Hunter, 2001). Indeed, an informal assessment by Hunter (2001) suggested that insect–root herbivory represents 2–5% of the literature published on insect herbivory, whereas plants from most ecosystems allocate 50–90% of their net primary production in root systems (Coleman, 1976). That the collective biomass of root–xylem-feeding cicadas is greater than any other animal in terms of biomass per unit area in certain North American forests (Karban, 1980) or that the weight of sheep per hectare is exceeded by the biomass of chafers in some Australian pastures (Britton, 1978) also illustrate well the importance of belowground insect herbivory. As hypothesised by Hunter (2001), soil-dwelling insects and roots are difficult to access, which likely contributes to most scientists' limited attention despite the clear linkages of below- and aboveground processes.

However, during the last decades, root–insect and root–microorganism interactions have drawn increasing consideration (e.g. Bonkowski et al., 2009; De Deyn et al., 2003; De La Peña et al., 2006; Gange and Brown, 2002; Rasmann and Agrawal, 2008; Rasmann et al., 2012b; Strong et al., 1999; van Dam, 2009; Van Der Putten, 2003). Because of their evident impact on human societies, most of the rhizophagous research has been conducted on insect pests of crops (Johnson and Murray, 2008). However, not all root herbivory is detrimental to human interests. Indeed, a total of 49 root-herbivore insect species were released in nearly 100 attempts to control 19 different target plant pest species in 10 different countries (Blossey and Hunt-Joshi, 2003 and references therein). The weevil *Cyphocleonus achates* (Coleoptera: Curculionidae) reduces the biomass production and fitness of spotted knapweed *Centaurea stoebe* (Asterales: Asteraceae) (Knochel and Seastedt, 2010), an invasive weed in North America. The effectiveness of biological control by the weevil was, however, compromised by biotic (plant competition) and abiotic (soil resources) factors that ultimately favoured the targeted pest plant (Knochel and Seastedt, 2010). Yet, the application of root herbivory in weed control still requires optimisation (Ortega et al., 2012) and with respect to environmental safety, in-depth studies have to be conducted before a root herbivore can be mass-released to control invasive weeds such as knapweed. Nonetheless, Blossey and Hunt-Joshi (2003) concluded that root feeders are more likely to contribute to weed control than aboveground-dwelling insect herbivores.

Just as with aboveground plant–insect interactions, plants can also interact with rhizospheric insects without deleterious effects. Such commensal association has been described in the interaction between the broadleaf cattail (*Typha latifolia*, Typhales: Typhaceae) and *Coquillettidia* species (Diptera: Culicidae) (Sérandour et al., 2008). Indeed, the aquatic larvae of these insects rely on root aerenchymes as a source of oxygen in benthic anoxic substrate (Bosak and Crans, 2002) and therefore respond to water-soluble secondary metabolites released by its host plant to locate an adequate site of establishment (Bosak and Crans, 2002). This nondestructive relationship between roots and insects might provide the insect not only with oxygen but also with better sources of food. For instance, the extensive rhizospheric exudation of organic matter by the plant could favour the abundance and diversity of microorganisms on which the mosquito larvae feed (Merritt et al., 1992). This increase in nutrient turnover might also be beneficial to the plant (Bonkowski, 2004 and references therein), but this has not been tested in this particular context.

The theoretical and evolutionary concepts pertaining to how aboveground chemically mediated plant–insect interactions are structured (Karban and Baldwin, 1997; Price et al., 1980; Vet and Dicke, 1992) also apply to subterranean systems (e.g. Rasmann et al., 2011a; Strong et al., 1999; Van Der Putten et al., 2009; van Tol et al., 2001). Indeed, roots impact the behaviour, composition and abundance of belowground communities (e.g. Bonkowski et al., 2009) and their trophic links (Coleman, 1976; Monroy and van der Putten, 2009; Strong et al., 1999). However, conspicuous differences (physical and mechanistic) appear between these two broad ecological niches.

1.1. Soil versus air

Differences between soil and air characteristics are, to some extent, easy to conceive but imply dramatic divergences in their respective physical and chemical properties. At a glance, air is mainly composed of one phase of matter (gas), whereas soil is constituted of all three (gas, liquid and solid) (Fig. 3.1). Despite its apparent simplicity, atmospheric air composition plays an important role in chemically mediated interactions between insects and plants aboveground. Oxygen is highly chemically reactive and can therefore oxidise volatile organic compounds emitted by plants. Differences in temperature, water vapour levels and pressures result in diverse volatile organic compound diffusion characteristics on a micro- or macroscale.

However, the troposphere (75–80% of the atmosphere total mass over a height of 10–15 km) can be considered homogenous as compared to soils covering the surface of the globe. Indeed, adding two extra phases of matter

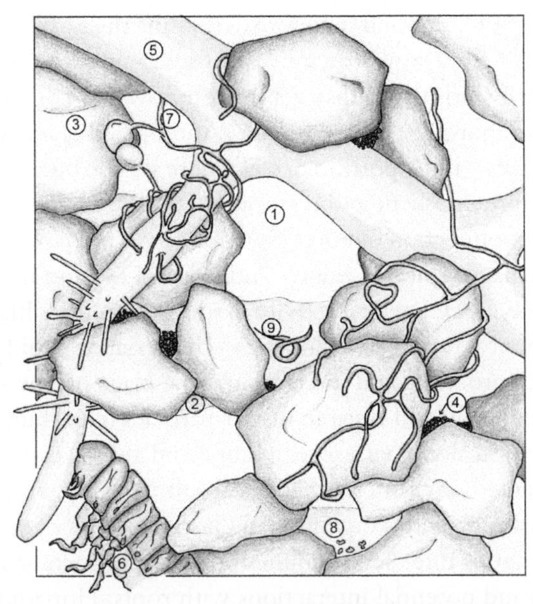

Figure 3.1 Despite an apparent homogeneity, soil is a very complex medium. Besides atmospheric air (20–30% of the total composition) filling up pores (1), water is present either in the gaseous phase or as a liquid ((2), 20–30% of the total composition), influencing both biotic and abiotic conditions. The mineral portion of soil ((3), 45% of the total composition) varies not only with the basal rock characteristics it originated from but also with external outcomes brought along by wind or water. Organic matter (4) represents a portion of about 8% of the total composition of soils and plays a major role in the formation of the clay–humic complexes that are unique to soils and ultimately allow life to proliferate. The soil biota is spanning from microorganisms to mammals. Plant roots (5) are an important component of soil biota, as they constitute a sink of carbon for insect herbivory (6) or other interaction with fungi (7) or microorganisms (8) such as nematodes (9). The diversity of the encountered organisms and the diversity of the holding media make interactions in soil very complex and extensively diverse. Chemical cues between soil-dwelling organisms are an important component of their communication arsenal. *Drawing: I. Hiltpold.* (For colour version of this figure, the reader is referred to the online version of this chapter.)

offers considerably more variation and therefore an increased diversity in soil types and properties (see also Chapter 1). Soil diversity is characterised by several properties. (1) The soil texture is probably the soil's physical property that has the most profound impact on other soil properties and on soil biota. Defined as the proportion of the three mineral particles, *sand*, *silt* and *clay*, texture results from basal mineral breakdown and exogenous deposition of material by wind and/or water. (2) Soil structure is the arrangement of the soil particles in aggregates. The structure of a soil will impact its stability, porosity and water regime and therefore determine the possible uses of a particular portion of land. Soil structure is classified in seven different types, such as granular or prismatic, that depend on how soil components clump together. (3) Soil porosity, defined as air-/water-filled spaces between particles and aggregates, is the portion of soil where several biotic processes take place. The two previously described properties influence the size, the number and the interconnection of pores. Soil porosity is not constant and can be modified by management procedures and/or land use and has an influence on the soil density. (4) Chemical activity in soils is extremely high. Indeed, in addition to chemical reactions occurring in the aerial or liquid phase of soil, several chemical processes and interactions take place on the surface of the colloids, the finest clay and organic matter particles (clay–humic complex). All these properties, along with several biotic and abiotic processes, result in a highly fragmented environment. Vertical soil diversity ranges from loose litter to the basal rock and is organised in "horizons". As a result of the variability, soil horizons offer several different niches impacting species diversity, behaviour and potential interactions with roots. Horizontally, soil can be relatively homogeneous over a particular surface (i.e. agricultural land) or can dramatically vary depending on local parameters such as topography, water regime and vegetation. These variations affect both the diffusion of chemical signals (Hiltpold and Turlings, 2008) and the related behaviour of the insects (see Chapter 1). As pointed out by Johnson and Gregory (2006), a volatile diffuses more rapidly through 1 m of air than through 1 mm film of water covering soil pores (Payne and Gregory, 1988).

These various soil parameters impact behaviour, mobility and signalling, therefore altering interactions between soil-dwelling organisms in a dramatically different manner than air would in aboveground ecosystems. Indeed, soil organisms (excluding those living in litter) usually do not explore more than a 1 m^2 during their lifespan, whereas above-ground, some insects can wander through several times that surface within 1 day (Hedlund et al., 2004). Chemical properties of soil suggest higher metabolite retention capacity and, therefore, more stable chemical gradient establishment allowing slow-motion soil organisms to locate conspecifics or food resources. As compared to plant shoot physiology, roots evolved particular physiology, potentially leading to very distinct interactions with herbivores in soils (Erb et al., 2012).

Despite these differences between above- and belowground environments, these two ecospheres should not be seen as two separate and independent milieus. Indeed, besides the obvious physical link made by plants crossing the "border" between both spheres, many heterophagous insects spend different parts of their life stages in both environments (e.g. notably Coleoptera and Diptera species) or inhabit both as adults (e.g. Isoptera species) (Brown and Gange, 1990). A growing literature describes aboveground–belowground interactions (e.g. Johnson et al., 2012, 2013; Rasmann and Turlings, 2007; Soler et al., 2007; van Dam et al., 2003, 2004, 2005). As compared to aboveground insects that rely on olfactory cues for long-distance orientation and visual stimuli at a closer range (e.g. Otálora-Luna et al., 2013), soil-dwelling insects possibly do not use visual cues at all and rather chemical or tactile/gustatory cues affect their behaviour and interactions with their surrounding environment (Johnson and Gregory, 2006; Johnson and Nielsen, 2012).

1.2. Root exudation of chemical compounds in the rhizosphere

Thousands of papers report on plant metabolites and their impact on insect ecology and evolution. Indeed, it is commonly accepted that plants produce over 200,000 different compounds (Dixon and Strack, 2003), historically known as "secondary metabolites" but more recently redefined as "specialised metabolites" (Pichersky et al., 2006; Schilmiller et al., 2008; Yonekura-Sakakibara and Saito, 2009). Since the pioneer work by Fraenkel (1959), a wide diversity of compounds and biosynthetic pathways have been described and elucidated (e.g. Pichersky et al., 2006). This striking number of known plant-produced compounds is likely an underestimate. Only a few plant species have been investigated for their production of specialised metabolites, and some compounds are present only in minute quantities that prevent their isolation or detection. However, new methods and specialised equipment, which facilitate in identifying new chemicals, further help to elucidate new biochemical pathways and the ecological relevance of plant metabolites (e.g. Glauser et al., 2013). Care must be taken with these techniques, however, as sample preparation and the instability of some compounds can erroneously produce molecules that are not originally present in plants. Currently, most studies evaluating how root exudates shape their environment have been on agricultural pests such as root-feeding insects (Johnson and Gregory; 2006, Johnson and Nielsen, 2012), plant parasitic nematodes (Rasmann et al., 2012a) or other microorganisms (e.g. Akiyama et al., 2005). Nevertheless, roots synthesise a similar diversity of specialised metabolites as compared to plant shoots (Kaplan et al., 2008). This metabolite diversity is reflected in traditional Chinese medicine where more than one-quarter of the pharmacopoeia were derived from belowground plant material (Bensky and Gamble, 1986).

As discussed in the previous section, soils simultaneously offer two phases for root exudates to diffuse through connected pores: a gaseous phase composed of air and a liquid phase composed of water and water-soluble molecules and ions. This provides the opportunity for soil-dwelling insects, much as other belowground organisms, to exploit volatile organic compounds or water-soluble organic compounds for long-distance detection, local orientation and food acceptance. The role of root-exuded metabolites on soil biota and biotic interactions has been discussed in seminal reviews (Badri et al., 2009; Bais et al., 2006; Bonkowski et al., 2009; Flores et al., 1999; Hiltpold and Turlings, 2012; Rasmann et al., 2012b; van Dam, 2009; Wenke et al., 2010), and growing interest in such interactions will likely lead to a rapid increase of our knowledge. Here, we focus on the interaction between insect and roots, their evolution and potential applications in an agricultural context.

1.3. Root-mediated chemical identification and selection by insects

Plants can release up to 20% of their fixed carbon via exudation through roots (Barber and Martin, 1976). This release of organic metabolites can affect abiotic and biotic soil conditions, and there is increasing evidence that insect behaviour is partially mediated by root exudates. In a comprehensive review, Johnson and Gregory (2006) proposed a seminal schematic model of the importance of plant metabolites in the location and selection of root by insects. Further discussed by Hiltpold and Turlings (2012) and Chapter 2, this model offers a detailed illustration of the different steps affecting insect behaviour during root foraging. First, foraging insects shift from a random to biased-random movement. Such a shift is most likely triggered by carbon dioxide (CO_2) (Johnson et al., 2006). This ubiquitous compound plays a similar role aboveground (Dekker et al., 2005; Turner et al., 2011) as it does with soil-dwelling organisms (Dillman et al., 2012b; Turlings et al., 2012). Certain insects can detect very low variations in CO_2 concentrations (Bernklau and Bjostad, 1998a; Doane et al., 1975; Klinger, 1958) enabling them to follow gradients. As a result, CO_2-biased movement will likely bring the insect into contact with more specific plant cues such as plant-emitted volatile organic compounds and/or water-soluble organic compounds (Johnson and Nielsen, 2012). Either the volatile organic compounds or the CO_2 gradient eventually leads the foraging insect to the plant tissue.

On a finer scale, insects may rely on water-soluble organic compounds to locate their plant host (Bernklau and Bjostad, 2008; Hibbard et al., 1994). Microbes interacting with roots also shape interactions in the rhizosphere (e.g. Cipollini et al., 2012) and therefore can affect insect behaviour (Vannette and Rasmann, 2012). Insects themselves carry chemical-producing symbiotic microbes (e.g. Dematheis et al., 2012) and can additionally facilitate such interactions.

Even though this does not well fit the scope of the present discussion, it is interesting to mention that vibrations are important physical cues for subterranean insects. This growing field of research increases our fundamental knowledge in insect ecology and potential applications in pest control (e.g. De Groot et al., 2011; Ennis et al., 2010; Eriksson et al., 2012; Kojima et al., 2012; Laumann et al., 2013; Mankin, 2012; Mukai et al., 2012; Tokuda et al., 2010), in plant physiology (Gagliano et al., 2012b) and in root ecology (Gagliano et al., 2012a). The conjugated effects of both acoustic and chemical signals on the behaviour of soil-dwelling insects require further research to document the respective and/or cumulative effect of both cues. Since plants can detect vibrations (Gagliano, 2013), those caused by insects in the rhizosphere might result in shifts of specialised metabolite synthesis and exudation to directly defend the plant tissues or to trigger indirect defence pathways.

1.4. Aims and scope

The present discussion aims to provide an overview of the insect physiology of chemical perception and environment-specific adaptation in subterranean insects. The evolution of belowground insect chemotaxis will be discussed. Whereas insects rely on chemical cues to locate their food source, they also emit volatiles that can be used as location cues by natural enemies (e.g. Dillman et al., 2012b) or they could induce shifts in root volatile organic compound profiles resulting in the attraction of such natural enemies (e.g. Hiltpold et al., 2011; Rasmann et al., 2005). Evolutionary aspects of these interactions are still scarce but could help in drawing general hypotheses and theories on soil community interactions, even though there is a strong bias towards agricultural insect pests in our actual knowledge. An update on newly released methodologies to study belowground insect behaviour and physiology in relation to roots is also proposed. Indeed, recent developments are offering root biologists new powerful approaches to disentangle root interactions with their surrounding belowground communities. Because of their economic and social impact on human societies, most of the belowground research focuses on root pests mainly from the Coleoptera and Diptera orders. Therefore, a detailed example of root pest chemical ecology is finally discussed, using Western corn rootworm (WCR) (*Diabrotica virgifera virgifera* LeConte, Coleoptera: Chrysomelidae) as a model species, and we discuss the potential of chemical ecology in managing soil-dwelling insect pests.

2. OLFACTION IN SOIL-DWELLING INSECTS

As for any other organism, insects rely on a wide range of sensory systems not only to locate and evaluate food but also to shelter, mate or avoid threats such as predation and harmful abiotic environments. Smelling their surrounding direct and distal environment is of pivotal importance for insects (Dethier, 1947) even though it can be sometimes subordinate to vision in aboveground species such as *Diaprepes abbreviatus* (Coleoptera: Curculionidae) (Otálora-Luna et al., 2013). Yet, the necessity of olfaction reflects the wide range of antennal structures that both aboveground and, to some moderate extent, belowground insects possess. In addition to their antennae, insects also evolved other organs to detect chemical compounds such as maxillary or labial palps (Schneider, 1964). However, the necessity for multiple types and locations of olfactory organs still remains unclear. It can be hypothesised that both antennae and palps are designed to sense different sorts of compounds, as in the malaria vector *Anopheles gambiae* (Diptera: Culicidae), which has palpal sensillum extremely sensitive to CO_2 and 1-octen-3-ol (Lu et al., 2007), two volatile organic compounds usually emitted by its hosts. However, some other species exhibit overlapping in odour detection by antennae and palps. The vinegar fly *Drosophila melanogaster* (Diptera: Drosophilidae) accomplishes CO_2 detection with both organs as their respective olfactory sensory neuron activity overlaps (De Bruyne et al., 1999). The reason for this remains unclear but is very likely to result from selection for high sensitivity and phylogenetic and/or developmental limits and abiotic constraints (Hansson and Stensmyr, 2011). This ability to accurately smell the environment and the evolutional plasticity of this trait may explain why insects are so abundant and diverse in heterogeneous ecological niches, like soils.

2.1. Molecular mechanisms of odour detection by insects

Sensory organs in insects come in multiple shapes and sizes and are located in several places on the insect body. Nonetheless, mechanisms of odour capture and further identification are very similar. Molecular responses of insects to odorant cues are characterised by two large and exclusive gene families coding for (1) the odorant-binding proteins (Vogt and Riddiford, 1981) and (2) odorant receptors (Clyne et al., 1999; Vosshall et al., 1999). Odorant-binding proteins are produced and secreted in large quantities in the lymph surrounding the olfactory sensory neuron (Swarup et al., 2011). The exact role of these proteins still remains to be elucidated, but it is widely hypothesised that odorant-binding proteins are involved in the transport of odour ligands to the receptor site on the neural dendrites (Hansson and Stensmyr, 2011) when volatiles reach the sensillum lymph via pores or slits in the sensillum cuticle (Steinbrecht, 1997). Similar proteins have also been described in mammals (Vieira et al., 2007); however, they are structurally dissimilar. Hence, insect odorant-binding proteins could be as old as insects themselves and might have evolved to face the dry land conquest constraints (Vieira et al., 2007). The contact between the odour stimuli and the odorant receptors on the surface of the olfactory sensory neuron membrane triggers the production of signals in the dendrite, which are then transmitted to the insect's brain. Similar to the odorant-binding protein gene family, odorant receptor genes, coding for the neural receptors, are also exclusive to insects (as well as insect gustatory receptors) (Clyne et al., 2000; Scott et al., 2001). Although odorant receptors are more documented in comparison to odorant-binding proteins, the evolution and diversification of odorant receptor gene family remains uncertain (Hansson and Stensmyr, 2011 and references therein). In some species, such as the *D. melanogaster* (Kondoh et al., 2003), a sexual brain dimorphism evolved to respond to the necessity of males to detect pheromones to identify reproductive conspecifics.

Even if many evolutionary and physiological aspects still need to be clarified, insects have the ability to detect odours using very specific and fine-tuned morphological and physiological adaptations. Olfactory machinery has evolved to specifically respond to ecological constraints encountered by root-feeding insects.

2.2. Smelling the dark

Most of the studies on insect olfaction focus on the olfactory perception of their environment by aboveground insects. Therefore, even though it is likely to exhibit similarities, little is known about the sensory appendages and olfactory physiology of soil-dwelling insects or their larvae. Yet, because of their economic impact, studies on different rhizophagous species have been conducted, mainly focusing on the detection by insects of the ubiquitous CO_2 signal (e.g. Cobb, 1999; Doane and Klingler, 1978; Giglio et al., 2003; Jepson, 1937; Keil, 1996; Klinger, 1957; Stange and Stowe, 1999; Sutherland and Hilier, 1974). A comprehensive study was recently published by Eilers et al. (2012) where they demonstrated that belowground insects are able to physiologically discriminate several volatile organic compounds. After an in-depth description of the sensory structure of the European cockchafer *Melolontha melolontha* (Coleoptera: Scarabaeidae) larva, the authors tested the olfactory responses of these appendixes to compounds likely to be present in the rhizosphere such as host plant kairomones (Eilers et al., 2012). In a series of elegant experiments, they demonstrated that the chafer larva has a highly developed chemosensory apparatus and responds to a wide range of compounds typically emitted by plants and potentially by plant roots. Moreover, each olfactory structure showed a very specific response profile to various stimuli with some discriminations down to the level of molecular chirality (Eilers et al., 2012), suggesting a very well-developed sense of smell, even though appendixes are more much discreet than similar ones on aboveground insect herbivores (Eilers et al., 2012). Narrowness of soil pores and cracks also limits the development of delicate olfactory appendixes such as those sometime encountered aboveground. In a similar study on the congeneric forest cockchafer *M. hippocastani* (Coleoptera: Scarabaeidae), Weissteiner et al. (2012) reported an antennal lobe composed of around 70 glomeruli. This brain complexity reflects the diversity of the odorant receptor proteins and consequently the olfactory sensory neuron (Stocker, 1994) and is comparable to the well-established complexity described in aboveground adult model insects (e.g. Grosse-Wilde et al., 2011; Stocker, 2001). Because of its extended development time spent in soil, the cockchafer could have evolved a very complex chemosensory apparatus that may not be shared with all subterranean insect larvae. Yet, other species possess well-developed olfactory appendixes (Cobb, 1999), and it is likely that the knowledge of the olfactory physiology of soil-dwelling insects will increase in the future.

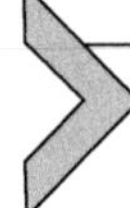

3. EVOLUTION OF INSECT CHEMOTAXIS IN ROOT–INSECT INTERACTIONS

3.1. Adopting an evolutionary approach

Chemotaxis is generally defined as a directed behaviour towards or away from a source of chemical stimuli. In this section, we summarise knowledge on major attractants and phagostimulants for foraging root–insect herbivores. Whereas CO_2 plays an important role in subterranean chemotaxis (e.g. Turlings et al., 2012, and Section 5 of the present discussion), it is a ubiquitous belowground cue that most organisms studied respond to (e.g. Johnson and Gregory, 2006; Johnson and Nielsen, 2012; Rasmann et al., 2012a and Section 5) and therefore was excluded from the following analyses.

Using phylogenetic tools and food web analytic approaches, evolutionary patterns emerged from the published literature on insect–root-positive chemotaxis or host acceptance, even though they were biased towards crop insect pests (Tables 3.1 and 3.2). Briefly, the phylogeny was based on gene sequences publicly available on GenBank for three nuclear DNA regions (18S rRNA (KC177301.1, KC177311.1, AF515703.1, AF308339.1, AY244865.1, AY988454.1), 28S rRNA (DQ198733.1, KC177741.1, KC177819.1, FJ867676.1, AF308387.1, JX263722.1, AY243741.2, GU226589.1, EU286372.1) and ace2 (AJ868345.1, AJ868312.1)). The three regions were initially aligned individually with ClustalX (Thompson et al., 1997) and thereafter manually adjusted with BioEdit (Hall, 1999) using the similarity criterion (Morrison, 2006). SequenceMatrix (Vaidya et al., 2011) was used to construct a supermatrix. ML analyses were performed using RAxML (Stamatakis et al., 2008) with 500 rapid bootstrap analyses followed by the search of the best-scoring ML tree in one single run. This analysis was done using the facilities offered by the Vital-IT portal in Lausanne, Switzerland (http://phylobench.vital-it.ch/raxml-bb/). Classical systematics was used for the second set of data; indeed, the phylogeny of the present taxon was not accurate enough as the overlap between gene alignments across species was too low. Identified molecules were sorted in broad chemical classes and the relation between a particular insect and a particular class of compound was quantified as a percentage of the overall response of this insect to all chemicals identified and having a similar function. Food webs were further analysed using the package "bipartite" (Dormann et al., 2008) in R (R Core Team, 2012). The insect chemotaxic orientation towards roots and root acceptance as suitable food by insects were not nested ($p=0.33$ and $p=0.67$, respectively). Nestedness is often related to the robustness of a given ecological network (Bascompte et al., 2003; Bastolla et al., 2009), but a possible explanation why no such networks are observed here can be the very limited amount of data available, mainly focusing on a restricted number of root pests instead of representing the actual complexity of soil food webs. However, the insect

Table 3.1 Chemical cues emitted by plants in the rhizosphere that enable soil-dwelling insects to locate host roots

Insect order	Insect species	Plant specificity	Chemical compound	Class	References
Coleoptera	*Hylastes nigrinus*	S	α-Pinene	te	Rudinsky (1966) and Rudinsky and Zethner-Møller (1967)
			β-Pinene	te	
			Camphene	te	
	Hylastinus obscurus	S	Pentadecanal	ad	Kamm and Buttery (1984)
			Hexadecanal	ad	
			(*E*)-2-Hexenal	ad	Tapia et al. (2007)
			Hexanoic acid	ca	Kamm and Buttery (1984)
			Ethyl laurate	es	
			Ethyl benzoate	es	
			Methyl benzoate	es	Tapia et al. (2007)
			Estragole	o	Kamm and Buttery (1984)
	Dibrotica v. virgifera	S	(*E*)-β-Caryophyllene	te	Robert et al. (2012a)
			Oleic acid	li	Hibbard et al. (1994)
			Stearic acid	li	
			DIMBOA	o	Robert et al. (2012b)
	Hylobius abietis	G	α-Pinene	te	Nordenhem and Nordlander (1994)
			Ethanol	al	

Table 3.1 Chemical cues emitted by plants in the rhizosphere that enable soil-dwelling insects to locate host roots—cont'd

Insect order	Insect species	Plant specificity	Chemical compound	Class	References
	Cylas formicarius	S	Geraniol	al	Wang and Kays (2002)
			Gurjunene	te	
			Humulene	te	
			Ylangene	te	
	Melolontha melolontha	G	Benzaldehyde	ad	Eilers et al. (2012)
			Ethyl acetate	es	
			Limonene	te	
			Linalool	te	
			α-Pinene	Te	
			β-Pinene	te	
			β-Myrcene	te	
			(±)-Camphene	te	
			Farnesene	te	
			β-Caryophyllene	te	
			β-Cedrene	te	
			Octanol	al	
			Hexyl acetate	es	
	Agriotes spp.	G	Ethyl acetate	es	Morgan and Crumb (1929)
			Nitrobenzene	o	
			Aspartic acid	aa	Thorpe et al. (1946)
			Asparagine	aa	
			Malic acid	ca	
			Succinic acid	ca	
			Glutamine	aa	
			Glutamic acid	aa	

Continued

Table 3.1 Chemical cues emitted by plants in the rhizosphere that enable soil-dwelling insects to locate host roots—cont'd

Insect order	Insect species	Plant specificity	Chemical compound	Class	References
Diptera	*Psila rosae*	S	Bornyl acetate	es	Guerin and Ryan (1984) and Ryan and Guerin (1982)
			2,4-Dimethyl styrene	hy	
			Biphenyl	hy	
			α- and β-Ionone	ke	
			Falcarinol	o	Maki and Ryan (1989) and Maki et al. (1989)
			Falcarinodiol	o	
			Falcarinodiol monoacetate	o	
	Delia antiqua	S	*n*-Propyl disulphide	es	Matsumoto (1970) and Ross and Anderson (1992)
			Methyl disulphide	es	
			Ethyl acetate	es	
			21 Esters[a]	es	Mochizuki et al. (1989)
			n-Propyl mercaptan	mc	Matsumoto (1970) and Ross and Anderson (1992)
			n-Heptanal	ad	Mochizuki et al. (1989)
			Pentanal	ad	
			Hexanal	ad	
			Heptanal	ad	
			Propanol	al	
			Butanol	al	
			Pentanol	al	

Table 3.1 Chemical cues emitted by plants in the rhizosphere that enable soil-dwelling insects to locate host roots—cont'd

Insect order	Insect species	Plant specificity	Chemical compound	Class	References
			Hexanol	al	
			Heptanol	al	
			Valeric acid	ca	
			Caproic acid	ca	
			Enanthic acid	ca	
			Allyl isothiocyanate	ie	Ross and Anderson (1992)
			Sulphur compounds[b]	su	Soni and Finch (1979)
			Ethyl sulphide	su	Matsumoto (1970)
			n-Butyl sulphide	su	
			iso-Butyl sulphide	su	
			n-Butyl methyl sulphide	su	
			n-Butyl ethyl sulphide	su	
			Isopentyl sulphide	su	
			Allyl sulphide	su	
			n-Propyl disulphide	su	Ross and Anderson (1992)
			Tetramethylpyrazine	o	Matsumoto (1970)
	Delia floralis	S	Allyl isothiocyanate	ie	Ross and Anderson (1992)
			n-Dipropyl disulphide	su	
			Allyl alcohol	al	
			Methyl eugenol	o	

Continued

Table 3.1 Chemical cues emitted by plants in the rhizosphere that enable soil-dwelling insects to locate host roots—cont'd

Insect order	Insect species	Plant specificity	Chemical compound	Class	References
	Delia radicum	S	Isothiocyanates[c]	ie	Finch and Skinner (1974)
			Allyl isothiocyanate	ie	Košt'ál (1992) and Ross and Anderson (1992)
			Ethyl isothiocyanate	ie	
			Hexanol	al	Košt'ál (1992)
			cis-3-Hexen-1-ol	al	
			Linalool	al	
			Hexanal	ad	
			n-Dipropyl disulphide	su	Ross and Anderson (1992)
			Allyl alcohol	o	
			Methyl eugenol	o	
	Coquillettidia spp.		Glycerol	al	Sérandour et al. (2008)
			Uracil	py	
			Thymine	py	
			Uridine	nu	
			Thymidine	nu	

[a]Mochizuki et al. (1989) listed 21 esters (not listed here) that are attractive to *Delia antiqua*.
[b]These compounds reported by Soni and Finch (1979) were either attractive or repellent to *Delia antiqua* depending on the concentration.
[c]Finch and Skinner (1974) reported unspecified isothiocyanates attractive to *Delia radicum*.
"Plant specificity" refers to the host range of the insect; (S) specialist (mono- and oligophagous) and (G) generalist (polyphagous). "Class" describes the broad nature of each compound; (aa) amino acids, (ad) aldehydes, (al) alcohols, (ca) carboxylic acids, (es) esters, (hy) hydrocarbons, (ie) isothiocyanates, (ke) ketones, (li) lipids, (mc) mercaptans, (nu) nucleosides, (py) pyrimidines, (su) sulphides, (te) terpenes and (o) others. Links between insects and these classes are further analysed and presented in Fig. 3.2.
Adapted from Johnson and Nielsen (2012).

Table 3.2 Chemical cues emitted by plants in the rhizosphere that enable soil-dwelling insects to accept host roots as a suitable source of food

Insect order	Insect species	Plant specificity	Chemical compound	Class	References
Diptera	*Delia floralis*	S	Allyl isothiocyanate	ie	Rygg and Sömme (1972)
			Glucose	sg	Hopkins et al. (1993)
			Fructose	sg	
	Psila rosae	S	Chlorogenic acid	o	Cole (1985)
	Delia antiqua	S	Fructose	sg	Honda and Ishikawa (1987) and Mochizuki et al. (1985)
			Sucrose	sg	
Coleoptera	*Popillia japonica*	G	Sucrose	sg	Ladd (1988)
			Maltose	sg	
			Fructose	sg	
			Glucose	sg	
			Trehalose	sg	
	Costelytra zealandica	S	Sucrose	sg	Sutherland (1971)
			Aspartic acid	aa	Sutherland and Hillier (1974)
			Glutamic acid	aa	
			Serine	aa	
			Ascorbic acid	o	
	Acalymma blomorum	S	Cucurbitacins	te	Eben et al. (1997)
	Diabrotica balteata	S	Cucurbitacins	te	Eben et al. (1997)
	D. v. virgifera	S	Glucose	sg	Bernklau and Bjostad (2008)
			Fructose	sg	
			Sucrose	sg	
			Cucurbitacins	te	Tallamy et al. (2005)

Continued

Table 3.2 Chemical cues emitted by plants in the rhizosphere that enable soil-dwelling insects to accept host roots as a suitable source of food—cont'd

Insect order	Insect species	Plant specificity	Chemical compound	Class	References
	Diabrotica undecimpunctata	S	Cucurbitacins	te	Eben et al. (1997)
	Heteronychus arator	G	Alanine	aa	Sutherland et al. (1980)
			Aspartic acid	aa	
			Glutamic acid	aa	
			Maltose	sg	Sutherland and Hillier (1976)
			Sucrose	sg	
			Fructose	sg	
			Glucose	sg	
	Sericesthis geminata		Alanine	aa	Wensler and Dudzinski (1972)
			Leucine	aa	
			Isoleucine	aa	
			Sucrose	sg	
			Glucose	sg	
			Maltose	sg	
	Lepidiota negatoria	G	Fructose	sg	Allsopp (1992)
			Mannose	sg	
			Sucrose	sg	
			Raffinose	sg	
			Trehalose	sg	
			Melezitose	sg	
			Cysteine	aa	
			Glutamic acid	aa	
			Histidine mono-hydrochloride	o	

Table 3.2 Chemical cues emitted by plants in the rhizosphere that enable soil-dwelling insects to accept host roots as a suitable source of food—cont'd

Insect order	Insect species	Plant specificity	Chemical compound	Class	References
	Antitrogus parvulus	G	Sucrose	sg	Allsopp (1992)
			Raffinose	sg	
			Tyrosine	aa	
	Agriotes spp.	G	Glucose	sg	Thorpe et al. (1946)
			Sucrose	sg	
			Fructose	sg	
			Galactose	sg	
			Maltose	sg	
			Stachyose	sg	

"Plant specificity" refers to the host range of the insect; (S) specialist (mono- and oligophagous) and (G) generalist (polyphagous). "Class" describes the broad nature of each compound; (aa) amino acids, (ie) isothiocyanates, (sg) sugars, (te) terpenes and (o) others. Links between insects and these classes are further analysed and presented in Fig. 3.3.
Adapted from Johnson and Gregory (2006).

chemotaxic orientation towards roots and root acceptance as suitable food by insects are not random but, as often when nestedness is not observed (Fortuna et al., 2010), appear to have a modular topology, indicating that particular groups of insects were responding to particular groups of chemical classes, therefore reflecting a certain stability (Krause et al., 2003; Teng and McCann, 2004). Modularity was not significant either, but again, these kinds of tests often require a more diverse and documented data set in order to be interpreted correctly. However, strong trends are still present and are discussed in the following sections.

3.2. Plant location

To date, belowground insects have demonstrated positive chemotaxis to over 70 different chemical compounds exuded by healthy or damaged host roots (Table 1). Even though limited to a relatively small number of insect species and chemical classes, the chemotaxic web resulting from the available literature appears highly complex. Indeed, several species from several orders respond to a number of classes of compounds (Fig. 3.2A and B). However,

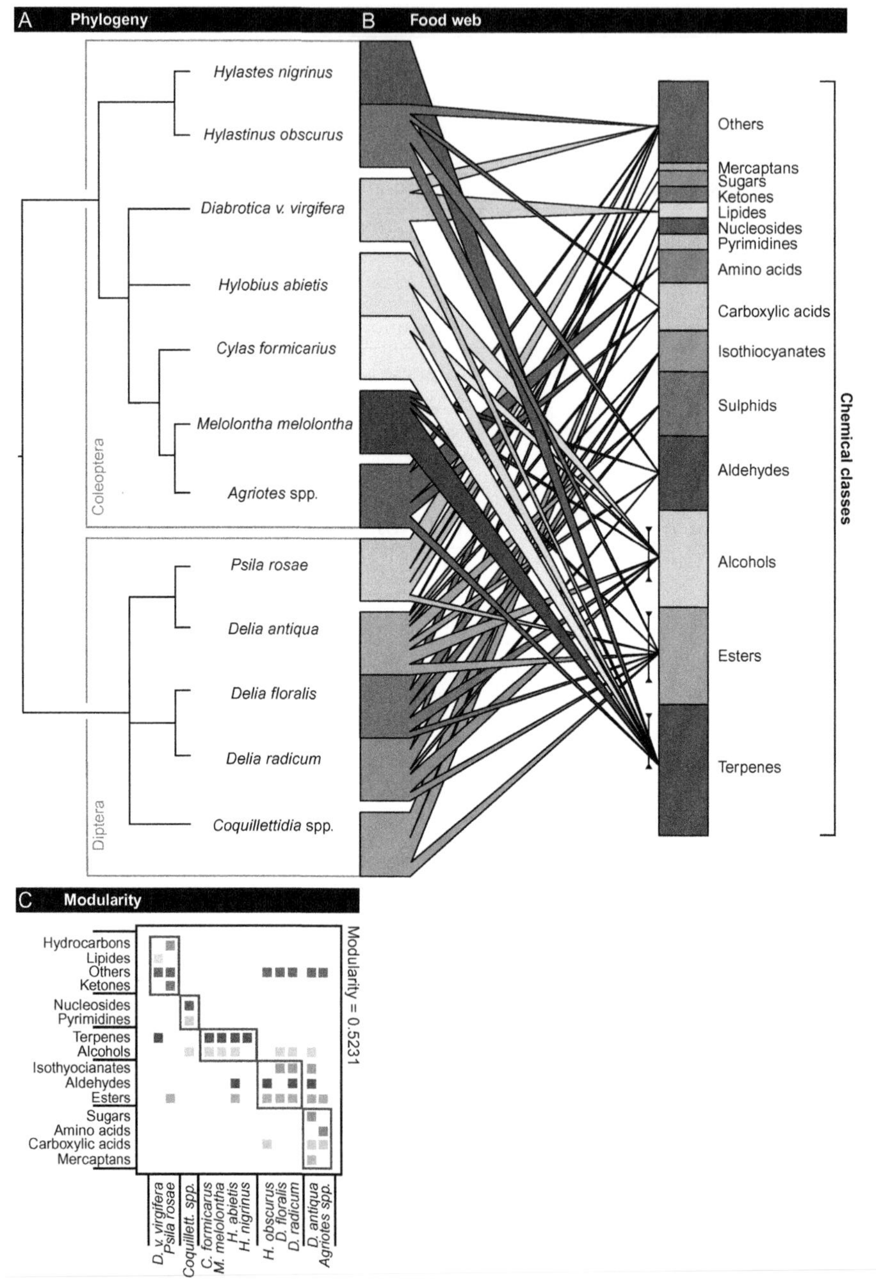

Figure 3.2 Evolution and chemical web of the location of roots by insects. (A) Phylogenetic tree of insects showing positive chemotaxis to certain compounds emitted by roots. (B) Chemical web representing the interactions between each insect species and each chemical class. The width at the origin of the links represents the strength of the relationship as a percentage of the total interaction. Arrows are discussed in the text (Section 3.2). (C) Modular distribution of the insect response to the root-emitted chemicals. Groups are discussed in the text (Section 3.2). (For colour version of this figure, the reader is referred to the online version of this chapter.)

patterns, to some extent supported by the analysis of modularity (Fig. 3.2C), can be highlighted in Fig. 3.2. Indeed, three classes of plant chemical compounds appear to be pivotal in the attraction of insects towards roots and constitute 44% of the chemicals playing such a role. Terpenes are the major class of compounds that insects respond to (17% of total); compounds from this class act as a chemoattractant to insect species in the order Coleoptera, whereas no Dipteran species have been shown to exhibit positive chemotaxis to that same class. Terpenes are volatile organic compounds and act as potent long-distance signals for some foraging insects. Even though their solubility in water is low, moist conditions are necessary for an optimal diffusion of terpenes in soil (Hiltpold and Turlings, 2008), as they may interact significantly more with soil colloids or evaporate when water is lacking. The next two chemical classes with the most documented influence on foraging soil-dwelling insects are esters and alcohols (13.5% of the total, respectively) (Fig. 3.2B). Small esters have relatively high solubility in water and therefore would not be of much help for foraging insects trying to locate roots from a distance but rather support a root acceptance process (see Section 3.3). However, esters' solubility dramatically falls with the length of their carbon chain. As compared to terpenes, the response to esters is more widespread across the current list of soil insect species documented to have interactions with plants based on chemistry (Fig. 3.2B). This general response might be the result of the extensive presence of ester metabolites in plants and roots. Indeed, some major plant hormones belong to this chemical class (Erb et al., 2008; Xie et al., 2013) and several fruit and flower smells perceived as sweet are esters. Equally important as esters, alcohols represent the third main class of chemicals that insects rely on to locate potential root hosts (Fig. 3.2B). Similar to esters, the water solubility of alcohols dramatically drops with the increasing number of carbon atoms (and/or their special organisation) attached to the hydroxyl group. As most of the molecules belonging to this class are rather complex (Table 3.1) and their water solubility is low enough, they can be classified as volatile organic compounds. Whereas these three classes of chemical compounds all diffuse mainly in the subterranean atmosphere and may act as distance cues, several of the remaining classes have various water solubility and can be considered as either water-soluble or not. However, they may still trigger foraging behaviour in certain insects or constitute a component of blends that induce such behaviour. It is also interesting to note that subterranean insects that are considered to have a specific relationship with their host plant are more likely to respond to very special classes of compounds that would be either avoided by generalist feeders (isothiocyanate; e.g. Ranger et al. (2011)) or considered neutral (pyrimidines and nucleosides; Sérandour et al. (2008)).

Yet, besides CO_2, soil-dwelling insects also rely on some commonly found classes of compounds in order to adopt the biased-random movement defined by Johnson and Gregory (2006). However, they also use more specific compounds such as water-soluble organic compounds or compounds involved in particular root functions (such as defences) to refine their search and target their appropriate host. Most of the molecules used by insects to locate their host are volatile compounds, suggesting that this type of chemicals play a major role in the long-range foraging behaviour of root-herbivore insects.

3.3. Plant acceptance or rejection

Once the insects have found roots in the soil matrix, they again rely on various chemicals to accept them as suitable food. Acting as contact cues, these molecules are likely to be more soluble in water, therefore less motile in soil (Johnson and Gregory, 2006; Payne and Gregory, 1988), than volatile organic compounds used by insect to locate their potential host root over a longer distance. Johnson and Gregory's hypothesis is supported by peer-reviewed literature where a large majority of the compounds promoting host acceptance or further feeding are molecules that are soluble in water (Fig. 3.3B; sugars 53% and amino acids 12%). Despite the presence of several systematic groups (Fig. 3.3A), most of the tested insect species rely on either sugars or amino acids (sometimes both) to accept root material as suitable food. Scarabaeid larvae, in particular, use these two classes of compounds as phagostimulants (Fig. 3.3A and B). As scarab larvae are rather sedentary in soil, they might have evolved a more accurate method in detecting water-soluble organic compounds as compared to using long-range cues, such as volatile organic compounds. More motile species, such as *Delia* spp. or *Diabrotica* spp., do rely solely on not only water-soluble organic compounds but also less soluble molecules such as terpenes or isothiocyanates as phagostimulants. Because they are longer-range cues, relying on volatile organic compounds as a phagostimulant could allow specialised foraging insects to not only detect the presence of roots but also detect the suitability of the potential food and either to move towards a particular resource or to seek another potent signal. These insects could subsequently use water-soluble organic compounds for shorter-range food acceptance.

Patterns in web structure (Fig. 3.3C) suggest a strong influence of terpenes on the studied chrysomelid larvae. Indeed, the chrysomelid species

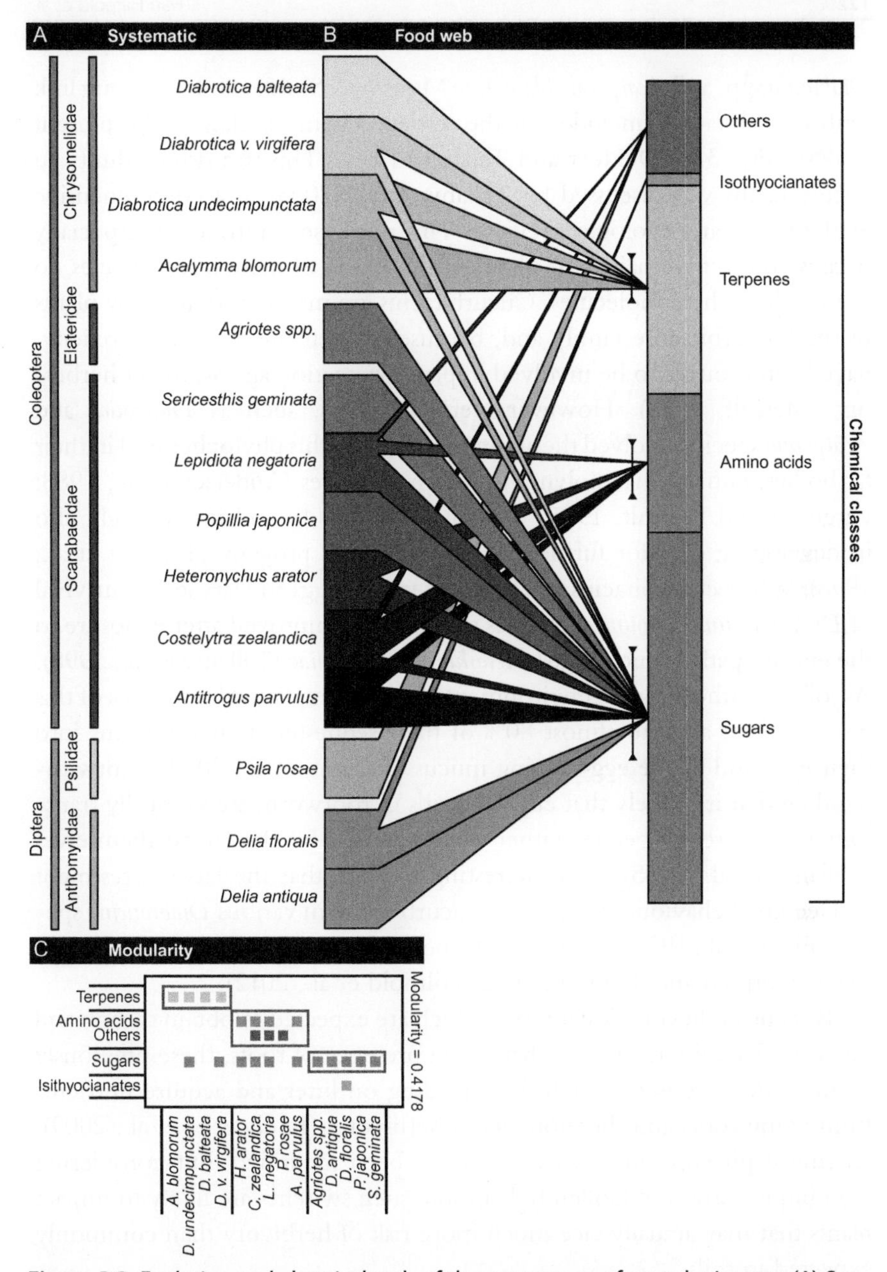

Figure 3.3 Evolution and chemical web of the acceptance of roots by insects. (A) Systematic groups of insect exploiting certain compounds emitted by roots to accept the plant material as suitable food. (B) Chemical web representing the interactions between each insect species and each chemical class. The width at the origin of the links represents the strength of the relationship as a percentage of the total interaction. Arrows are discussed in the text (Section 3.3). (C) Modular distribution of the insect response to the root-emitted chemicals. Groups are discussed in Section 3.3. (For colour version of this figure, the reader is referred to the online version of this chapter.)

Diabrotica spp. and *Acalymma blomorum* Munroe & Smith exhibit a strong link with cucurbitacins (included in the terpene chemical class in the present analysis, Fig. 3.3) (DeHeer and Tallamy, 1991). This surprising affinity to cucurbitacins was discussed by Tallamy et al. (2005) in a comprehensive evolutionary and ecological review. This response is intriguing, especially because the actual main host plants of the larvae of these insect species do not produce these molecules. Cucurbitacins are mainly produced by plants of the Cucurbitaceae family and, because of their bitterness and toxicity, have been thought to be involved in plant protection against insect herbivory (Metcalf, 1985). However, certain insects, such as *Diabrotica* and *Acalymma* species, evolved the ability to sequester this phytochemical in their fat bodies, cuticles, haemolymphs, spermatophores (Andersen et al., 1988; Ferguson and Metcalf, 1985; Tallamy et al., 2000) and interestingly also in developing eggs for further protection of the progeny (Tallamy et al., 1998). When cucurbitacins were sequestered in eggs and larvae, the survival of *Diabrotica undecimpunctata* L. was significantly improved after exposure to the entomopathogenic fungus *Metarhizium anisopliae* (Tallamy et al., 1998). As soil is a pathogen-rich environment, this may explain why females of this insect species allocate almost 80% of their sequestered cucurbitacins into their eggs and in the egg-coating mucus (Tallamy et al., 2000). Notwithstanding that it is likely that cucurbitacins in rootworm are vertically transmitted by parental beetles rather than acquired by the larvae themselves (Tallamy et al., 2005), it is interesting to note that the larval arrestment and feeding behaviour triggered by cucurbitacins in various *Diabroticina* species (Eben et al., 1997) can potentially be useful in rootworm larval management (Hiltpold and Turlings, 2012; Hiltpold et al., 2012).

Recent studies on Collembola, which are expected to obtain carbon and nitrogen from litter, showed that in the presence of roots, these previously assumed decomposers switch from feeding on litter and acquire nutrients from living roots and therefore act as herbivores (Endlweber et al., 2009). Chemical phagostimulants have not yet been identified, but considering the ubiquitousness of Collembola in soil, such switches are likely to impact plants that may actually face much more risk of herbivory than commonly expected in soils.

Insects also rely on exudates to identify unsuitable resources. Interestingly, most of the identified deterrent compounds belong to the isoflavonoid class of molecules (Johnson and Gregory, 2006 and references therein). Isoflavonoids are widely found in several plant groups, particularly in legumes, and are involved in several biological processes in the shoots and the roots.

Interestingly, their solubility in water varies with the number of bonds with sugar residues, making them either local water-soluble organic compound cues or more distal volatile organic compound signals, possibly allowing mobile insects to discriminate (e.g. accept/reject) at a certain distance from the root. While those examples listed by Johnson and Gregory (2006) were mainly repellent, others such as formononetin were actually attractive to larvae of the clover root weevil, *Sitona lepidus* (Coleoptera: Curculionidae) (Johnson et al., 2005).

Several questions on how insects use and respond to root exudates while in the process of accepting or rejecting plant material as suitable food remain unsolved. However, the present discussion highlights that water-soluble organic compounds are crucial for various insect species in the process of host acceptance. While this apparent homogeneity in insect response to sugars and amino acids is unlikely to explain insect specificity to a certain diet, this is important to note that ratios between these chemicals can mediate a specific response (e.g. Bernklau and Bjostad, 2008), therefore allowing insects to specialise on particular plant species. With the realisation that root herbivores use several other classes of compounds as phagostimulants, a broad field of research is now open, an undertaking that is sure to lead knowledge applicable to applied science, especially in the context of plant breeding for natural insect resistance. This could be achieved by selecting plants overexpressing deterrent compounds or roots lacking one or several key phagostimulants (Hiltpold and Turlings, 2012) or in the development of baited soil insecticides (Bernklau and Bjostad, 2005).

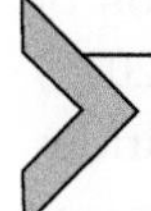

4. METHODOLOGY IN STUDYING ROOT–INSECT INTERACTIONS

Developing experimental procedures to study subterranean interactions between plants and insects is challenging and studying insect behaviour without seeing the insect necessitates inventiveness. In the context of interactions mediated by root volatile organic compounds and water-soluble organic compounds, researchers also face the challenge of collecting and isolating chemicals from the soil for identification purposes. Several peer-reviewed papers, book chapters and handbooks describe proven and innovative methods to study the rhizosphere (e.g. Dawson and Byers, 2008; Johnson et al., 2007; Luster and Finlay, 2006; Mankin et al., 2008; Rasmann et al., 2012b). Therefore, in the following sections, we only focus on some recently published approaches to study root–insect interactions and the chemicals involved in these particular trophic links. Details on methods that are not discussed in the succeeding text were reviewed by Neumann et al. (2009) or described in the publications cited earlier.

4.1. Insect behaviour in soil

4.1.1 What did the insect feed on?

This apparently trivial question might be hard to tackle when a new or barely known species is the focus of a research programme. For instance, Hill and O'Malley (2010) conducted a series of biological studies on the Mount Hermon June beetle (*Polyphylla barbata*, Coleoptera: Scarabaeidae). This beetle is endemic to the Zayante Sandhills of the Santa Cruz Mountains in the San Francisco Bay Area (California, USA) and is listed as an endangered species. The species was believed to be specialist on tree roots but very little was known about its host plant range, making the selection of appropriate conservation practices problematic. During a two-year study, Hill and O'Malley (2010) captured beetle larvae and collected their frass. After visual microscopic analyses of the fragments contained in the frass, they could only identify fungal and plant structures (Hill and O'Malley, 2010). However, the authors described in detail the vegetation in the habitats from which they collected the larvae (Hill and O'Malley, 2010). Therefore, specific primers could be designed according to the surrounding vegetative composition and more detailed analyses could be conducted. Indeed, molecular tools allow *a priori* and *a posteriori* identification of food sources from digested material and could be a very potent approach in the qualification and quantification of belowground food webs. In an *a priori* approach, Wallinger et al. (2012) designed a diagnostic PCR (polymerase chain reaction) to identify various plant taxa and species from the click beetle (*Agriotes* spp.) gut content or faeces. After having designed primers to amplify specific fragments of chloroplastic DNA from the tmL-F region, Wallinger et al. (2012) distinguished plant DNA from the entire insect body DNA and other DNAs from the gut contents. The optimised diagnostic PCR estimated that 30% of *Agriotes* tested positive for DNA of at least one of the 15 targeted plant species (Wallinger et al., 2012). Using a similar approach, plant material from various species could be detected in the guts of click beetles over a period of 72 h postconsumption (Wallinger et al., 2013). *Agriotes* species seemed to prefer rhizome- and rosette-forming plants to those species that grew tap or fine roots (Wallinger et al., 2013). These studies, as well as an extensive number of others conducted in various ecosystems (e.g. Chapter 5; Campos-Herrera et al., 2011, 2012; Chapman et al., 2012; Gariepy and Messing, 2012; Gariepy et al., 2008; Lundgren et al., 2009a), support the use of molecular approaches as a powerful tool in describing *a priori* trophic links between known species. In an *a posteriori* approach, next-generation sequencing is a highly potent tool to assess the diet of several animals from various groups (Pompanon et al., 2012 and references therein). This approach seems to be much more consistent and comprehensive than other techniques used so far, such as protein electrophoresis (e.g. Paill et al., 2002; Walrant and Loreau, 1995), the use of antibodies (e.g. Harwood et al., 2001) or stable isotopes (e.g. Sasakawa, 2011; Traugott et al., 2007). High-throughput sequencing results in the sequencing of many of the species' DNA present in the guts of a particular root–insect feeder. Sequences can then be *a posteriori* compared to existing databases and the dietary regime of an animal determined. The costs of such analyses decline as new products and technologies evolve, and together with the ever-expanding publicly available DNA libraries, this approach is very potent when identifying trophic linkages within subterranean food webs. Next-generation sequencing may also be a very powerful tool for understanding how insects adapt to plant direct defences (e.g. Pauchet et al., 2010).

To summarise, molecular tools offer great potential for ecologists to study the considerable complexity of subterranean food webs (also see Chapter 5). In soil, where direct observation is seldom possible, this indirect estimation of the dietary behaviour/preference of insects will certainly provide useful applied information in various contexts such as agroecology, conservation, alien species monitoring or more fundamental knowledge, such as energy flow between/within ecosystem niches.

4.1.2 What is the foraging behaviour of a soil-dwelling pest?

Once the dietary range of a particular insect is described, one could wonder how the insect got to its food source. Again, the lack of an easy direct observation of a subterranean insect foraging for its root makes the description of this biological process difficult. Several methods in laboratory, greenhouse and field setups are well established and often used to address behavioural questions on belowground insects (Dawson and Byers, 2008). Recently, host plant selection by WCR larvae was assessed (Robert et al., 2012a) in a modified belowground olfactometer (Rasmann et al., 2005). Allowed to move in an empty glass tube, the pest larvae were attracted towards roots already infested by conspecifics in addition to performing better in comparison to larvae fed on uninfested roots (Robert et al., 2012a). While practical, this approach has two major general drawbacks: (1) the insects are moving in an environment far different from their natural niche, lacking any aggregates or particles to crawl on and around, and (2) they may be affected by direct light exposure, which could dramatically disturb their behaviour. In an interesting attempt to solve this first issue, Downie et al. (2012) developed a synthetic transparent, porous and heterogeneous substrate. The water retention of this "transparent" soil was similar to vermiculite and the polymer used absorbed nutrient that would subsequently be available for growing root systems (Downie et al., 2012). Various root growth parameters measured in transparent soil were more similar to those in soil or sand than to those grown in gels, indicating that this synthetic transparent media could accurately mimic natural conditions while allowing direct observation in a complex matrix (Downie et al., 2012). Designing experiments with transparent soil in an environment with low-intensity light would certainly still allow direct observation while reducing the impact of light on the subterranean organisms and on the root physiology. In order to completely isolate the insect from light, there are not many other solutions than using soil itself, which prevents direct observation.

To assess the behaviour of WCR larvae in response to synthetic blends of attractant and feeding stimulants, Hiltpold et al. (2012) designed a destructive laboratory experiment where the insect was monitored by digging out racks previously burrowed in soil microcosms. Visually checking for the presence of larvae in the individual racks, they evaluated the attractiveness of the offered blends of volatile organic compounds and water-soluble organic compounds and further attempted to develop novel biological control strategies (Hiltpold et al., 2012). However, such destructive approaches are not appropriate for long-term observations of a particular process. Therefore, Schumann et al. (2013) designed a more sophisticated experiment to evaluate WCR behaviour. The larvae were allowed to move in a rhizotron (Neumann et al., 2009) thin enough so that their position could be recorded over time. Using GIS (geographic information system) analyses, their behaviour in response to the offered volatiles was mapped (Schumann et al., 2013). Adopting a similar approach, Wilson et al. (1995) built up a large-scale observation chamber dug in the field to examine the effect of pesticides on soil communities.

Recent technological improvements make use of acoustic detection or 3D X-ray reconstruction feasible. For instance, acoustic detection of white grub species (Coleoptera: Scarabaeidae) damaging sugarcane and further computer processing enabled the identification of the species *Dermolepida albohirtum* individuals and of the members of the genus *Antitrogus* in the field. Indeed, the grubs showed notably distinctive acoustic patterns of activity, therefore potentially helping in the monitoring effort of these melolonthine pests in Australian sugarcane fields (Mankin et al., 2009). Further development is yet still needed to relate acoustic measures with economic thresholds or to improve bioinformatic analyses (Mankin et al., 2009), but such approaches, if also applicable to other species, could be of significant value, enabling nondestructive sampling and monitoring of the insects in their natural belowground environment. While acoustic methods enable researchers to detect active soil insects, the visualisation of their 3D environment still remains very challenging. Root digging and further imaging have been a widely used approach to quantify root properties (e.g. Pierret et al., 2005; Trachsel et al., 2011). Also known as shovelomic, this technique is still valid to assess particular traits and new bioinformatics models and software are currently available to evaluate root systems in both 2D (e.g. Clark et al., 2013) or in 3D (e.g. Grift et al., 2011). Nevertheless, this destructive approach results in a snapshot of the root phenology in time and, therefore, is very limiting when studies aim to address natural process occurring over longer time periods, like root growth or root colonisation of particular niches in soil. Therefore, other systems have been imagined. For instance, Clark et al. (2011) grew rice (*Oryza sativa* L.) in gellan gum and developed a high-throughput platform for 3D imaging. From 40 2D images of a single root system, their software (RootReader3D) generates a high-resolution 3D reconstruction of a root system that can be further analysed for various root traits such as length, width or (substrate) exploitation (Clark et al., 2011). However, as recently demonstrated by Downie et al. (2012), the material used to grow plants can significantly affect root architecture; therefore, preliminary assessment of architectural parameters should be conducted before using such growth material. Yet, the best material for realistic root measures would again ideally be soil itself. Using X-ray computed tomography (Hounsfield, 1973), pioneer studies could describe several processes occurring around the roots without disturbing the system (Aylmore, 1993; Crestana et al., 1986). Indeed, several studies have used this approach to characterise the behaviour of root-feeding insects in relation to their host plants, albeit in simplified systems (Harrison et al., 1993; Johnson et al., 2004a,b). Since the first use of the X-ray tomography in soil sciences three decades ago, this technology has continuously improved and is now very accurate (Mooney et al., 2012). During several development phases, limits have been significantly pushed further (Mooney et al., 2012 and references therein) and new instruments are now sensitive enough to detect microscopic variation (down to spatial resolution of 0.5 μm) in soil organisation and root structures (Tracy et al., 2010). This very precise imaging procedure offers various opportunities to unravel complex interactions in the rhizosphere and the ongoing development to visualise much larger portions of soil will definitely scale up the possibilities offered by this technology.

4.2. Collection of water-soluble organic compounds and volatile organic compounds in the rhizosphere

The next question that can be asked is "What triggers this behaviour in subterranean insects?". As already mentioned in the previous sections, insects are very likely to respond to root exudates, along with other belowground cues. On average, 1000–2200 $kgha^{-1}\ y^{-1}$ of organic carbon is exuded by roots (Kuzyakov and Domanski, 2000); it is therefore crucial to be able to sample this enormous fraction of organic compounds and to identify their ecological functions. Root exudates can roughly be sorted in two chemical classes: (1) water-soluble molecules that will be mainly sampled from the water phase in soil (water-soluble organic compounds) and (2) volatile molecules, which are substantially less soluble in water and are more likely to be detected in the soil atmosphere. Rather than two separate groups, root exudates are spread over a continuum of water solubility, according to their chemical properties and the abiotic conditions of a particular soil. Rhizodeposition is defined as all organic compounds released both from healthy root systems and from damaged or senescent roots, whereas root exudates comprise organic compounds released by healthy root systems only (Neumann and Römheld, 2007).

4.2.1 Sampling the water phase in a soil matrix

Since exudation is not homogeneous along a root, the first question to address is where to sample for organic rhizodeposition. This can be achieved using gels (i.e. agar, agarose or polyacrylamide) containing chemical reagents that will stain zones with high exudation activity of a particular class of chemical in a root system. The gels can contain dyes that will react with exudates and change in colour, such as a decolouration of the Al–aluminon complex dye by root exudates (Dinkelaker et al., 1993) or the reduction/oxidation of iron ions (Dinkelaker et al., 1993). pH variations resulting from root exudation can also be visualised while using pH indicators as chemical reagents (Plassard et al., 1999).

Once the optimal location for exudate sampling has been determined, localised sampling techniques can be set up, such as establishing a contact between exudation regions and sorbent materials. Depending on the material used, this sampling can be very generalist or highly selective on a particular category of compounds. Microsuction devices have been used to collect the soil solution for further analyses (Dessureault-Rompré et al., 2006) and have been shown to be consistent in contrasting soil textures (Shen and Hoffland, 2007). Besides an accurate sampling of exudates, microsuction offers the possibility to filter the rhizodeposition and therefore remove most of the microorganisms (Shen and Hoffland, 2007). Indeed, microbes are likely to degrade part of the rhizodeposition while and after sampling, therefore making further chemical analyses inaccurate. Filtration has been shown to significantly reduce microbial degradation (Shen and Hoffland, 2007), as well as formaldehyde (Dessureault-Rompré et al., 2006), even though this treatment has to be carefully realised to prevent contamination of the samples with exogenous chemicals or significant changes in pH. Suction can also be used to sample a larger portion of the soil solution (Neumann et al., 2009). In an elegant *in situ* study, pine tree roots were dug out, carefully washed from remaining soil and inserted into a syringe filled with glass beads (Phillips et al., 2008). A fine mesh was placed in the syringe to prevent the glass beads from clogging the outlet while sampling the exudates with a vacuum pump (Phillips et al., 2008). The collection device was then placed back in the ground. *In situ* spatial and seasonal variations were measured resulting in high exudation at the root tips, which was greater in spring than in fall (Phillips et al., 2008). As with any method, this not only produces experimental artefacts but also underpins the importance of temporal variation in the plant physiology, which should be taken into account while studying such biological processes. It also has the advantage of sampling root exudates without much contamination from microbes present in the rhizosphere.

4.2.2 Sampling volatiles emitted from the roots

This field of research has only recently developed and therefore, methodologies to sample volatiles from the gaseous phase of soil are still scarce. In 2005, Rasmann et al. (2005) isolated and identified volatiles from WCR-damaged corn roots using solid-phase microextraction (SPME) fibre. Briefly, they ground root material in liquid nitrogen and placed the resulting powder in sealed glass vials, from which they collected the volatile organic compounds (Rasmann et al., 2005). SPMEs are made of adsorbent material fused to a silica fibre and, therefore, allow sampling of volatiles without the use of solvent. Adsorbed molecules can be thermally desorbed in the injection device of a gas chromatography (GC)–mass spectrometer and possibly identified. This method is rapid, easy to set up and, in under some conditions, highly sensitive as SPME fibres can adsorb parts per trillion. A comprehensive book on this volatile sampling approach has been recently published and describes in detail the use of SPME fibres (Pawliszyn, 2011). Although SPME has proven its great potential in sampling root volatiles (Rasmann et al., 2005) and monitoring their diffusion in soil (Hiltpold and Turlings, 2008), the original method was destructive, therefore limiting time-related measures. It has to be noted that the method developed by Rasmann et al. (2005) results in the collection of volatiles within the entire root material and not only compounds exuded by the root system. To overcome this drawback, Ali et al. (2010) subsequently developed an *in vivo* non-destructive method to sample volatile organic compounds of citrus roots induced by the root-herbivore *D. abbreviatus* (Coleoptera: Curculionidae). By vacuuming air from a glass chamber containing citrus plants through an adsorbent trap, they could sample rhizospheric volatiles (Ali et al., 2010). After extraction from the traps, mass-spectrometric (MS) analyses led to the identification of volatiles emitted by induced citrus trees (Ali et al., 2010). In addition to the *in situ* collection, this approach offers the possibility to test the collected volatiles for further bioactivity on soil-dwelling micro-fauna (Ali et al., 2010, 2011) and to monitor for potential temporal variation in volatile organic compound release. Using the same approach, Hiltpold et al. (2011) could confirm that the volatile organic compounds earlier identified by Rasmann et al. (2005) were indeed emitted from maize roots fed upon by WCR (Hiltpold et al., 2010c; Köllner et al., 2008; Rasmann et al., 2005) and were not artefacts from the sampling method (Hiltpold et al., 2011).

Recently, Ali et al. (2012) developed a probe to sample volatiles emitted by citrus tree roots in the field. Stainless steel probes, coupled to adsorbent traps, were burried in the field 20 cm deep at 1 and 10 m from the tree trunk; soil air was sucked through the traps with a vacuum pump and samples were subsequently extracted in solvent and analysed in GC–MS (Ali et al., 2012). The detection of the most abundant volatile organic compound (pregeijerene; Ali et al., 2010) emitted by damaged citrus roots was still possible 10 m from the source (Ali et al., 2012) underpinning the long range of influence of such exudates *in situ* and their potential impact in shaping biological interaction close and relatively far from the root systems.

Danner et al. (2012) demonstrated the great potential of proton-transfer-reaction mass spectrometry in capturing dynamic processes of root volatile emission. This noninvasive approach allows real-time volatile sampling and is highly sensitive; however, the results from the spectrometry can be difficult to link actual compounds as they provide only masses that could be related to several volatiles (Danner et al., 2012).

4.3. The rise of soil chemical ecology

Recent development in both molecular biology and chemical sampling and analyses will certainly allow rapid progresses in the field of soil chemical ecology and chemically mediated interactions in this environment. All described techniques have not only their limitations but also many advantages. Several new techniques or adaptations of existing approaches are currently developed and soil chemical ecology will certainly become, in a close future, a hub in our global understanding of ecosystems.

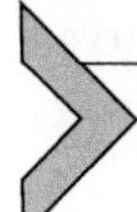

5. THE CHEMICAL ECOLOGY OF THE WCR: A CASE STUDY OF A SOIL-DWELLING PEST

The WCR is one of the most studied belowground pests, averaging nearly 50 refereed publications per year recently as documented in SCOPUS and more than 85 publications per year as documented in CAB Direct over the four-year period from 2009 to 2012. It is the most important insect pest of corn in the United States and is also expanding its range in parts of Europe. Both the larvae and adult stages have been relatively well studied, but since it is the larval stage that primarily causes economic damage, the ecology and chemical ecology of this stage have increasingly been studied (Moeser and Hibbard, 2005; Spencer et al., 2009). It is for this reason that we are using this organism as an example.

5.1. Larval studies

The larval stage of the WCR is the economically significant life stage of the insect (Pike et al., 1995). Given the economic importance of this pest (discussed in the preceding text), any new knowledge regarding the insect–host plant interactions is beneficial to corn agriculture. However, chemical and behavioural studies of WCR larvae are challenging because of the nature of the soil environment, in addition to the larvae being rather delicate. Unlike working with aboveground insects, which are generally easier to see and manipulate, it is difficult to design experiments that give accurate insight into behaviours of soil-dwelling insects (but see Section 4.1). Additional challenges are encountered when testing root exudates, extracts or other behaviourally active compounds, because it is difficult to evenly distribute such materials in the soil matrix. Experiments are often conducted employing techniques that have previously been used with aboveground larvae or adult insects (but see Section 4.2). While such studies can provide useful information (behavioural, physiological, toxicological, etc.), the results may not accurately reflect what occurs in the natural soil environment. Despite the challenges, we now have substantial knowledge of the specific host location behaviours of this root-feeding larva and the root infochemicals that elicit those behaviours.

Female WCR beetles lay their eggs in the soil and upon hatching, the burden of host location lies with the neonate larvae. The tiny (2 mm) larvae must travel through the soil to locate the roots of a host plant, a feat that must be accomplished within 24 h of hatching in order to avoid death (MacDonald and Ellis, 1990). The process of host location by WCR larvae involves three distinct, chemically mediated phases: attraction, host recognition and feeding.

5.2. Attraction

Larvae are initially attracted to corn roots by carbon dioxide (CO_2), which is produced by corn roots in the soil (Harris and Van Bavel, 1957). It is not surprising that WCR larvae and other soil-dwelling insects use CO_2 as an orientation cue. CO_2, which is a primary metabolite of plants, is a small molecule that is able to diffuse readily through soil-pore spaces and establish a concentration gradient in the soil that can serve as a "trail" back to the roots. Other soil-dwelling insects reported to use CO_2 in host location include the carrot fly larva, *Psila rosae* (Jones and Coaker, 1977, 1979); the black vine weevil, *Otiorhynchus sulcatus* F. (Klingler, 1957); cockchafers, *Melolontha* spp. (Klingler, 1957); wireworms (Klingler, 1965, 1966), *Agriotes* spp. (Doane et al., 1975; Klingler, 1957); and subterranean termites, *Reticulitermes* spp. (Bernklau, 2003). Table 3.1 and Fig. 3.2 summarise the response of soil-dwelling insects to metabolites other than CO_2 (the table and figure are discussed in Section 3). In early experiments involving an olfactometer, WCR larvae were attracted to CO_2 gas at a low rate of 0.3 ml/h (Strnad and Bergman, 1987). In a later study, Bernklau and Bjostad (1998a) established a detailed dose–response of neonate larvae to CO_2. This key study employed a vertical "Y" tube filled with glass beads. This particular bioassay apparatus was designed to accommodate the geotropic tendency of the larvae, and the glass beads simulated large soil particles providing thigmotactic cues that larvae normally encounter in the soil (Bernklau, 1997). This study demonstrated that larvae are attracted to CO_2 concentrations as low as 1.25 mmol/mol (0.125%) but respond most strongly to 0.5% CO_2. The reduced response of larvae to higher concentrations (10% and above) was not due to a lack of attraction, but rather to toxic effects of CO_2. The same body of work also showed that WCR larvae are able to detect differences in CO_2 concentrations as low as 12% (Bernklau and Bjostad, 1998a). The remarkable sensitivity of WCR larvae to very small changes in CO_2 concentration may benefit the insects by enabling them to detect a CO_2 gradient while still some distance from the root of a host plant and then accurately follow the trail back to its source.

The experiments that established the details of attraction to CO_2 were conducted in laboratory bioassays in order to allow accurate gas measurements and, hence, these tests do not accurately reflect what actually happens in the soil environment. Soil atmosphere contains CO_2 from root respiration and respiration of soil-dwelling organisms and CO_2 produced from the breakdown of organic material by microorganisms (Clinton and Vose, 1999; Frank et al., 2002). A portion of soil CO_2 is released into the atmosphere during a natural soil–air exchange of CO_2 and oxygen, and a share of the CO_2 that is then retained in the soil reacts with soil–water to form carbonic acid ($CO_2 = H_2O \rightarrow H_2CO_3$). The movement and concentration of remaining soil CO_2 is influenced by factors such as soil type and texture, compaction, depth and the amount of organic material in the soil (Brady, 1990). CO_2 concentrations in surface soils range from approximately 0.3% in well-aerated soil to 10% in poorly drained soils (Brady, 1990).

Although only a small-scale representation of the field environment, lab experiments that include the use of a soil medium can provide an indication of the abilities and limitations that WCR larvae might have in the soil. In tub assays, CO_2 concentrations measured in soil within 2 cm of the roots of a small (6-day-old) corn seedling were 0.44%, and levels in the same soil (well-aerated top soil amended with 25% compost) without the corn were 0.2% (Bernklau and Bjostad, 1998a). In this scenario, up to 100% of neonate larvae are able to locate the roots of a single corn plant (V2; Ritchie et al., 1992) within 24 h from a distance of 13 cm (Bernklau and Bjostad, unpublished data). When a smaller germinating corn seed (4 days old) was used in this same setup, the CO_2 concentrations were much lower, measuring 0.13%

near the corn seed and 0.10% 13 cm away, but this 25% gradient was sufficient for more than 80% of larvae to locate the corn within 6 hours (Bernklau, 2003).

Under normal conditions, the larva's ability to detect concentration differences causes them to orient to and then follow the correct gradient produced by the host plant (higher) and to ignore other incorrect CO_2 sources in the soil (lower). This instinctive response to CO_2 can also be used to the detriment of the insect. For example, Bernklau and Bjostad (1998b) reported that larvae are attracted away from corn roots by a higher concentration of CO_2 alone. This effect was first demonstrated in choice tests using a glass bead bioassay. In this test, larvae overwhelmingly chose a 5 mmol/mol (0.5%) concentration of CO_2 alone over a collection of volatiles produced by corn roots (which contained 2 mmol/mol of CO_2 (Bernklau and Bjostad, 1998b)). When the same concept was applied in soil bioassays, larvae crawled away from the roots of a growing corn seed and towards an artificial source of CO_2 alone (Bernklau, 2003; Bernklau et al., 2004). In subsequent tests, larvae were attracted away from corn roots by a variety of CO_2-generating materials placed in the soil, including yeast granules, effervescing tablets and sucrose pellets. These experiments were conducted using small plastic tubs filled with moist soil in which a 2-day-old corn seedling was placed in the soil at one end and the CO_2-generating formulation was placed at the opposite end (13 cm away). When granules comprising yeast and corn syrup (with added corn meal and yeast nutrient agar) proved to be the most consistent treatment, additional trials were conducted with the granules, increasing the size of the container and the size of the corn seedling and varying the placement of the granules and/or the larvae. Two arrangements were particularly effective in preventing larvae from locating corn roots. In the first, the CO_2-generating granules were concentrated in a band in the soil halfway between the larvae and the corn seedling. In the second, the granules were concentrated in a band on one side of the container, the corn seed was placed on the other end, and the larvae were introduced midway between the granules and the corn. One interpretation of these results is that, in these two "attraction" arrangements, larvae were attracted to the CO_2 produced by the band of granules and either stayed in this region or remained in this region long enough that they were then too weak to continue on and reach the corn root. The third and least effective arrangement was an attempt to "disrupt" host location by dispersing the CO_2 granules throughout the entire soil volume. This technique only prevented larvae from locating the corn root when a very large amount of granules (18 g) were applied to the soil. The fact that most of the larvae were able to locate the corn root despite the lack of a CO_2 gradient (no difference in the CO_2 concentration throughout the soil) suggests that larvae may revert to reliance on other cues for host location when they cannot detect a sufficient CO_2 gradient.

Larvae have previously been reported to be attracted to 6-methoxy-2-benzoxazolinone (Bjostad and Hibbard, 1992) and also to specific long-chain fatty acids when carbon dioxide levels were similar on both sides of the choice (Hibbard et al., 1994) (Table 3.1 and Fig. 3.2B). More recently, when (*E*)-β-caryophyllene or ethylene was added to healthy roots, these plants were more attractive to WCR larvae than healthy roots alone (Robert et al., 2012a). Robert et al. (2012b) conducted a series of studies with maize lines with and without 2,4-dihydroxy-7-methoxy-1,4-benzoxazin-3-one (DIMBOA) and concluded that WCR larvae also utilise this compound for cues to "optimally forage". Blending chemicals previously identified as attractants or phagostimulants to WCR larvae into a capsule shell, Hiltpold et al. (2012) recorded a similar number of second instar larvae attracted to a maize plant or to capsules over a distance of 25 cm in soil under laboratory conditions.

5.3. Host recognition

The second phase of rootworm host location, which we term "host recognition", involves a behaviour that is unique to this insect. Strnad and Dunn (1990) first reported that the search behaviour exhibited by larvae after being exposed for a short time (only 5 min) to the roots of corn plants is distinctly different from the search behaviour exhibited after exposure to the roots of a nonhost plant. The response to nonhost roots, which these investigators termed "long-range search behaviour", is characterised by a rapid rate of crawling in a fairly straight path that incorporates few turns and path crossings. In contrast, the "localised search behaviour" is best described as a "tight-turning" movement because it involves slow, minimal movement, with constant sharp turns and path crossings. When exhibiting tight-turning behaviour, larvae are observed to be constantly, yet slowly, moving their heads back and forth and frequently biting on the neutral substrate (Bernklau et al., 2009; Strnad and Dunn, 1990).

The most unusual aspect of this behaviour is that it does not occur while the insect is on the root surface but rather after the larva has been removed from the root and placed on a neutral surface (untreated filter paper).

The host recognition behaviour of the potato tuber moth larva, *Phthorimaea operculella* Zeller, is somewhat similar in that they reduce their rate of movement, make fewer turns and take more frequent and longer pauses in movement (Varela and Bernays, 1988). However, this behaviour is reported to only occur while the larvae are in contact with an extract from the host plant and does not continue when they are removed from the substrate. This "priming" aspect of the rootworm's tight-turning behaviour raises the question of whether or not this is indeed a host recognition behaviour. Host recognition by insect larvae more commonly involves behaviours such as biting (Bernays et al., 1997; Eigenbrode and Espelie, 1995), palpating (Chapman and Sword, 1993; Harrison, 1987) or arresting behaviour (Heisswolf et al., 2007), and no other insect larva has been reported to exhibit a specific type of search behaviour after being exposed to the host plant. The nature of this unusual behaviour might best be defined by how it benefits the organism in the soil environment. If a larva that has located a host plant root were to lose contact with the root surface, execution of the localised (tight-turning) search behaviour would keep the insect in the immediate vicinity of the root and would likely result in the larva once again contacting the root surface. On the other hand, a larva that has come into contact with an incorrect (nonhost) root would have a better chance of eventually locating an acceptable host root if it were to exhibit the long-range search behaviour. Based on this logic, as well as the original observations made by Strnad and Dunn (1990), it could be argued that a larva exhibits the tight-turning behaviour because it recognised a root as belonging to a host plant and a larva exhibits the long-range search behaviour because it did not recognise a root as a host.

Work is currently being conducted towards the identification of the host recognition factors. In earlier studies, tight-turning by neonate larvae was elicited using a solvent extract of corn roots, demonstrating that the behavioural cues are chemical in nature (Bernklau et al., 2009). However, subsequent experiments showed that the active compounds are not volatile (Bernklau et al., 2009). Recent studies demonstrated that the host recognition behaviour is elicited by a blend of compounds that includes both polar and nonpolar components. In further tests, tight-turning was elicited when a synthetic blend of four small sugars identified from the polar fraction was combined with the active nonpolar fraction. The sugar blend consists of glucose–fructose–sucrose–myoinositol at 30:8:8:2 mg/ml in the active corn root extract (E. Bernklau, unpublished data). The polar fraction contained many other compounds that are not required for behaviour, including diacids, amino acids and inorganic compounds. It is possible that these and other compounds have an enhancement effect that has not yet become apparent. While the blend of sugars was shown to be active, it has not yet been determined whether all four sugars are necessary to elicit a larval response, and the activity of the sugars individually has not yet been determined. The active nonpolar fraction consists of large lipids containing fatty acyl groups (Bernklau et al., 2013) and recent chromatographic purification and subsequent chemical analysis of the nonpolar fraction provided evidence that the active component is a large (>700 *m/z*) glycolipid compound that is present in the corn root at a concentration of approximately 2 mg/g (dry weight) of root (E. Bernklau, unpublished data).

5.4. Feeding behaviour

The third phase of host location behaviour by WCR larvae is intense feeding. When a larva reaches a root in the soil, it takes several "test bites", and if the root is determined to be an appropriate host, the insect typically begins vigorously feeding immediately (Clark et al., 2006). A specific blend of compounds isolated from corn roots serves as feeding stimulants for WCR larvae (Bernklau and Bjostad, 2008). The feeding stimulant blend consists of 30:4:4 mg/ml glucose–sucrose–fructose, plus at least one free fatty acid (FFA) (either linoleic acid (18:2 FA) or oleic acid (18:1 FA), 0.3 mg/ml). Chemical analysis of the liquid pressed from fresh corn roots revealed a blend of small sugars, amino acids, diacids, FFAs and inorganic compounds. In feeding bioassays, the sugar blend alone elicited feeding by approximately 50% of the larvae tested, but none of the other groups of compounds elicited feeding. The number of larvae feeding and the intensity of feeding were significantly increased (>90% of larvae feeding) with the addition of one of the FFA (linoleic or oleic acid). FFAs also impact other belowground insect pests (see Chapter 2). Branson (1982) previously reported that a 50% sucrose solution elicited feeding by second instar larvae. Although the blend of 20 amino acids identified in the corn roots did not elicit feeding, it is possible that one or more specific amino acids could enhance feeding on the sugar–fatty acid blend. Amino acids alone (Hollister and Mullin, 1998; Kim and Mullin, 1998), as well as blends of sugars and amino acids (Kim and Mullin, 2007), serve as phagostimulatory cues for adult WCRs, and amino acids are documented as chemosensory cues for several root-feeding insects (Johnson and Gregory, 2006), including grass grub, *Costelytra zealandica* (Sutherland and Hilier, 1974); black beetle, *Heteronychus arator* (Sutherland et al., 1980); Australian scarab, *Sericesthis geminate* (Wensler and Dudzinski, 1972); and scarabaeid beetle, *Lepidiota negatoria* (Melolonthinae) (Allsopp, 1992).

Although the exact host recognition blend has not yet been identified, results to this point suggest that there is a relationship between the feeding and the host recognition cues for WCR larvae. In previous experiments, the identified feeding stimulant blend did not elicit the host recognition behaviour (Bernklau et al., 2009), but the most recent study does suggest some overlap in the two active blends and it is possible that one or more of the compounds in the feeding stimulant blend are also contained in the host recognition blend.

5.5. Application of behavioural cues in agriculture

Identification of specific larval host location cues may have practical implications in rootworm control. In laboratory bioassays, the synthetic feeding stimulant blend of sugars and FFAs increased the efficacy of thiamethoxam insecticide (Bernklau et al., 2011). When larvae were exposed to treatments for 30 min, the LD_{50} of thiamethoxam insecticide was 568 pg/ml, but this was decreased to 0.045 pg/ml when the feeding stimulant blend was added, causing a 10,000-fold increase in the efficacy of the insecticide. It might be possible to use feeding stimulants, or even the host recognition cues, to enhance the effectiveness of granular insecticide formulations and provide control of WCR damage with a reduced amount of insecticide. As mentioned earlier, the basic concept has already been proven in bioassays with insecticide/behavioural cue combinations (Bernklau et al., 2011) or with capsules releasing these compounds and luring the pest (Hiltpold et al., 2012), but further work is needed to develop formulations that can be applied in the field that would endure for a period of weeks in the soil environment. A second approach might be to reduce or alter amounts of the behavioural cues contained in corn roots through plant breeding or genetic engineering in order to make the roots less recognisable or less palatable to the larvae and thereby resistant to WCR damage.

5.6. Natural enemies, chemical ecology and pest management

WCR not only is a major pest on maize but also has to face a tremendous number of threats in the ground. As larvae are likely to encounter several pathogens and predators around the root system, understanding these interactions better can have dramatic effects on rootworm populations and the damage they inflict (Lundgren and Fergen, 2010; Pilz et al., 2009). Just as insects forage for plant roots, so too do soil-dwelling entomopathogens use chemical cues to locate and assess their host. Early researchers in this field observed that entomopathogenic nematodes were attracted to roots of *Thuja occidentalis* L. and *Fragaria* × *ananassa* 'Elsanta' that were damaged by root–insect herbivores, whereas the root or insect alone was not attractive to the nematodes (Boff et al., 2002; van Tol et al., 2001). Entomopathogenic nematodes are obligate parasites of insects that usually kill their host with 24–48 h postinfection (Dillman et al., 2012a). Because they are easy to mass-produce (Ehlers, 2001) and effective against most WCR larval stages (Kurzt et al., 2009), nematodes were thought to be very potent biological control agents against this root pest (Kuhlmann and van der Brugt, 1998; Pilz et al., 2009). It was subsequently shown (Rasmann et al., 2005) that herbivory by WCR larvae on the roots of several maize varieties resulted in root production of sesquiterpene (*E*)-β-caryophyllene; this chemical is highly attractive to the entomopathogenic nematode *Heterorhabditis megidis*. Not all maize varieties produce this compound (Hiltpold et al., 2010c; Köllner et al., 2008; Rasmann et al., 2005), but (*E*)-β-caryophyllene can be restored via genetic engineering of the plant to create genetically modified maize compatible with nematode biocontrol (Degenhardt et al., 2009). This ability of the maize root to call for help when damaged significantly reduces WCR populations in the field, and nematode-attractive plants generally suffered relatively less damage when nematodes were applied to the soil (Degenhardt et al., 2009; Hiltpold et al., 2010a,c; Rasmann et al., 2005). As earlier suggested by Hiltpold et al. (2010a), the strain-specific response of entomopathogenic nematodes to root-emitted volatiles has been confirmed (Laznik and Trdan, 2013), underpinning the necessity to carefully examine the behaviour of the entomopathogenic nematodes before their use in biological control exploiting such tritrophic interactions. Examples of such influence of roots on upper trophic levels remain scarce (but see Ali et al., 2010, 2011, 2012, 2013; Rasmann et al., 2011b), but they illustrate the potential of chemical ecology in the management of subterranean pests such as WCR, especially since entomopathogenic nematodes respond to a wide range of common plant volatile organic compounds (Hallem et al., 2011).

Entomopathogenic nematodes are attracted to nonplant chemical compounds as well. Insect hosts also emit volatiles that recruit several species of entomopathogenic nematodes (Dillman et al., 2012b). Volatile organic compounds emitted by WCR remain poorly explored but could produce additional chemical attractants useful in nematode biological control programmes (Hiltpold et al., 2010a,b).

Not all chemically mediated signals favour the natural enemies of subterranean herbivores; WCR larvae are protected against predation by a potent haemolymph defence that is at least partially chemical in nature. Arthropod predator communities are abundant and diverse within soil systems and can have important effects on subterranean herbivores. WCR larvae are no exception, and each preimaginal life stage is exposed to its own predator community (Lundgren et al., 2009c). Recent work has revealed that the haemolymph of WCR larvae coagulates quickly on the mouthparts of arthropod predators, causing them to abandon their attack and vigorously clean their mouthparts (Lundgren et al., 2009b). In addition to the physical defensive components of the defence, Lundgren et al. (2010) discovered that there is an ethanol extractable fraction of the haemolymph that is repellent to some predators. The chemistry underlying this defence is actively being pursued, and preliminary results by J.G. Lundgren suggest that hydroxamic acids (DIMBOA, MBOA, BOA, etc.) used in WCR larval foraging behaviour have little bioactivity against predators (Fig. 3.4) or are not found in the rootworm haemolymph. It is notable that the chemical aspect of the defence is separate from the well-established cucurbitacin-derived haemolymph defences of Diabroticines (Tallamy et al., 2005). Research seems to indicate that there is a "dose" component to the defence; first and second instars are apparently undefended against larger predators like some ants, carabid beetles, wolf spiders and crickets (J.G.Lundgren personal observation), but first instars effectively repel predatory mites (Prischmann et al., 2011). WCR is not the only *Diabrotica* sp. to display this type of haemolymph defence; at least four other related species have predator-defended larvae (Wallace and Blum, 1971, J.G.Lundgren, personal observation). To summarise, this haemolymph defence is multifaceted and the physical and chemical facets function with varying effectiveness against the predators of WCR larvae (Lundgren and Fergen, 2010; Lundgren et al., 2009a). This is not to say that WCR larvae cannot be effectively managed using generalist predators. Indeed, research shows that conserving healthy soil arthropod predator communities in agroecosystems can overcome the larval defence of WCR and lead to reduced damage to maize roots (Lundgren and Fergen, 2010, 2011, in press).

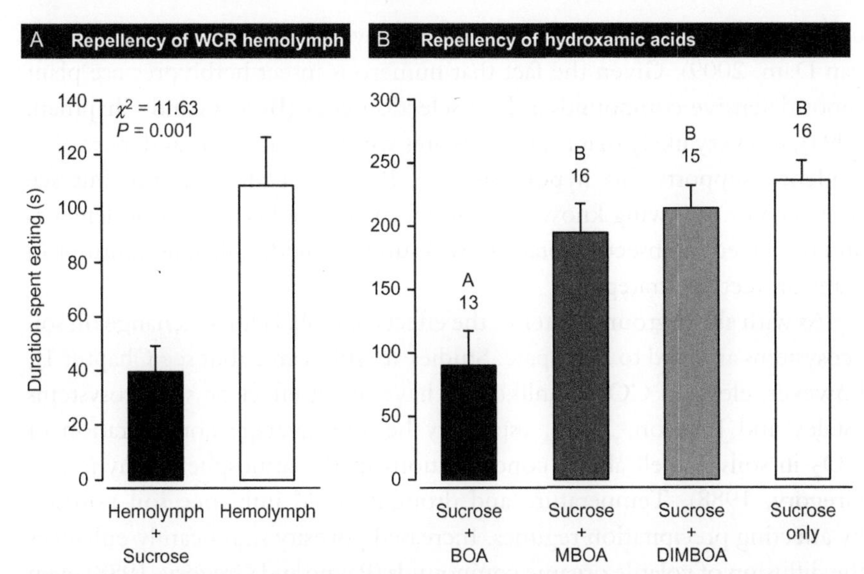

Figure 3.4 Predator response to defensive compounds. (A) Chemical repellency of ethanol-extracted haemolymph of WCR to the predator *Harpalus pensylvanicus* (Coleoptera: Carabidae). (B) Repellency of hydroxamic acids to this WCR predator. Bars represent the mean duration (SEM) spent eating the solutions; above each bar is the number of observation and bars capped with different letters are significantly different from each other (Tukey's test, $\alpha = 0.05$). In 3.4.A, the concentration of haemolymph in the sucrose solution was equivalent to the quantity bled from the third instar WCR. In 3.4.B, predatory beetles were fed 0.24 M sucrose solutions containing 0.01 M concentrations of DIMBOA, MBOA or BOA. Although there was some antipredator bioactivity of BOA, neither this nor any of the other hydroxamic acids could be identified in the haemolymph of WCR.

6. FUTURE CHALLENGES AND CONCLUSIONS

Belowground chemical ecology is a vast but barely explored field of research. The chapter mainly focuses on insects interacting with roots; however, there are hundreds of other types of possible chemical interactions in soil (e.g. Bonkowski et al., 2009; Rasmann et al., 2012a,b) and therefore challenges in this area of research are tremendous.

Yet, some aspects are likely to receive more attention in the future. As a main research avenue, root defences against herbivores are still mostly unknown, even though progress has been seen recently (e.g. Johnson et al., 2011; Robert et al., 2012b; Chapter 2). In order to effectively use chemical ecology in pest management strategies, it is crucial to better understand how insects are interacting and coping with root exudates and chemical defences (comprehensive reviews by Hiltpold and Turlings, 2012; van Dam, 2009). Given the fact that numerous insect herbivores use plant shoot defensive compounds as host selection cues (Bernays and Chapman, 1994), it is very likely that root herbivores would do the same and increasing evidence supports this hypothesis (e.g. Robert et al., 2012a,b, but see Section 5.6). Growing knowledge on the arms race between root defences and root-feeding insect adaptations will surely provide valuable solutions in plant protection strategies.

As with aboveground systems, the effects of global climate change on soil ecosystems are hard to anticipate. Studies are still scarce (but see Chapter 1); however, elevated CO_2 is unlikely to have direct effect on soil ecosystems (Staley and Johnson, 2008), especially because average concentration of CO_2 in soils is well above concentrations in the atmosphere (Payne and Gregory, 1988). Temperature and drought could influence soil porosity by affecting precipitation regimes. Increased porosity significantly enhances the diffusion of volatile organic compounds (Payne and Gregory, 1988) even though some moisture is necessary for volatile organic compounds to diffuse efficiently in soil (Hiltpold and Turlings, 2008). Water-soluble organic compound diffusion and availability for rhizospheric organisms are likely to be impaired under dryer conditions. Consequently, subterranean chemical signalling might be affected by global climate change, subsequently impacting several soil-dwelling organisms and ecosystem's functions.

This chapter aims to give an overview of chemical interactions occurring in the rhizosphere between roots, insects and, to some extent, organisms in the upper trophic levels. As primary producers, plants profoundly shape their surrounding environment in many ways. In response, not only insects but also other organisms such as nematodes, bacteria or fungi evolved very complex machinery to perceive and, subsequently, benefit from root chemical exudation of volatile, as long-distance signals, and water-soluble organic compounds, during short-distance foraging or plant acceptance/rejection. Whether beneficial, deleterious or neutral for plants, expanding our knowledge on these interactions is likely to play a major role in future plant protection strategies and will give us tools to guarantee sustainable functionality of this unique and pivotal ecosphere.

ACKNOWLEDGEMENTS

We thank Dr. S.N. Johnson for the fruitful discussion and his help to improve the chapter. We are grateful for the invitation to write this report and to the reviewers for their constructive comments and suggestions. We thank Prof. M. Traugott (University of

Innsbruck, Austria) for his help in chemical web analyses and Philip Smith for proofreading this chapter. This work was funded by the Swiss National Science Foundation Grant PBNEP3-134385 awarded to IH and by the NSF grant 1045322 partially supporting NM-S.

REFERENCES

Akiyama, K., Matsuzaki, K., Hayashi, H., 2005. Plant sesquiterpenes induce hyphal branching in arbuscular mycorrhizal fungi. Nature 435, 824–827.

Ali, J.G., Alborn, H.T., Stelinski, L.L., 2010. Subterranean herbivore-induced volatiles released by citrus roots upon feeding by *Diaprepes abbreviatus* recruit entomopathogenic nematodes. J. Chem. Ecol. 36, 361–368.

Ali, J.G., Alborn, H.T., Stelinski, L.L., 2011. Constitutive and induced subterranean plant volatiles attract both entomopathogenic and plant parasitic nematodes. J. Ecol. 99, 26–35.

Ali, J.G., Alborn, H.T., Campos-Herrera, R., Kaplan, F., Duncan, L.W., Rodriguez-Saona, C., Koppenhöfer, A.M., Stelinski, L.L., 2012. Subterranean, herbivore-induced plant volatile increases biological control activity of multiple beneficial nematode species in distinct habitats. PLoS One 7, e38146.

Ali, J.G., Campos-Herrera, R., Alborn, H.T., Duncan, L., Stelinski, L.L., 2013. Sending mixed messages: a trophic cascade produced by a belowground herbivore-induced cue. J. Chem. Ecol. 39, 1140–1147.

Allsopp, P.G., 1992. Sugars, amino-acids, and ascorbic acid as phagostimulants for larvae of *Antitrogus parvulus* and *Lepidiota negatoria* (Coleoptera: Scarabaeidae). J. Econ. Entomol. 85, 106–111.

Andersen, J.F., Plattner, R.D., Weisleder, D., 1988. Metabolic transformations of cucurbitacins by *Diabrotica virgifera virgifera* LeConte and *D. undecimpunctata howardi* Barber. Insect Biochem 19, 71–78.

Aylmore, L.A.G., 1993. Use of computer-assisted tomography in studying water movement around plant roots. Adv. Agron. 49, 1–54.

Badri, D.V., Quintana, N., El Kassis, E.G., Kim, H.K., Choi, Y.H., Sugiyama, A., Verpoorte, R., Martinoia, E., Manter, D.K., Vivanco, J.M., 2009. An ABC transporter mutation alters root exudation of phytochemicals that provoke an overhaul of natural soil microbiota. J. Plant Physiol. 151, 2006–2017.

Bais, H.P., Weir, T.L., Perry, L.G., Gilroy, S., Vivanco, J.M., 2006. The role of root exudates in rhizosphere interactions with plants and other organisms. Annu. Rev. Plant Biol. 57, 233–266.

Barber, D.A., Martin, J.K., 1976. The release of organic substances by cereal roots into soil. New Phytol. 76, 60–80.

Bascompte, J., Jordano, P., Melián, C.J., Olesen, J.M., 2003. The nested assembly of plant–animal mutualistic networks. Proc. Natl. Acad. Sci. U.S.A. 16, 9383–9387.

Bastolla, U., Fortuna, M.A., Pascual-García, A., Ferrera, A., Luque, B., Bascompte, J., 2009. The architecture of mutualistic networks minimizes competition and increases biodiversity. Nature 458, 1018–1020.

Bensky, D., Gamble, A., 1986. Chinese Herbal Medicine: Materia Medica. Eastland Press, Seattle, USA.

Bernays, E.A., Chapman, R.F., 1994. Host-Plant Selection by Phytophagous Insects. Chapman & Hall, New York.

Bernays, E.A., Blaney, W.M., Chapman, R.F., Cook, A.G., 1997. The ability of *Locusta migratoria* L. to perceive plant surface waxes. In: Jermy, T. (Ed.), The Host-Plant in Relation to Insect Behaviour and Reproduction. Akadémiai Kiadó, Budapest, Hungary, pp. 35–40, Plenum Publishing Corporation New York, NY, USA.

Bernklau, E.J., 1997. Infochemicals and Host Location by Western Corn Rootworm Larvae. Colorado State University, Fort Collins, CO, USA.

Bernklau, E.J., 2003. Behavioural Effects of Carbon Dioxide on Western Corn Rootworm and Subterranean Termites with Implications for Pest Management. Colorado State University, Fort Collins, CO, USA, PhD Dissertation.

Bernklau, E.J., Bjostad, L.B., 1998a. Behavioural responses of first-instar western corn rootworm (Coleoptera: Chrysomelidae) to carbon dioxide in a glass bead bioassay. J. Econ. Entomol. 91, 444–456.

Bernklau, E.J., Bjostad, L.B., 1998b. Re-investigation of host location by western corn rootworm larvae (Coleoptera: Chrysomelidae): CO_2 is the only volatile attractant. J. Econ. Entomol. 91, 1331–1340.

Bernklau, E.J., Bjostad, L.B., 2005. Insecticide enhancement with feeding stimulants in corn for western corn rootworm larvae (Coleoptera: Chrysomelidae). J. Econ. Entomol. 98, 1150–1156.

Bernklau, E.J., Bjostad, L.B., 2008. Identification of feeding stimulants in corn roots for western corn rootworm (Coleoptera: Chrysomelidae) larvae. J. Econ. Entomol. 101, 341–351.

Bernklau, E.J., Fromm, E.A., Bjostad, L.B., 2004. Disruption of host location of western corn rootworm larvae (Coleoptera: Chrysomelidae) with carbon dioxide. J. Econ. Entomol. 97, 330–339.

Bernklau, E.J., Bjostad, L.B., Meihls, L.N., Coudron, T.A., Lim, E., Hibbard, B.E., 2009. Localized search cues in corn roots for western corn rootworm (Coleoptera: Chrysomelidae) larvae. J. Econ. Entomol. 102, 558–562.

Bernklau, E.J., Bjostad, L.B., Hibbard, B.E., 2011. Synthetic feeding stimulants enhance insecticide activity against western corn rootworm larvae, *Diabrotica virgifera virgifera* (Coleoptera: Chrysomelidae). J. Appl. Entomol. 135, 47–54.

Bernklau, E.J., Bjostad, L.B., Hibbard, B.E., 2013. Isolation and characterization of host recognition cues in corn roots for larvae of the western corn rootworm (Coleoptera: Chrysomelidae). J. Econ. Entomol. Accepted for publication.

Bjostad, L.B., Hibbard, B.E., 1992. 6-Methoxy-2-benzoxazolinone—a semiochemical for host location by western corn rootworm larvae. J. Chem. Ecol. 18, 931–944.

Blossey, B., Hunt-Joshi, T.R., 2003. Belowground herbivory by insects: influence on plants and aboveground herbivores. Annu. Rev. Entomol. 48, 521–547.

Boff, M.I.C., Van Tol, R., Smits, P.H., 2002. Behavioural response of *Heterorhabditis megidis* towards plant roots and insect larvae. Biocontrol 47, 67–83.

Bonkowski, M., 2004. Protozoa and plant growth: the microbial loop in soil revisited. New Phytol. 162, 617–631.

Bonkowski, M., Villenave, C., Griffiths, B., 2009. Rhizosphere fauna: the functional and structural diversity of intimate interactions of soil fauna with plant roots. Plant Soil 321, 213–233.

Bosak, P.J., Crans, W.J., 2002. The structure and function of the larval siphon and spiracular apparatus of *Coquillettidia perturbans*. J. Am. Mosq. Control. Assoc. 18, 280–283.

Brady, N.C., 1990. The Nature and Properties of Soils. MacMillan Publishing, Inc., New York, NY, USA.

Branson, T.F., 1982. Olfactory response of larvae of *Diabrotica virgifera virgifera* to plant roots. Entomol. Exp. Appl. 31, 303–307.

Britton, E., 1978. A revision of the Australian chafers (Coleoptera: Scarabaeidae: Melolonthinae). Vol. 2. Tribe Melolonthini. Aust. J. Zool 26, 1–150, Supplementary Series.

Brown, V.K., Gange, A.C., 1990. Insect herbivory insect below ground. In: Begon, M., Fitter, A.H., Macfadyen, A. (Eds.), Advances in Ecological Research. Elsevier, The Netherlands, pp. 1–58.

Campos-Herrera, R., El-Borai, F.E., Stuart, R.J., Graham, J.H., Duncan, L.W., 2011. Entomopathogenic nematodes, phoretic *Paenibacillus* spp., and the use of real time

quantitative PCR to explore soil food webs in Florida citrus groves. J. Invertebr. Pathol. 108, 30–39.
Campos-Herrera, R., El-Borai, F.E., Duncan, L.W., 2012. Real-time PCR as an effective technique to assess the impact of phoresy by *Paenibacillus* sp. bacteria on *Steinernema diaprepesi* nematodes in nature. Mol. Ecol. Resour. 12, 885–893.
Chapman, R.F., Sword, G., 1993. The importance of palpation in food selection by a polyphagous grasshopper (Orthoptera: Acrididae). J. Insect Behav. 6, 79–91.
Chapman, E.G., Schmidt, J.M., Welch, K.D., Harwood, J.D., 2012. Molecular evidence for dietary selectivity and pest suppression potential in an epigeal spider community in winter wheat. Biol. Control. 65, 72–86.
Cipollini, D., Rigsby, C.M., Barto, E.K., 2012. Microbes as targets and mediators of allelopathy in plants. J. Chem. Ecol. 38, 714–727.
Clark, P.L., Vaughn, T.T., Meinke, L.J., Molina-Ochoa, J., Foster, J.E., 2006. *Diabrotica virgifera virgifera* (Coleoptera: Chrysomelidae) larval feeding behaviour on transgenic maize (MON 863) and its isoline. J. Econ. Entomol. 99, 722–727.
Clark, R.T., MacCurdy, R.B., Jung, J.K., Shaff, J.E., McCouch, S.R., Aneshansley, D.J., Kochian, L.V., 2011. Three-dimensional root phenotyping with a novel imaging and software platform. Plant Physiol. 156, 455–465.
Clark, R.T., Famoso, A.N., Zhao, K., Shaff, J.E., Craft, E.J., Bustamante, C.D., McCouch, S.R., Aneshansley, D.J., Kochian, L.V., 2013. High-throughput two-dimensional root system phenotyping platform facilitates genetic analysis of root growth and development. Plant Cell Environ. 36, 454–466.
Clinton, B.D., Vose, J.M., 1999. Fine root respiration in mature eastern white pine (*Pinus strobus*) in situ: the importance of CO_2 in controlled environments. Tree Physiol. 19, 475–479.
Clyne, P.J., Warr, C.G., Freeman, M.R., Lessing, D., Kim, J., Carlson, J.R., 1999. A novel family of divergent seven-transmembrane proteins: candidate odorant receptors in *Drosophila*. Neuron 22, 327–338.
Clyne, P.J., Warr, C.G., Carlson, J.R., 2000. Candidate taste receptors in *Drosophila*. Science 287, 1830–1834.
Cobb, M., 1999. What and how do maggots smell? Biol. Rev. Camb. Philos. 74, 425–459.
Cole, R.A., 1985. Relationship between the concentration of chlorogenic acid in carrot roots and the incidence of carrot fly larval damage. Ann. Appl. Biol. 106, 211–217.
Coleman, D.C., 1976. A review of root production processes and their influence on soil biota in terrestrial ecosystems. In: Andersen, J.M., Macfadyen, A. (Eds.), The Role of Terrestrial and Aquatic Organisms in Decomposition Processes. Blackwell, Oxford, UK, pp. 417–434.
Crestana, S., Casaero, R., Mascarenhas, S., 1986. Using a computer assisted tomography miniscanner in soil science. Soil Sci. 142, 56–61.
Danner, H., Samudrala, D., Cristescu, S.M., Van Dam, N.M., 2012. Tracing hidden herbivores: time-resolved non-invasive analysis of belowground volatiles by proton-transfer-reaction mass spectrometry (PTR-MS). J. Chem. Ecol. 38, 785–794.
Dawson, L.A., Byers, R.A., 2008. Methods for studying root herbivory. In: Johnson, S.N., Murray, P.J. (Eds.), Root Feeders: An Ecosystem Perspective. CABI Publishing, Wallingford, UK, pp. 3–19.
De Bruyne, M., Clyne, P.J., Carlson, J.R., 1999. Odor coding in a model olfactory organ: the *Drosophila maxillary* palp. J. Neurosci. 19, 4520–4532.
De Deyn, G.B., Raaijmakers, C.E., Zoomer, H.R., Berg, M.P., De Ruiter, P.C., Verhoef, H.A., Bezemer, T.M., Van Der Putten, W.H., 2003. Soil invertebrate fauna enhances grassland succession and diversity. Nature 422, 711–713.
De Groot, M., Fåokl, A., Virant-Doberlet, M., 2011. Species identity cues: possibilities for errors during vibrational communication on plant stems. Behav. Ecol. 22, 1209–1217.

De La Peña, E., Echeverría, S.R., Van Der Putten, W.H., Freitas, H., Moens, M., 2006. Mechanism of control of root-feeding nematodes by mycorrhizal fungi in the dune grass *Ammophila arenaria*. New Phytol. 169, 829–840.

Degenhardt, J., Hiltpold, I., Köllner, T.G., Frey, M., Gierl, A., Gershenzon, J., Hibbard, B.E., Ellersieck, M.R., Turlings, T.C.J., 2009. Restoring a maize root signal that attracts insect-killing nematodes to control a major pest. Proc. Natl. Acad. Sci. U.S.A. 106, 13213–13218.

Deheer, C.J., Tallamy, D.W., 1991. Cucumber beetle larval affinity to cucurbitacins. Environ. Entomol. 20, 775–788.

Dekker, T., Geier, M., Carde, R.T., 2005. Carbon dioxide instantly sensitizes female yellow fever mosquitoes to human skin odours. J. Exp. Biol. 208, 2963–2972.

Dematheis, F., Kurtz, B., Vidal, S., Smalla, K., 2012. Microbial communities associated with the larval gut and eggs of the western corn rootworm. PLoS One 7, e44685.

Dessureault-Rompré, J., Nowack, B., Schulin, R., Luster, J., 2006. Modified micro suction cup/rhizobox approach for the in-situ detection of organic acids in rhizosphere soil solution. Plant Soil 286, 99–107.

Dethier, V.G., 1947. Chemical Insect Attractants and Repellents. Blakiston Co, Philadelphia, PA, USA.

Dillman, A.R., Chaston, J.M., Adams, B.J., Ciche, T.A., Goodrich-Blair, H., Stock, S.P., Sternberg, P.W., 2012a. An entomopathogenic nematode by any other name. PLoS Pathog. 8, e1002527.

Dillman, A.R., Guillermin, M.L., Lee, J., Kim, B., Sternberg, P.W., Hallem, E.A., 2012b. Olfaction shapes host-parasite interactions in parasitic nematodes. Proc. Natl. Acad. Sci. U.S.A. 109, E2324–E2333.

Dinkelaker, B., Hahn, G., Römheld, V., Wolf, G.A., Marschner, H., 1993. Non-destructive methods for demonstrating chemical changes in the rhizosphere I. Description of methods. Plant Soil 155–156, 67–70.

Dixon, R.A., Strack, D., 2003. Phytochemistry meets genome analysis, and beyond. Phytochemistry 62, 815–816.

Doane, J.F., Klingler, J., 1978. Location of CO_2-receptive sensilla on larvae of wireworms *Agriotes lineatus-obscurus* and *Limonius californicus*. Ann. Entomol. Soc. Am. 71, 357–363.

Doane, J.F., Lee, Y.W., Klinger, J., Westcott, N.D., 1975. The orientation response of *Ctenicera destructor* and other wireworms (Coleoptera: Elateridae) to germinating grain and to carbon dioxide. Can. Entomol. 107, 1233–1251.

Dormann, C.F., Gruber, G., Fruend, J., 2008. Introducing the bipartite package: analysing ecological networks. R news 8, 8–11.

Downie, H., Holden, N., Otten, W., Spiers, A.J., Valentine, T.A., Dupuy, L.X., 2012. Transparent soil for imaging the rhizosphere. PLoS One 7, e44276.

Eben, A., Barbercheck, M.E., Aluja, S.M., 1997. Mexican diabroticite beetles: I. Laboratory test on host breadth of *Acalymma* and *Diabrotica* spp. Entomol. Exp. Appl. 82, 53–62.

Ehlers, R.U., 2001. Mass production of entomopathogenic nematodes for plant protection. Appl. Microbiol. Biotechnol. 56, 623–633.

Eigenbrode, S.D., Espelie, K.E., 1995. Effects of plant epicuticular lipids on insect herbivores. Annu. Rev. Entomol. 40, 171–194.

Eilers, E.J., Talarico, G., Hansson, B.S., Hilker, M., Reinecke, A., 2012. Sensing the underground— ultrastructure and function of sensory organs in root-feeding *Melolontha melolontha* (Coleoptera: Scarabaeinae) larvae. PLoS One 7, e41357.

Endlweber, K., Ruess, L., Scheu, S., 2009. Collembola switch diet in presence of plant roots thereby functioning as herbivores. Soil Biol. Biochem. 41, 1151–1154.

Ennis, D.E., Dillon, A.B., Griffin, C.T., 2010. Simulated roots and host feeding enhance infection of subterranean insects by the entomopathogenic nematode *Steinernema carpocapsae*. J. Invertebr. Pathol. 103, 140–143.

Erb, M., Ton, J., Degenhardt, J., Turlings, T.C.J., 2008. Interactions between arthropod-induced aboveground and belowground defenses in plants. Plant Physiol. 146, 867–874.

Erb, M., Glauser, G., Robert, C.A.M., 2012. Induced immunity against belowground insect herbivores-activation of defenses in the absence of a jasmonate burst. J. Chem. Ecol. 38, 629–640.

Eriksson, A., Anfora, G., Lucchi, A., Lanzo, F., Virant-Doberlet, M., Mazzoni, V., 2012. Exploitation of insect vibrational signals reveals a new method of pest management. PLoS One 7, e32954.

Ferguson, J.E., Metcalf, R.L., 1985. Cucurbitacins: plant derived defense compounds for Diabroticina (Coleoptera: Chrysomelidae). J. Chem. Ecol. 11, 311–318.

Finch, S., Skinner, G., 1974. Studies of the Cabbage Root Fly. 24th Annual Report for 1973, National Vegetable Research Station, Wellesbourne, Warwick, pp. 84–85.

Flores, H.E., Vicanco, J.M., Loyola-Vargas, V.M., 1999. 'Radicle' biochemistry: the biology of root-specific metabolism. Trends Plant Sci. 4, 220–226.

Fortuna, M.A., Stouffer, D.B., Olesen, J.M., Jordano, P., Mouillot, D., Krasnov, B.R., Poulin, R., Bascompte, J., 2010. Nestedness versus modularity in ecological networks: two sides of the same coin? J. Anim. Ecol. 79, 811–817.

Fraenkel, G.S., 1959. The raison d'être of secondary plant substances. Science 129, 1466–1470.

Frank, A.B., Liebig, M.A., Hanson, J.D., 2002. Soil carbon dioxide fluxes in northern semi-arid grasslands. Soil Biol. Biochem. 34, 1235–1241.

Gagliano, M., 2013. Green symphonies: a call for studies on acoustic communication in plants. Behav. Ecol. 24, 789–796.

Gagliano, M., Mancuso, S., Robert, D., 2012a. Towards understanding plant bioacoustics. Trends Plant Sci. 17, 323–325.

Gagliano, M., Renton, M., Duvdevani, N., Timmins, M., Mancuso, S., 2012b. Acoustic and magnetic communication in plants: is it possible? Plant Signal. Behav. 7, 1346–1348.

Gange, A.C., Brown, V.K., 2002. Multitrophic Interactions in Terrestrial Systems. Cambridge University Press, Cambridge, UK.

Gariepy, T.D., Messing, R.H., 2012. Development and use of molecular diagnostic tools to determine trophic links and interspecific interactions in aphid-parasitoid communities in Hawaii. Biol. Control. 60, 26–38.

Gariepy, T.D., Kuhlmann, U., Gillott, C., Erlandson, M., 2008. Does host plant influence parasitism and parasitoid species composition in *Lygus rugulipennis*? A molecular approach. B. Entomol. Res. 98, 217–221.

Giglio, A., Ferrero, E.A., Perrotta, E., Tripepi, S., Zetto Brandmayr, T., 2003. Ultrastructure and comparative morphology of mouth-part sensilla in ground beetle larvae (insecta, coleoptera, carabidae). Zool. Anz. 242, 277–292.

Glauser, G., Boccard, J., Wolfender, J.L., Rudaz, S., 2013. Metabolomics: application in plant sciences. In: Laemmerhofer, M., Weckewerh, W. (Eds.), Metabolomics in Practice: Successful Strategies to Generate and Analyze Metabolic Data. Wiley-VCH, Weinheim, Germany, pp. 311–341.

Grift, T.E., Novais, J., Bohn, M., 2011. High-throughput phenotyping technology for maize roots. Biosyst. Eng. 110, 40–48.

Grosse-Wilde, E., Kuebler, L.S., Bucks, S., Vogel, H., Wicher, D., Hansson, B.S., 2011. Antennal transcriptome of *Manduca sexta*. Proc. Natl. Acad. Sci. U.S.A. 108, 7449–7454.

Guerin, P.M., Ryan, M.F., 1984. Relationship between root volatiles of some carrot cultivars and their resistance to the carrot fly, *Psila rosae*. Entomol. Exp. Appl. 36, 217–224.

Hall, T.A., 1999. Bioedit: a user-friendly biological sequence alignment editor and analysis program for Windows 95/98/NT. Nucl. Acids Symp. Ser. 41, 95–98.

Hallem, E.A., Dillman, A.R., Hong, A.V., Zhang, Y.J., Yano, J.M., Demarco, S.F., Sternberg, P.W., 2011. A sensory code for host seeking in parasitic nematodes. Curr. Biol. 21, 377–383.

Hansson, B., Stensmyr, M., 2011. Evolution of insect olfaction. Neuron 72, 698–711.

Harris, D.G., Van Bavel, C.H.M., 1957. Root respiration of tobacco, corn and cotton plants. Agron. J. 49, 182–184.

Harrison, G.D., 1987. Host-plant discrimination and evolution of feeding preference in the Colorado potato beetle *Leptinotarsa decemlineata*. Physiol. Entomol. 12, 407–415.

Harrison, R.D., Gardner, W.A., Tollner, W.E., Kinard, D.J., 1993. X-ray computed tomography studies of the burrowing behavior of 4th-instar pecan weevil (Coleoptera, Curculionidae). J. Econ. Entomol. 86, 1714–1719.

Harwood, J.D., Phillips, S.W., Sunderland, K.D., Symondson, W.O.C., 2001. Secondary predation: quantification of food chain errors in an aphid-spider-carabid system using monoclonal antibodies. Mol. Ecol. 10, 2049–2057.

Hedlund, K., Griffiths, B., Christensen, S., Scheu, S., Setälä, H., Tscharntke, T., Verhoef, H., 2004. Trophic interactions in changing landscapes: responses of soil food webs. Basic Appl. Ecol. 5, 495–503.

Heisswolf, A., Gabler, D., Obermaier, E., Muller, C., 2007. Olfactory versus contact cues in host plant recognition of a monophagous Chrysomelid beetle. J. Insect Behav. 20, 247–266.

Hibbard, B.E., Bernklau, E.J., Bjostad, L.B., 1994. Long-chain free fatty-acids: semiochemicals for host location by western corn rootworm larvae. J. Chem. Ecol. 20, 3335–3344.

Hill, K.E., O'Malley, R., 2010. A picky palate? The host plant selection of an endangered June beetle. J. Insect Conserv. 14, 277–287.

Hiltpold, I., Turlings, T.C.J., 2008. Belowground chemical signalling in maize: when simplicity rhymes with efficiency. J. Chem. Ecol. 34, 628–635.

Hiltpold, I., Turlings, T.C.J., 2012. Manipulation of chemically mediated interactions in agricultural soils to enhance the control of crop pests and to improve crop yield. J. Chem. Ecol. 38, 641–650.

Hiltpold, I., Baroni, M., Toepfer, S., Kuhlmann, U., Turlings, T.C.J., 2010a. Selection of entomopathogenic nematodes for enhanced responsiveness to a volatile root signal helps to control a major root pest. J. Exp. Biol. 213, 2417–2423.

Hiltpold, I., Baroni, M., Toepfer, S., Kuhlmann, U., Turlings, T.C.J., 2010b. Selective breeding of entomopathogenic nematodes for enhanced attraction to a root signal did not reduce their establishment or persistence after field release. Plant Signal. Behav. 5, 1450–1452.

Hiltpold, I., Toepfer, S., Kuhlmann, U., Turlings, T.C.J., 2010c. How maize root volatiles influence the efficacy of entomopathogenic nematodes against the western corn rootworm? Chemoecology 20, 155–162.

Hiltpold, I., Erb, M., Robert, C.A.M., Turlings, T.C.J., 2011. Systemic root signalling in a belowground, volatile-mediated tritrophic interaction. Plant Cell Environ. 34, 1267–1275.

Hiltpold, I., Hibbard, B.E., French, D.W., Turlings, T.C.J., 2012. Capsules containing entomopathogenic nematodes as a Trojan horse approach to control the western corn rootworm. Plant Soil 358, 11–25.

Hollister, B., Mullin, C.A., 1998. Behavioural and electrophysiological dose-response relationships in adult western corn rootworm (*Diabrotica virgifera virgifera* LeConte) for host pollen amino acids. J. Insect Physiol. 44, 463–470.

Honda, I., Ishikawa, Y., 1987. Electrophysiological studies on the dorsal and anterior organs of the onion fly larva, *Hylemya antiqua* Meigen (Diptera: Anthomyiidae). Appl. Entomol. Zool. 22, 410–416.

Hopkins, R.J., Griffiths, D.W., Birch, A.N.E., McKinlay, R.G., Hall, J.E., 1993. Relationships between turnip root fly (*Delia floralis*) larval development and the sugar content of swede (*Brassica napus* ssp. *rapifera*) roots. Ann. Appl. Biol. 122, 405–415.

Hounsfield, G.N., 1973. Computerized transverse axial scanning (tomography): I. Description of system. Brit. J. Radiol. 46, 1016–1022.

Hunter, M.D., 2001. Out of sight, out of mind: the impacts of root-feeding insects in natural and managed systems. Agric. For. Entomol. 3, 3–9.

Jepson, W.F., 1937. The morphology of the larva of *Serica brunnea* L. Bull. Entomol. Res. 28, 149–165.

Johnson, S.N., Gregory, P.J., 2006. Chemically-mediated host-plant location and selection by root-feeding insects. Physiol. Entomol. 31, 1–13.

Johnson, S.N., Murray, P.J., 2008. Root Feeders: An Ecosystem Perspective. CABI Publishing, Wallingford, UK.

Johnson, S.N., Nielsen, U.N., 2012. Foraging in the dark—chemically mediated host plant location by belowground insect herbivores. J. Chem. Ecol. 38, 604–614.

Johnson, S.N., Gregory, P.J., Murray, P.J., Zhang, X., Young, I.M., 2004a. Host plant recognition by the root feeding clover weevil, *Sitona lepidus* (Coleoptera: Curculionidae). Bull. Entomol. Res. 94, 433–439.

Johnson, S.N., Read, D.B., Gregory, P.J., 2004b. Tracking larval insect movement within soil using high resolution X-ray microtomography. Ecol. Entomol. 29, 117–122.

Johnson, S.N., Gregory, P.J., Greenham, J.R., Zhang, X.X., Murray, P.J., 2005. Attractive properties of an isoflavonoid found in white clover root nodules on the clover root weevil. J. Chem. Ecol. 31, 2223–2229.

Johnson, S.N., Zhang, X.X., Crawford, J.W., Gregory, P.J., Hix, N.J., Jarvis, S.C., Murray, P.J., Young, I.M., 2006. Effects of carbon dioxide on the searching behaviour of the root-feeding clover weevil *Sitona lepidus* (Coleoptera: Curculionidae). Bull. Entomol. Res. 96, 361–366.

Johnson, S.N., Crawford, J.W., Gregory, P.J., Grinev, D.V., Mankin, R.W., Masters, G.J., Murray, P.J., Wall, D.H., Zhang, X., 2007. Non-invasive techniques for investigating and modelling root-feeding insects in managed and natural systems. Argic. Forest Entomol. 9, 39–46.

Johnson, S.N., Barton, A.T., Clark, K.E., Gregory, P.J., McMenemy, L.S., Hancock, R.D., 2011. Elevated atmospheric carbon dioxide impairs the performance of root-feeding vine weevils by modifying root growth and secondary metabolites. Glob. Chang. Biol. 17, 688–695.

Johnson, S.N., Clark, K.E., Hartley, S.E., Jones, T.H., McKenzie, S.W., Koricheva, J., 2012. Aboveground-belowground herbivore interactions: a meta-analysis. Ecology 93, 2208–2215.

Johnson, S.N., Mitchell, C., McNicol, J.W., Thompson, J., Karley, A.J., 2013. Downstairs drivers—root herbivores shape communities of above-ground herbivores and natural enemies via changes in plant nutrients. J. Anim. Ecol. 82, 1021–1030.

Jones, O.T., Coaker, T.H., 1977. Oriented responses of carrot fly larvae *Psila rosae* to plant odours, carbon dioxide and carrot root volatiles. Physiol. Entomol. 2, 189–197.

Jones, O.T., Coaker, T.H., 1979. Responses of carrot fly larvae, *Psila rosae*, to the odorous and contact-chemostimulatory metabolites of host and non-host plants. Physiol. Entomol. 4, 353–360.

Kamm, J.A., Buttery, R.G., 1984. Root volatile components of red clover identification and bioassay with the clover root borer (Coleoptera, Scolytidae). Environ. Entomol. 13, 1427–1430.

Kaplan, I., Halitschke, R., Kessler, A., Sardanelli, S., Denno, R.F., 2008. Constitutive and induced defenses to herbivory in above- and belowground plant tissues. Ecology 89, 392–406.

Karban, R., 1980. Periodical cicada nymphs impose periodical oak tree wood accumulation. Nature 287, 326–327.

Karban, R., Baldwin, I., 1997. Induced responses to herbivory. University Press of Chicago, Chicago, IL, USA.

Keil, T.A., 1996. Sensilla on the maxillary palps of *Helicoverpa armigera* caterpillars: in search of the CO_2-receptor. Tissue Cell 28, 703–717.

Kim, J.H., Mullin, C.A., 1998. Structure–phagostimulatory relationships for amino acids in adult western corn rootworm, *Diabrotica virgifera virgifera*. J. Chem. Ecol. 24, 1499–1511.

Kim, J.H., Mullin, C.A., 2007. An isorhamnetin rhamnoglycoside serves as a costimulant for sugars and amino acids in feeding responses of adult western corn rootworms (*Diabrotica virgifera virgifera*) to corn (*Zea mays*) pollen. J. Chem. Ecol. 33, 501–512.

Klinger, J., 1958. Die Bedeutung der Kohlendioxyd-Ausscheidung der Würzeln für die Orientierung der Larven von *Otiorrhynchus sulcatus* F. und anderer bodenbewohnender phytophager Insektenarten. Mitt. Schweiz. Entomol. Ges. 31, 205–269.

Klingler, J., 1957. Über die Bedeutung des Kohlendioxyds für die Orientierung der Larven von *Otiorrhynchus sulcatus* F., *Melolontha* und *Agriotes* (Col.) im Boden (Vorläufige Mitteilung). Mitt. Schweiz. Entomol. Ges. 30, 317–322.

Klingler, J., 1965. On the orientation of plant nematodes and of some other soil animals. Nematologica 11, 4–18.

Klingler, J., 1966. Uber den Sitz der CO_2-Rezeptoren bei der Larve von *Otiorrhynchus sulcatus*. Entomol. Exp. Appl. 9, 271–277.

Knochel, D.G., Seastedt, T.R., 2010. Reconciling contradictory findings of herbivore impacts on spotted knapweed (*Centaurea stoebe*) growth and reproduction. Ecol. Appl. 20, 1903–1912.

Kojima, W., Ishikawa, Y., Takanashi, T., 2012. Deceptive vibratory communication: pupae of a beetle exploit the freeze response of larvae to protect themselves. Biol. Lett. 8, 717–720.

Köllner, T.G., Held, M., Lenk, C., Hiltpold, I., Turlings, T.C.J., Gershenzon, J., Degenhardt, J., 2008. A maize (E)-β-caryophyllene synthase implicated in indirect defense responses against herbivores is not expressed in most American maize varieties. Plant Cell 20, 482–494.

Kondoh, Y., Kaneshiro, K.Y., Kimura, K.I., Yamamoto, D., 2003. Evolution of sexual dimorphism in the olfactory brain of Hawaiian *Drosophila*. Proc. R. Soc. B Biol. Sci. 270, 1005–1013.

Košt'ál, V., 1992. Orientation behavior of newly hatched larvae of the cabbage maggot, *Delia radicum* (L.) (Diptera: Anthomyiidae), to volatile plant metabolites. J. Insect Behav. 5, 61–70.

Krause, A.E., Frank, K.J., Mason, D.M., Ulanowicz, R.E., Taylor, W.W., 2003. Compartments revealed in food web structure. Nature 426, 282–285.

Kuhlmann, U., Van Der Brugt, W.A.C.M., 1998. Possibilities for biological control of the western corn rootworm, *Diabrotica virgifera virgifera* LeConte, in Central Europe. Biocontrol 19, 59–68.

Kurzt, B., Hiltpold, I., Turlings, T.C.J., Kuhlmann, U., Toepfer, S., 2009. Comparative susceptibility of larval instars and pupae of the western corn rootworm to infection by three entomopathogenic nematodes. Biocontrol 54, 255–262.

Kuzyakov, Y., Domanski, G., 2000. Carbon input by plants into the soil. Rev. J. Plant Nutr. Soil Sci. 163, 421–431.

Ladd Jr., T.L., 1988. Japanese beetle (Coleoptera: Scarabaeidae): influence of sugars on feeding response of larvae. J. Econ. Entomol. 81, 1390–1393.

Laumann, R.A., Kavčič, A., Moraes, M.C.B., Borges, M., Čokl, A., 2013. Reproductive behaviour and vibratory communication of the neotropical predatory stink bug *Podisus nigrispinus*. Physiol. Entomol. 38, 71–80.

Laznik, Ž., Trdan, S., 2013. An investigation on the chemotactic responses of different entomopathogenic nematode strains to mechanically damaged maize root volatile compounds. Exp. Parasitol. 134, 349–355.

Lu, T., Qiu, Y.T., Wang, G., Kwon, J., Rutzler, M., Kwon, H.W., Pitts, R.J., Van Loon, J.J.A., Takken, W., Carlson, J.R., Zwiebel, L.J., 2007. Odor coding in the maxillary palp of the Malaria vector mosquito *Anopheles gambiae*. Curr. Biol. 17, 1533–1544.

Lundgren, J.G., Fergen, J.K., 2010. The effects of a winter cover crop on *Diabrotica virgifera* (Coleoptera: Chrysomelidae) populations and beneficial arthropod communities in no-till maize. Environ. Entomol. 39, 1816–1828.

Lundgren, J.G., Fergen, J.K., 2011. Enhancing predation of a subterranean insect pest: a conservation benefit of winter vegetation in agroecosystems. Appl. Soil Ecol. 51, 9–16.

Lundgren, J.G., Fergen, J.K., in press. Predator community structure and trophic linkage strength to a focal prey: the influence of the prey's anti-predator defense. Mol. Ecol.

Lundgren, J.G., Ellsbury, M.E., Prischmann, D.A., 2009a. Analysis of the predator community of a subterranean herbivorous insect based on polymerase chain reaction. Ecol. Appl. 19, 2157–2166.

Lundgren, J.G., Haye, T., Toepfer, S., Kuhlmann, U., 2009b. A multifaceted hemolymph defense against predation in *Diabrotica virgifera virgifera* larvae. Biocontrol Sci. Technol. 19, 871–880.

Lundgren, J.G., Nichols, S., Prischmann, D.A., Ellsbury, M.M., 2009c. Seasonal and diel activity patterns of generalist predators associated with *Diabrotica virgifera* immatures (Coleoptera: Chrysomelidae). Biocontrol Sci. Technol. 19, 327–333.

Lundgren, J.G., Toepfer, S., Haye, T., Kuhlmann, U., 2010. Haemolymph defence of an invasive herbivore: its breadth of effectiveness against predators. J. Appl. Entomol. 134, 439–448.

Luster, J., Finlay, R.D., 2006. Handbook of Methods Used in Rhizosphere Research. Swiss Federal Research Institute WSL, Birmensdorf.

Macdonald, P.J., Ellis, C.R., 1990. Survival time of unfed, first-instar western corn rootworm (Coleoptera: Chrysomelidae) and the effects of soil type, moisture, and compaction on their mobility in soil. Environ. Entomol. 19, 666–671.

Maki, A., Ryan, M.F., 1989. Root-mediated effects in carrot resistance to the carrot fly, *Psila rosae*. J. Chem. Ecol. 15, 1867–1882.

Maki, A., Kitajima, J., Abe, F., Stewart, G., Ryan, M.F., 1989. Isolation, identification, and bioassay of chemicals affecting nonpreference carrot-root resistance to carrot-fly larva. J. Chem. Ecol. 15, 1883–1897.

Mankin, R.W., 2012. Applications of acoustics in insect pest management. CAB Rev. 7, 1–7.

Mankin, R.W., Johnson, S.N., Grinev, D.V., Gregory, P.J., 2008. New experimental techniques for studying root herbivores. In: Johnson, S.N., Murray, P.J. (Eds.), Root Feeders: An Ecosystem Perspective. CABI Publishing, Wallingford, UK, pp. 20–32.

Mankin, R.W., Samson, P.R., Chandler, K.J., 2009. Acoustic detection of melolonthine larvae in Australian sugarcane. J. Econ. Entomol. 102, 1523–1535.

Matsumoto, Y., 1970. Volatile organic sulfur compounds as insect attractants with special reference to host selection. In: Wood, D.L., Nakajima, M. (Eds.), Control of Insect Behavior by Natural Products. Academic Press, New York, pp. 133–160.

Merritt, R.W., Dadd, R.H., Walker, E.D., 1992. Feeding behavior, natural food, and nutritional relationships of larval mosquitoes. Annu. Rev. Entomol. 37, 349–376.

Metcalf, R.L., 1985. Plant kairomones and insect pest control. Bull. Ill. Nat. Hist. Surv. 35, 175.

Mochizuki, A., Ishikawa, Y., Matsumoto, Y., 1985. Sugars as phagostimulants for larvae of the onion fly, *Hylemya antiqua* Meigen (Diptera, Anthomyiidae). Appl. Entomol. Zool. 20, 465–469.

Mochizuki, A., Ishikawa, Y., Matsumoto, Y., 1989. Olfactory response of the larvae of the onion fly, *Hylemya antiqua* Meigen (Diptera: Anthomyiidae) to volatile compounds. Appl. Entomol. Zool. 24, 29–35.

Moeser, J., Hibbard, B.E., 2005. A synopsis of the nutritional ecology of larvae and adults of *Diabrotica virgifera virgifera* (LeConte) in the New and Old World— Nouvelle cuisine for the invasive maize pest *Diabrotica virgifera virgifera* in Europe? In: Vidal, S., Kuhlmann, U., Edwards, R. (Eds.), Western Corn Rootworm: Ecology and Management. CABI Publishers, Wallingford, pp. 45–61.

Monroy, F., Van Der Putten, W.H., 2009. Local variation in belowground multitrophic interactions. Soil Biol. Biochem. 41, 1689–1695.

Mooney, S.J., Pridmore, T.P., Helliwell, J., Bennett, M.J., 2012. Developing X-ray computed tomography to non-invasively image 3-D root systems architecture in soil. Plant Soil 352, 1–22.

Morgan, A.C., Crumb, S.E., 1929. Notes on the chemotrophic responses of certain insects. J. Econ. Entomol. 21, 913–920.

Morrison, D.A., 2006. Multiple sequence alignment for phylogenetic purposes. Aust. Syst. Bot. 19, 476–539.

Mukai, H., Hironaka, M., Tojo, S., Nomakuchi, S., 2012. Maternal vibration induces synchronous hatching in a subsocial burrower bug. Anim. Behav. 84, 1443–1448.

Neumann, G., Römheld, V., 2007. The release of root exudates as affected by the plant physiological status. In: Pinton, R., Varanini, Z., Nannipieri, Z. (Eds.), The Rhizosphere: Biochemistry and Organic Substances at the Soil-Plant Interface. CRC Press, Boca Raton, pp. 23–72.

Neumann, G., George, T.S., Plassard, C., 2009. Strategies and methods for studying the rhizosphere-the plant science toolbox. Plant Soil 321, 431–456.

Nordenhem, H., Nordlander, G., 1994. Olfactory oriented migration through soil by root-living *Hylobius abietis* (L.) larvae (Col, Curculionidae). J. Appl. Entomol. 117, 457–462.

Ortega, Y.K., Pearson, D.E., Waller, L.P., Sturdevant, N.J., Maron, J.L., 2012. Population-level compensation impedes biological control of an invasive forb and indirect release of a native grass. Ecology 93, 783–792.

Otálora-Luna, F., Lapointe, S.L., Dickens, J.C., 2013. Olfactory cues are subordinate to visual stimuli in a neotropical generalist weevil. PLoS One 8, e53120.

Paill, W., Backeljau, T., Grimm, B., Kastberger, G., Kaiser, H., 2002. Isoelectric focusing as a tool to evaluate carabid beetles as predatory agents of the pest slug *Arion lusitanicus*. Soil Biol. Biochem. 34, 1333–1342.

Pauchet, Y., Wilkinson, P., Vogel, H., Nelson, D.R., Reynolds, S.E., Heckel, D.G., Ffrench-Constant, R.H., 2010. Pyrosequencing the *Manduca sexta* larval midgut transcriptome: messages for digestion, detoxification and defence. Insect Mol. Biol. 19, 61–75.

Pawliszyn, J., 2011. Handbook of Solid Phase Microextraction. Elsevier, Chennai, India.

Payne, D., Gregory, P.J., 1988. The soil atmosphere. In: Wild, A. (Ed.), Russell's Soil Conditions and Plant Growth. Longman, Harlow, pp. 298–314.

Phillips, R.P., Erlitz, Y., Bier, R., Bernhardt, E.S., 2008. New approach for capturing soluble root exudates in forest soils. Funct. Ecol. 22, 990–999.

Pichersky, E., Noel, J.P., Dudareva, N., 2006. Biosynthesis of plant volatiles: nature's diversity and ingenuity. Science 311, 808–811.

Pierret, A., Moran, C.J., Doussan, C., 2005. Conventional detection methodology is limiting our ability to understand the roles and functions of fine roots. New Phytol. 166, 967–980.

Pike, D.R., Steffey, K.L., Gray, M.E., Kirby, H.W., Edwards, D.I., Hornbaker, R.H., 1995. Biological and Economic Assessment of Pesticide Use on Corn and Soybeans. USDA-National Agricultural Pesticide Impact Assessment Program Program, Washington, DC.

Pilz, C., Keller, S., Kuhlmann, U., Toepfer, S., 2009. Comparative efficacy assessment of fungi, nematodes and insecticides to control western corn rootworm larvae in maize. Biocontrol 54, 671–684.

Plassard, C., Meslem, M., Souche, G., Jaillard, B., 1999. Localization and quantification of net fluxes of H+ along maize roots by combined use of pH-indicator dye videodensitometry and H+-selective microelectrodes. Plant Soil 211, 29–39.

Pompanon, F., Deagle, B.E., Symondson, W.O.C., Brown, D.S., Jarman, S.N., Taberlet, P., 2012. Who is eating what: diet assessment using next generation sequencing. Mol. Ecol. 21, 1931–1950.

Price, P.W., Bouton, C.E., Gross, P., McPheron, B.A., Thompson, J.N., Weis, A.E., 1980. Interactions among three trophic levels—influence of plants on interactions between insect herbivores and natural enemies. Annu. Rev. Ecol. Syst. 11, 41–65.

Prischmann, D.A., Knutson, E.M., Dashiell, K.E., Lundgren, J.G., 2011. Generalist-feeding subterranean mites as potential biological control agents of immature corn rootworms. Exp. Appl. Acarol. 55, 233–248.

Ranger, C.M., Reding, M.E., Oliver, J.B., Schultz, P.B., Moyseenko, J.J., Youssef, N., 2011. Comparative efficacy of plant-derived essential oils for managing *Ambrosia* beetles (Coleoptera: Curculionidae: Scolytinae) and their corresponding mass spectral characterization. J. Econ. Entomol. 104, 1665–1674.

Rasmann, S., Agrawal, A.A., 2008. In defense of roots: a research agenda for studying plant resistance to belowground herbivory. Plant Physiol. 146, 875–880.

Rasmann, S., Turlings, T.C.J., 2007. Simultaneous feeding by aboveground and belowground herbivores attenuates plant-mediated attraction of their respective natural enemies. Ecol. Lett. 10, 926–936.

Rasmann, S., Köllner, T.G., Degenhardt, J., Hiltpold, I., Toepfer, S., Kuhlmann, U., Gershenzon, J., Turlings, T.C.J., 2005. Recruitment of entomopathogenic nematodes by insect-damaged maize roots. Nature 434, 732–737.

Rasmann, S., Bauerle, T.L., Poveda, K., Vannette, R., 2011a. Predicting root defence against herbivores during succession. Funct. Ecol. 25, 368–379.

Rasmann, S., Erwin, A.C., Halitschke, R., Agrawal, A.A., 2011b. Direct and indirect root defences of milkweed (*Asclepias syriaca*): trophic cascades, trade-offs and novel methods for studying subterranean herbivory. J. Ecol. 99, 16–25.

Rasmann, S., Ali, J.G., Helder, J., Van Der Putten, W.H., 2012a. Ecology and evolution of soil nematode chemotaxis. J. Chem. Ecol. 38, 615–628.

Rasmann, S., Hiltpold, I., Ali, J.G., 2012b. The role of root-produced volatile secondary metabolites in mediating soil interactions. In: Montanaro, G., Dichio, B. (Eds.), Advances in Selected Plant Physiology Aspects, 269–290, InTech, Rijeka, Croatia.

Ritchie, S.W., Hanway, J.J., Benson, G.O., 1992. How a Corn Plant Develops. Iowa State University of Science and Technology Cooperative Extension Service Special Report No. 48

R Core Team (2012). R: A language and environment for statistical computing. R Foundation for Statistical Computing, Vienna, Austria. ISBN 3-900051-07-0, http://www.r-project.org/

Robert, C.A.M., Erb, M., Duployer, M., Zwahlen, C., Doyen, G.R., Turlings, T.C.J., 2012a. Herbivore-induced plant volatiles mediate host selection by a root herbivore. New Phytol. 194, 1061–1069.

Robert, C.A.M., Veyrat, N., Glauser, G., Marti, G., Doyen, G.R., Villard, N., Gaillard, M.D.P., Köllner, T.G., Giron, D., Body, M., Babst, B.A., Ferrieri, R.A., Turlings, T.C.J., Erb, M., 2012b. A specialist root herbivore takes advantage of defensive metabolites to locate nutritious tissues. Ecol. Lett. 15, 55–64.

Ross, K.T.A., Anderson, M., 1992. Larval responses of three vegetable root fly pests of the genus *Delia* (Diptera: Anthomyiidae) to plant volatiles. B. Entomol. Res. 82, 393–398.

Rudinsky, J.A., 1966. Scolytid beetles associated with Douglas fir: response to terpenes. Science 152, 218–219.
Rudinsky, J.A., Zethner-Møller, O., 1967. Olfactory responses of *Hylastes nigrinus* to various host materials. Can. Entomol. 99, 911–916.
Ryan, M.F., Guerin, P.M., 1982. Behavioral responses of the carrot fly larva, *Psila rosae*, to carrot root volatiles. Physiol. Entomol. 7, 315–324.
Rygg, T., Sömme, L., 1972. Oviposition and larval development of *Hylemya floralis* (Fallén) (Diptera, Anthomyiidae) on varieties of swedes and turnips. Norw. J. Entomol. 19, 81–90.
Sasakawa, K., 2011. Laboratory studies on larval food habits of two syntopic, related, granivorous ground beetles *Amara chalcites* and *A. congrua* (Coleoptera: Carabidae): a comparison with stable isotope analysis. Appl. Entomol. Zool. 46, 511–518.
Schilmiller, A.L., Last, R.L., Pichersky, E., 2008. Harnessing plant trichome biochemistry for the production of useful compounds. Plant J. 54, 702–711.
Schneider, D., 1964. Insect Antennae. Annu. Rev. Entomol. 9, 103–122.
Schoonhoven, L.M., Van Loon, J.J.A., Dicke, M., 2005. Insect-Plant Biology. Oxford University Press, Oxford.
Schumann, M., Patel, A., Vidal, S., 2013. Evaluation of an attract and kill strategy for western corn rootworm larvae. Appl. Soil Ecol. 64, 178–189.
Scott, K., Brady Jr., R., Cravchik, A., Morozov, P., Rzhetsky, A., Zuker, C., Axel, R., 2001. A chemosensory gene family encoding candidate gustatory and olfactory receptors in *Drosophila*. Cell 104, 661–673.
Sérandour, J., Reynaud, S., Willison, J., Patouraux, J., Gaude, T., Ravanel, P., Lempérière, G., Raveton, M., 2008. Ubiquitous water-soluble molecules in aquatic plant exudates determine specific insect attraction. PLoS One 3, e3350.
Shen, J., Hoffland, E., 2007. In situ sampling of small volumes of soil solution using modified micro-suction cups. Plant Soil 292, 161–169.
Soler, R., Harvey, J.A., Kamp, A.F.D., Vet, L.E.M., Van Der Putten, W.H., Van Dam, N. M., Stuefer, J.F., Gols, R., Hordijk, C.A., Bezemer, T.M., 2007. Root herbivores influence the behaviour of an aboveground parasitoid through changes in plant-volatile signals. Oikos 116, 367–376.
Soni, S.K., Finch, S., 1979. Laboratory evaluation of sulphur-bearing chemicals as attractants for larvae of the onion fly, *Delia antiqua* (Meigen) (Diptera: Anthomyiidae). Bull. Entomol. Res. 69, 291–298.
Spencer, J.L., Hibbard, B.E., Moeser, J., Onstad, D.W., 2009. Behaviour and ecology of the western corn rootworm (*Diabrotica virgifera virgifera* LeConte). Agric. For. Entomol. 11, 9–27.
Staley, J.T., Johnson, S.N., 2008. Climate change impacts on root herbivores. In: Johnson, S.N., Murray, P.J. (Eds.), Root Feeders: An Ecosystem Perspective. CABI Publishing, Wallingford, pp. 192–213.
Stamatakis, A., Hoover, P., Rougemont, J., 2008. A rapid bootstrap algorithm for the RAxML web servers. Syst. Biol. 57, 758–771.
Stange, G., Stowe, S., 1999. Carbon-dioxide sensing structures in terrestrial arthropods. Microsc. Res. Techniq. 47, 416–427.
Steinbrecht, R.A., 1997. Pores structures in insect olfactory sensilla: a review of data and concepts. Int. J. Insect Morphol. 26, 229–245.
Stocker, R.F., 1994. The organization of the chemosensory system in *Drosophila melanogaster*: a review. Cell Tissue Res. 275, 3–26.
Stocker, R.F., 2001. *Drosophila* as a focus in olfactory research: mapping of olfactory sensilla by fine structure, odor specificity, odorant receptor expression, and central connectivity. Microsc. Res. Techniq. 55, 284–296.

Strnad, S.P., Bergman, M.K., 1987. Movement of first-instar western corn rootworms (Coleoptera, Chrysomelidae) in soil. Environ. Entomol. 16, 975–978.
Strnad, S.P., Dunn, P.E., 1990. Host search behavior of neonate western corn rootworm (*Diabrotica virgifera virgifera*). J. Insect Physiol. 36, 201–205.
Strong, D.R., Whipple, A.V., Child, A.L., Dennis, B., 1999. Model selection for a subterranean trophic cascade: root-feeding caterpillars and entomopathogenic nematodes. Ecology 80, 2750–2761.
Sutherland, O.R., 1971. Feeding behaviour of the grass grub *Costelytra zealandica* (White) (Coleoptera: Melolonthinae). 1. The influence of carbohydrates. N. Z. J. Sci. 14, 18–23.
Sutherland, O.R.W., Hilier, J.R., 1974. Olfactory response of *Costelytra zealandica* (Coleoptera: Melolonthinae) to the roots of several pasture plants. N. Z. J. Sci. 1, 365–369.
Sutherland, O.R.W., Hillier, J.R., 1974. Feeding behaviour of the grass grub *Costelytra zealandica* (White) (Coleoptera: Melolonthinae). 3. The influence of amino acids, ascorbic acid, and inorganic salts. N. Z. J. Zool 1, 211–216.
Sutherland, O.R.W., Hillier, J.R., 1976. The influence of maltose and other carbohydrates on the feeding behaviour of *Heteronychus arator* (Scarabaeidae: Coleoptera). Experientia 32, 701–702.
Sutherland, O.R.W., Russell, G.B., Biggs, D.R., Lane, G.A., 1980. Insect feeding deterrent activity of phytoalexin isoflavonoids. Biochem. Syst. Ecol. 8, 73–75.
Swarup, S., Williams, T.I., Anholt, R.R.H., 2011. Functional dissection of odorant binding protein genes in *Drosophila melanogaster*. Genes Brain Behav. 10, 648–657.
Tallamy, D.W., Whittington, D.P., Defurio, F., Fontaine, D.A., Gorski, P.M., Gothro, P., 1998. The effect of sequestered cucurbitacins on the pathogenicity of *Metarhizium anisopliae* (Moniliales: Moniliaceae) on spotted cucumber beetle eggs and larvae (Coleoptera: Chrysomelidae). Environ. Entomol. 27, 366–372.
Tallamy, D.W., Gorski, P.M., Burzon, J.K., 2000. The fate of male-derived cucurbitacins in spotted cucumber beetle females. J. Chem. Ecol. 26, 413–427.
Tallamy, D.W., Hibbard, B.E., Clark, T.L., Gillespie, J.J., 2005. Western corn rootworm, cucurbits and cucurbitacins. In: Vidal, S., Kuhlmann, U., Edwards, R. (Eds.), Western Corn Rootworm: Ecology and Management. CABI publishers, Wallingford, pp. 67–93.
Tapia, T., Perich, F., Pardo, F., Palma, G., Quiroz, A., 2007. Identification of volatiles from differently aged red clover (*Trifolium pratense*) root extracts and behavioural responses of clover root borer (*Hylastinus obscurus*) (Marsham) (Coleoptera: Scolytidae) to them. Biochem. Syst. Ecol. 35, 61–67.
Teng, J., McCann, K.S., 2004. Dynamics of compartmented and reticulate food webs in relation to energetic flow. Am. Nat. 164, 85–100.
Thompson, J.D., Gibson, T.J., Plewniak, F., Jeanmougin, F., Higgins, D.G., 1997. The CLUSTAL X windows interface: flexible strategies for multiple sequence alignment aided by quality analysis tools. Nucleic Acids Res. 25, 4876–4882.
Thorpe, V.H., Crombie, A.C., Hill, R., Darrah, J.H., 1946. The behaviour of wireworms in response to chemical stimulation. J. Exp. Biol. 23, 234–266.
Tokuda, M., Tanaka, S., Maeno, K., Harano, K.I., Wakamura, S., Yasui, H., Arakaki, N., Akino, T., Fukaya, M., 2010. A two-step mechanism controls the timing of behaviour leading to emergence from soil in adult males of the scarab beetle *Dasylepida ishigakiensis*. Physiol. Entomol. 35, 231–239.
Trachsel, S., Kaeppler, S.M., Brown, K.M., Lynch, J.P., 2011. Shovelomics: high throughput phenotyping of maize (*Zea mays* L.) root architecture in the field. Plant Soil 341, 75–87.
Tracy, S.R., Roberts, J.A., Black, C.R., McNeill, A., Davidson, R., Mooney, S.J., 2010. The X-factor: visualizing undisturbed root architecture in soils using X-ray computed tomography. J. Exp. Bot. 61, 311–313.

Traugott, M., Pázmándi, C., Kaufmann, R., Juen, A., 2007. Evaluating $^{15}N/^{14}N$ and $^{13}C/^{12}C$ isotope ratio analysis to investigate trophic relationships of elaterid larvae (Coleoptera: Elateridae). Soil Biol. Biochem. 39, 1023–1030.
Turlings, T.C.J., Hiltpold, I., Rasmann, S., 2012. The importance of root-produced volatiles as foraging cues for entomopathogenic nematodes. Plant Soil 359, 51–60.
Turner, S.L., Li, N., Guda, T., Githure, J., Cardé, R.T., Ray, A., 2011. Ultra-prolonged activation of CO_2-sensing neurons disorients mosquitoes. Nature 474, 87–91.
Vaidya, G., Lohman, D.J., Meier, R., 2011. SequenceMatrix: concatenation software for the fast assembly of multi-gene datasets with character set and codon information. Cladistics 27, 171–180.
Van Dam, N.M., 2009. Belowground herbivory and plant defenses. Annu. Rev. Ecol. Evol. Syst. 40, 373–391.
Van Dam, N.M., Harvey, J.A., Wackers, F.L., Bezemer, T.M., Van Der Putten, W.H., Vet, L.E.M., 2003. Interactions between aboveground and belowground induced responses against phytophages. Basic Appl. Ecol. 4, 63–77.
Van Dam, N.M., Witjes, L., Svatos, A., 2004. Interactions between aboveground and below-ground induction of glucosinolates in two wild *Brassica* species. New Phytol. 161, 801–810.
Van Dam, N.M., Raaijmakers, C.E., Van Der Putten, W.H., 2005. Root herbivory reduces growth and survival of the shoot feeding specialist *Pieris rapae* on *Brassica nigra*. Entomol. Exp. Appl. 115, 161–170.
Van Der Putten, W.H., 2003. Plant defense belowground and spatiotemporal processes in natural vegetation. Ecology 84, 2269–2280.
Van Der Putten, W.H., Bardgett, R.D., De Ruiter, P.C., Hol, W.H.G., Meyer, K.M., Bezemer, T.M., Bradford, M.A., Christensen, S., Eppinga, M.B., Fukami, T., Hemerik, L., Molofsky, J., Schädler, M., Scherber, C., Strauss, S.Y., Vos, M., Wardle, D.A., 2009. Empirical and theoretical challenges in aboveground-belowground ecology. Oecologia 161, 1–14.
Van Tol, R.W.H.M., Van Der Sommen, A.T.C., Boff, M.I.C., Van Bezooijen, J., Sabelis, M.W., Smits, P.H., 2001. Plants protect their roots by alerting the enemies of grubs. Ecol. Lett. 4, 292–294.
Vannette, R.L., Rasmann, S., 2012. Arbuscular mycorrhizal fungi mediate below-ground plant-herbivore interactions: a phylogenetic study. Funct. Ecol. 26, 1033–1042.
Varela, L., Bernays, E.A., 1988. Behavior of newly hatched potato tuber moth larvae, *Phthorimaea operculella* Zell. (Lepidoptera: Gelechiidae), in relation to their host plants. J. Insect Behav 1, 261–275.
Vet, L.E.M., Dicke, M., 1992. Ecology of infochemical use by natural enemies in a tritrophic context. Annu. Rev. Entomol. 37, 141–172.
Vieira, F.G., Sánchez-Gracia, A., Rozas, J., 2007. Comparative genomic analysis of the odorant-binding protein family in 12 *Drosophila* genomes: purifying selection and birth-and-death evolution. Genome Biol. 8, R335.
Vogt, R.G., Riddiford, L.M., 1981. Pheromone binding and inactivation by moth antennae. Nature 293, 161–163.
Vosshall, L.B., Amrein, H., Morozov, P.S., Rzhetsky, A., Axel, R., 1999. A spatial map of olfactory receptor expression in the *Drosophila* antenna. Cell 96, 725–736.
Wallace, J.B., Blum, M.S., 1971. Reflex bleeding: a highly refined defensive mechanism in *Diabrotica* larvae (Coleoptera: Chrysomelidae). Ann. Entomol. Soc. Am. 64, 1021–1024.
Wallinger, C., Juen, A., Staudacher, K., Schallhart, N., Mitterrutzner, E., Steiner, E.M., Thalinger, B., Traugott, M., 2012. Rapid plant identification using species- and group-specific primers targeting chloroplast DNA. PLoS One 7, e29473.

Wallinger, C., Staudacher, K., Schallhart, N., Peter, E., Dresch, P., Juen, A., Traugott, M., 2013. The effect of plant identity and the level of plant decay on molecular gut content analysis in a herbivorous soil insect. Mol. Ecol. Resour. 13, 75–83.
Walrant, A., Loreau, M., 1995. Comparison of iso-enzyme electrophoresis and gut content examination for determining the natural diet of the groundbeetle species *Abax ater* (Coleoptera: Carabidae). Entomol. Gen. 19, 253–259.
Wang, Y., Kays, S.J., 2002. Sweetpotato volatile chemistry in relation to sweetpotato weevil (*Cylas formicarius*) behavior. J. Am. Soc. Hortic. Sci. 127, 656–662.
Weissteiner, S., Huetteroth, W., Kollmann, M., Weissübecker, B., Romani, R., Schachtner, J., Schütz, S., 2012. Cockchafer larvae smell host root scents in soil. PLoS One 7, e45827.
Wenke, K., Kai, M., Piechulla, B., 2010. Belowground volatiles facilitate interactions between plant roots and soil organisms. Planta 231, 499–506.
Wensler, R.J., Dudzinski, A.E., 1972. Gustation of sugars, amino acids and lipids by larvae of the scarabaeid *Sericethis geminata* (Coleoptera). Entomol. Exp. Appl. 15, 155–165.
Wilson, K., Gunn, A., Cherrett, J.M., 1995. The application of a rhizotron to study the subterranean effects of pesticides. Pedobiologia 39, 132–143.
Xie, R.J., Deng, L., Jing, L., He, S.L., Ma, Y.T., Yi, S.L., Zheng, Y.Q., Zheng, L., 2013. Recent advances in molecular events of fruit abscission. Biol. Plantarum 57, 201–209.
Yonekura-Sakakibara, K., Saito, K., 2009. Functional genomics for plant natural product biosynthesis. Nat. Prod. Rep. 26, 1466–1487.

The Biology, Physiology and Host–Plant Interactions of Grape Phylloxera *Daktulosphaira vitifoliae*

Kevin S. Powell[*], **Paul D. Cooper**[†], **Astrid Forneck**[‡]
[*]Biosciences Research Division, Department of Environment and Primary Industries, Rutherglen, Victoria, Australia
[†]Evolution, Ecology and Genetics, Research School of Biology, College of Medicine, Biology and Environment, Australian National University, Canberra, Australia
[‡]Department of Crop Sciences, Division of Viticulture and Pomology, University of Natural Resources and Applied Life Sciences, Vienna, Austria

Contents

Advances in Insect Physiology, Volume 45
ISSN 0065-2806
http://dx.doi.org/10.1016/B978-0-12-417165-7.00004-0

Abstract

Grape phylloxera *Daktulosphaira vitifoliae* (Fitch) is regarded as the most economically important insect pest of commercial grapevines *Vitis* worldwide. Root-feeding stages are the most economically damaging. Phylloxera has been managed effectively since the use of resistant rootstocks in the late nineteenth century. The more recent genetic characterisation and partial characterisation of the internal anatomy of *D. vitifoliae* has improved our knowledge of its life cycle, genetic diversity, physiology and geographic distribution of holocyclic and anholocyclic populations. Although the Phylloxeridae has a basal phylogenetic relationship to the Aphidoidea and provides an interesting comparative model, it has received far less research focus on its fundamental biological characteristics. Using existing techniques and recent advances in molecular, genomic and metabolomic techniques could address the knowledge gaps and lead to the development of improved management options.

1. INTRODUCTION

Grape phylloxera, *Daktulosphaira vitifoliae* (Fitch) (Hemiptera, Phylloxeridae), is a major pest of cultivated grapevines *Vitis* spp. (Vitaceae) worldwide (Powell, 2008). The root-feeding stages cause most economic damage and most research to date has focused on this part of the life cycle. The Phylloxeridae are a relatively small family, containing 75 species (Blackman and Eastop, 1984), which live on deciduous trees and perennial fruit crops. Grape phylloxera is the most economically destructive and geographically widespread pest species of commercial grapevines, on which it is an obligate biotroph of *Vitis* species, occurring in almost all viticultural regions around the world (Powell, 2012). Grape phylloxera is native to the NE Rockies where it coexists with native North American *Vitis* species (Wapshere and Helm, 1987).

Grape phylloxera gained economic pest status in the late nineteenth century when it spread to Europe causing significant damage to highly susceptible ungrafted European grapevine *V. vinifera* L. Considerable economic and social hardship occurred (Banerjee et al., 2007), initially in France then Western Europe, where the insect spread to over 2.5 million ha. of grapevines (Bournier, 1976). It then spread to Australia, Russia, New Zealand and South America (Crovetti and Rossi, 1987; Rossi, 1988) after it was inadvertently transferred on imported *Vitis* planting material (Buchanan, 1990; Mayet, 1890).

1.1. Host range

Grape phylloxera only feeds on *Vitis* species. It infests the root system and leaves of *Vitis* species and is cecidogenic (gall-forming). *Vitis vinifera* is particularly susceptible to radicole (root-galling) grape phylloxera, which induce nodosities on nonlignified roots and tuberosities on older lignified roots that disrupt water and nutrient uptake, resulting in loss of leaf surface area (Stellwaag, 1928), yield reduction, and even vine death within 4–7 years of infestation (Furkaliev, 1999). The use of resistant and partially resistant rootstocks bred from North American hybrids of *Vitis* species, onto which *V. vinifera* cultivars are grafted, is the most common form of phylloxera management worldwide and has been since the late nineteenth century (Carton et al., 2007). Gallicole (leaf-galling) grape phylloxera also occurs but is primarily associated with shooting resistant rootstocks in abandoned habitats and nursery plantings and rarely leads to economic loss. However, phylloxera-induced leaf galls have also been reported, albeit to a lesser extent, on *V. vinifera* (Strapazzon and Girolami, 1983; Koennecke et al., 2011; Vidart et al., 2013).

1.2. Biotype emergence

Worldwide phylloxera is largely managed through the use of *Vitis* rootstocks and to a lesser extent through the use of quarantine, although there have been some limited studies on other control options (see Benheim et al., 2012). Despite the success of rootstock management, since its introduction in the late nineteenth century, reliance on a single management approach has both economic consequences, due to the higher costs of grafted material, and certain associated risks. One of the emerging concerns is the potential risk of rootstock breakdown or failure due to the presence of phylloxera 'biotypes' (Printz, 1937), which had previously been described as 'races' (Börner, 1914; Kozhantschikov, 1928) and more recently described as xenotypes (Araya, 2001). Whilst there is some inconsistency and debate over the definition of 'biotypes' (Downie, 2010; Fergusson-Kolmes and Dennehy, 1991), the terminology used in the original citations will suffice for this chapter. A classic example of the economic consequences of 'biotype emergence' was that of biotype B (Granett et al., 1985), which caused the breakdown in resistance of the widely planted rootstock AXR#1 (*V. vinifera* 'Aramon' × *V. rupestris*) and cost the viticulture industry between US$ 1 and 6 billion (Gale, 2011; Sullivan, 1996).

Other rootstock hybrids with *V. vinifera* parentage, for example, 41B (*V. vinifera* × *V. berlandieri*) and 26G (*V. vinifera* × *V. riparia*) also have poor phylloxera resistance (Branas, 1963). There are also reports of rootstock decline, where there is no *V. vinifera* genetic background, associated with phylloxera biotypes, for example, in Germany and Hungary where 5C Teleki (*V. riparia* × *V. berlandieri*) succumbed to a combination of a virulent biotype and water stress (Walker et al., 1998). Biotypes have been reported and studied in Europe (Boubals, 1994; Song and Granett, 1990; Yvon and Peros, 2003), Australasia (Corrie et al., 1997; King and Rilling, 1985), Canada (Stevenson, 1970), South Africa (De Klerk, 1979) and the United States (De Benedictis and Granett, 1992; Williams and Shambaugh, 1987). In order to understand the potential implications of biotypes, along with the advancement in molecular techniques, there has subsequently been an increased focus on the genetics, life cycle, host–plant interactions and biology of the insect.

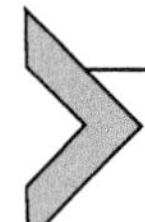

2. GRAPE PHYLLOXERA

2.1. Life cycle

Although grape phylloxera is monophagous, it has a complex life cycle, with a number of life cycle variations (see review Forneck and Huber, 2009). Its classic life cycle is holocyclic and includes sexual and asexual reproductive components. Cyclic parthenogenesis occurs on roots and leaves of *Vitis* species in spring and summer (Granett et al., 2001b) when temperature and host plant phenological conditions are optimal (Omer et al., 2002). In autumn, alate adults (winged sexuparae) are produced that have the potential, under suitable environmental conditions, to asexually produce nonfeeding male and female (oviparae) sexuales that can then mate and oviposit an overwintering egg (Forneck et al., 2001b). A female fundatrix can then hatch from the egg and in spring move onto *Vitis* leaves to recommence asexual parthenogenetic reproduction. First instars that hatch can then move down onto the root system in early spring and summer. This classic life cycle has a number of variations, and in some countries (e.g. Australia, China and New Zealand) and regions (e.g. California, USA), an anholocyclic life cycle is observed where asexual reproduction predominates and leaf galling and sexual reproduction, if it occurs at all, is relatively rare and first instar overwintering commonly occurs on the root system. In contrast in southern Europe leaf galling is more common than root-galling.

2.2. Genetic diversity

Molecular markers have been used to characterise genetic diversity, geographic distribution of holocyclic and anholocyclic populations and elucidate the life cycle of grape phylloxera using random amplified polymorphic DNA (RAPD) (Corrie et al., 1997; Fong et al., 1995; Kocsis et al., 1999; Yvon and Peros, 2003), amplified fragment length polymorphisms (AFLP) (Forneck et al., 2000, 2001b; Vorwerk and Forneck, 2007; Ritter et al., 2007) and microsatellite markers (Corrie et al., 2002). From the studies using microsatellite markers, asexual reproduction appears to predominate in Europe (Vorwerk and Forneck, 2006), Australia (Corrie and Hoffmann, 2004; Umina et al., 2007), China (Sun et al., 2009) and California, United States (Lin et al., 2006).

3. PHYLLOXERA AND THE PHYSIOLOGY OF GALLING

3.1. Cecidogenesis

Although of fundamental importance in understanding the basis of host plant resistance to grape phylloxera, there are relatively few studies that have focused on the process of cecidogenesis (gall induction) in *Vitis*.

Initiation of root galling occurs when a first instar radicole initially releases saliva that hydrolyses host tissue (Zotov et al., 1966); it then feeds and establishes on the meristem root tips of nonlignified young roots. Changes in cell structure and chemical composition, increased cell division and a subapical swelling or nodosity then occur just behind the root tip. Recent studies have highlighted that expansin genes are involved in the regulation of nodosity formation (Lawo et al., 2013). Once a feeding site has been established as phylloxera populations develop on nodosities on the fine nonlignified root system, first instars then migrate to older lignified roots and initiate tuberosities (Powell et al., 2000). Nodosities can occur both on American *Vitis* hosts and also on *V. vinifera*, whereas tuberosities predominantly, but not exclusively (see Korosi et al., 2007), occur on *V. vinifera*.

Leaf galls are initiated within 48 h (Rosen, 1916) by single first instar gallicole phylloxera feeding on the adaxial surface of *Vitis* leaves (Sterling, 1952). As the gall develops, the insect is effectively enclosed within and protected by the host plant tissue. Leaf galls predominantly occur on American *Vitis* hosts but have also been reported on *V. vinifera* in Europe and South America (Molnár et al., 2009; Vidart et al., 2013).

The galling response of different *Vitis* species roots to grape phylloxera is determined by both host plant and phylloxera genotype. For example, on a 'susceptible' host, such as *V. vinifera*, a typical reaction is cell hypertrophy (cell enlargement) and meristem hyperplasia (cell division) with a corresponding increase in starch, sugar, proline and amino acid concentrations (Du et al., 2008; Forneck et al., 2002; Hofman, 1957; Kellow et al., 2004). The gall becomes a major sink and nutrient reservoir drawing assimilates away from the rest of the plant. On resistant *Vitis* hybrids, the host plant response can range from galling to a hypersensitive response.

During galling, chemical defence reactions occur including activation of oxidising enzyme and hormones and accumulation of phenolic compounds. Early studies indicate that phenolic compounds occurring naturally in vine roots include chlorogenic acid and quercetin, which stimulate gall formation, whilst other phenolics such as caffeic acid, quinic acid and glycosylated quercetin appear to inhibit gall formation in resistant *V. riparia* × *V. rupestris* hybrids (Denisova, 1965). The level of defence response is dependent on the host *Vitis* species with the response being relatively weak in susceptible *V. vinifera* and strong in some resistant rootstocks (e.g. Börner (*V. riparia* × *V. cinerea*)).

Vitis root tips normally produce the endogenous hormones gibberellin, cytokinins, auxins and abscisic acid that regulate growth and development of the shoots and fruit (Freeman, 1983). Cytokinins in the root tip can promote cell division and elongation. Extreme stress in the root zone can interfere with cytokinin production causing premature leaf senescence and yellowing. Premature senescence is a common visual symptom expressed in *V. vinifera* foliage when radicole phylloxera are feeding on the roots and is sometimes used as part of routine visual surveying technique for phylloxera detection (Powell, 2008). Gibberellins promote stem elongation and abscisic acid inhibits transpiration, nutrient transport and stem and root elongation. The hypertrophy that occurs during leaf galling is associated with enhanced auxin levels, particularly indole acetic acid (IAA). However, it is unclear if IAA production is a causative or secondary effect of gall initiation. Some insects inject IAA into plants directly and salivary proteases can release IAA conjugates or IAA oxidase inhibitors. IAA has been detected in grape phylloxera saliva (Schäller, 1963a). Swellings similar to phylloxera-induced root galls have been observed following topical application of IAA to grapevine roots (Boubals, 1966a; Nysterakis, 1946). In more recent studies (Granett, 1990), topical applications of IAA to susceptible and resistant *Vitis* species result in root swellings in both, indicating that IAA may not be a major mechanism of resistance to grape phylloxera. Other endogenous plant hormones have been implicated in both leaf- and root-gall formation including abscisic acid, zeatin, gibberellic acid, cytokinins and kinetin, which all increase during gall formation (Du et al., 2011; Kislin et al., 1990; Zotov and Gadiyev, 1975).

At least three amino acids have been also implicated in gall formation. Topical applications of lysine, histidine and tryptophan to *Vitis* roots induce gall formation and have also been reported in phylloxera salivary excretions (Anders, 1958a,b), although some doubt has been cast on the validity of this claim (Miles, 1968). Moreover, some doubt also exists as to the potential role of salivary amino acids in leaf gall formation, because the host plant can produce the enzyme polyphenoloxidase that could deaminate free amino acids by oxidation (Henke, 1961). Rilling et al. (1975) suggested that although there are changes in the levels of some amino acids in root galls, notably a decrease in glutamine, this is related more to metabolic changes within the host plant rather than being due to cecidogenesis. In contrast, in leaf galls, glutamine, asparagine, histidine and arginine levels are increased, whilst other essential amino acids are decreased (Rilling et al., 1975). Kellow (2000) reported an increase in all amino acids including glutamine, during nodosity formation in *V. vinifera*. Other enzymes including quinone reductase, dehydrogenase, superoxide dismutase and phenylalanine ammonia lyase have also been shown to differ in concentration between resistant and susceptible *Vitis* species (Du et al., 2011; Henke, 1963; Qing et al., 2011), which may affect the galling response. Stylectomy is one approach that has not yet been used to analyse cell sap imbibed by grape phylloxera when feeding, although it has been used for other Hemiptera (e.g. Varkonyi-Gasic et al., 2010). If successfully applied, this approach could clarify the insect's basal dietary requirements and provide a qualitative analysis of salivary enzyme composition; however, the volumes of sap available for collection by parenchymal feeders are likely to be smaller than for phloem feeders. Another experimental approach would be to collect salivary enzymes using an *in vitro* feeding system.

3.1.1 Root-galling phylloxera

Because the root-feeding form of grape phylloxera is the most economically important in terms of damage to cultivated grapevines, it is important to focus research on this group. In Australia, the first evidence of genetically diverse phylloxera strains was obtained using RAPD markers to compare six phylloxera strains from three geographically distinct grape-growing regions (Corrie et al., 1997). Three radicole strains were initially described (Corrie et al., 1998). Subsequently, more detailed studies have shown that genetic diversity is more extensive and parthenogenetic reproduction predominates in Australia (Corrie et al., 2002; Corrie and Hoffmann, 2004). Eighty-three genetic strains, including both radicole and gallicole forms, have been characterised in a study using six microsatellite mitochondrial DNA markers (Umina et al., 2007). Although gallicole forms exist in Australia, they are less prevalent and geographically isolated to a single phylloxera infested zone (PIZ) quarantine region in northeast Victoria. The distribution and frequency of phylloxera strains vary within infested regions of Australia, suggesting the introduction of separate strains of *D. vitifoliae* with host associations. The greatest genetic diversity, representing all the 83 strains of phylloxera, is present in the northeast Victoria PIZ, which also represents one of the oldest state quarantine regions.

Two radicole genotypes, G1 and G4, predominate in Australia and have been described as 'superclones' (Umina et al., 2007). These superclones have a broader geographic distribution and comparatively higher virulence levels and have been identified in most of the phylloxera outbreaks occurring in Australia since 1994 (K.S. Powell, personal observation). Virulence levels of particular phylloxera strains may vary on susceptible *V. vinifera* as recently shown in laboratory, glasshouse and field trials that compared the performance of radicole phylloxera strains (Herbert et al., 2008, 2010; Trethowan and Powell, 2007). The strains differed in fecundity and survivorship and damage levels (Herbert et al., 2010). The two 'superclones' displayed the highest virulence and damage levels. These results point to only some phylloxera clones being highly virulent on *V. vinifera*, helping to explain differences in patterns of field damage, phylloxera distribution and continued survival and production of *V. vinifera* vines in some infested areas (Herbert et al., 2006; Powell et al., 2003).

3.1.2 Leaf-galling phylloxera

Leaf galling, particularly in Europe and South America, does in some instances cause economic damage to *V. vinifera* (Koennecke et al., 2011; Molnár et al., 2009; Quaglia and Rossi, 1987; Raspi and Antonelli, 1987; Vidart et al., 2013) but more commonly occurs in abandoned vineyards, on nursery plantings of rootstocks or on resistant rootstock suckers that emerge from below the graft union of grafted vines (Hellman, 2003). Phylloxera populations reach vast numbers in leaf galls and as such are primary sources for phylloxera infestations on leaves in neighbouring economic vineyards, moving to the roots and overwintering as hibernating morphs. It has been reported that leaf galling over subsequent years may lead to severe defoliation of *V. vinifera* cultivars and finally vine death (Botton and Walker, 2009). The resistance among *V. vinifera* host plants changes by genotype (Könnecke et al., 2009; Vidart et al., 2013; Wegner-Kiss, 2007) and is also affected by environmental conditions and inoculum pressure.

No information on field studies of leaf-feeding phylloxera strains exists to our knowledge. The population dynamics of the gallicole phylloxera may be affected through insect antagonists or coinhabitants within the leaf galls (Herrmann and Forneck, 2001).

3.1.3 Physiology of the nodosity

A study analysing genetic traits for nodosity and tuberosity formation, using quantitative trait loci, conducted on segregating populations parented by *V. vinifera* Aramon × *V. rupestris* indicated two loci being involved in the nodosity formation trait in *Vitis*. However, the traits for tuberosity formation did not indicate clear linkage to a genetic marker (Roush et al., 2007). Other studies analysing the genetic basis of nodosity formation using a transcriptomic approach indicate that the insect changes patterns of carbohydrate partitioning and suppresses plant defences for their own benefit in partially resistant hosts (M. Griesser et al., personal communication). There is interest in the identification of genes involved in the process of gall induction to understand the biological roles of these genes and their use for genetically engineering nonhost resistance in rootstocks.

Nodosities contain starch molecules, which occur as large (1–100 μm), water insoluble semicrystalline granules that stain blue in colour with iodine and most possibly serve as storage molecules (Forneck et al., 2002). Starch metabolism involves the combined actions of many enzymes including soluble granule-bound starch synthases, branching enzymes, the debranching enzymes isoamylase, α-glucan phosphorylases, α-amylases and β-amylases

(Koetting et al., 2010). Genomic approaches to understand the expression of genes that may provide data on the physiological networks and signalling behind gall induction and the effects of fluctuating mineral concentrations and carbohydrate partitioning in plants attacked by gall-forming insects are scarce (Marini-Filho and Fernandes, 2012; Raman, 2011). With the sequencing of the *Vitis* genome (Jaillon et al., 2007; Velasco et al., 2007), studies of the phylloxera–*Vitis* interaction provide an opportunity to understand how host plant genetics affects fundamental sink–source relationships and susceptibility to galling insects using microarray or qPCR approaches (Compson et al., 2011; Lawo et al., 2013). A recent study shows that the biosynthesis of starch for storage and to buffer sugar metabolites is not the only effect altering the gall-bound carbohydrate metabolism. Massive import of sucrose from source leaves is symplastically transported towards the feeding site, transcending the endodermis and Casparian strip where phylloxera ingests sucrose directly. The sink activity is enhanced through sucrose synthase activity, activity of expansin genes and the insect feeding activity by effectively tapping into the symplast (M. Griesser, personal communication).

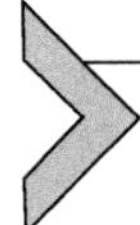

4. PHYLLOXERA ANATOMY AND PHYSIOLOGY

4.1. Digestive system morphology

In early studies, the digestive system of grape phylloxera was characterised, and most authors concluded, as an incomplete digestive system lacking a functional anus for waste excretion (Dreyfus, 1889; Federov, 1959; Schäller, 1960; Sobetsky and Derzhavina, 1973). Federov (1959) suggested the midgut ended 'blindly' and Dreyfus (1889) indicated there was no anal opening. More recent studies conclude the presence of a complete digestive system with an anal opening in both gallicole (Breider, 1952) and radicole phylloxera (Ponsen, 1997). Recent light and electron microscope studies have shown that radicole phylloxera have an atypical complete digestive system with a compartmentalised midgut. The midgut of radicole *D. vitifoliae* has anterior and posterior regions, separated by a hindgut connection (Fig. 4.1). The posterior region has a storage role prior to digestion activities in the anterior region (Kingston et al., 2009) and an anal opening with limited functionality (Andrews et al., 2012). In the apterous adult during oogenesis, the midgut becomes compressed. A midgut posterior chamber may be an essential nutrient storage area for the continual supply of energy

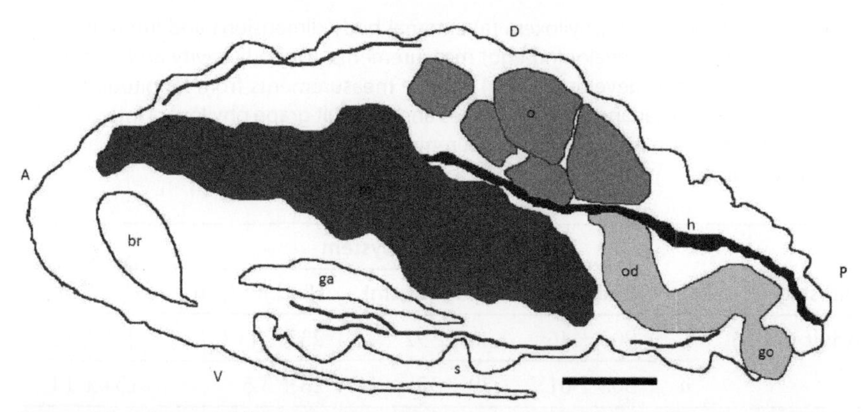

Figure 4.1 Schematic diagram of the digestive and reproductive features of radicole grape phylloxera. Scale bar 100 μm. Abbreviations: s, stylet; br, brain; ga, ganglion; m, midgut; o, ovaries; h, hindgut; od, oviduct; go, gonopore; insect orientation indicated: A, anterior; P, posterior; D, dorsal; V, ventral. Source: *Kingston et al. (2009)*. (For colour version of this figure, the reader is referred to the online version of this chapter.)

during periods of reduced food intake. The storage capacity may aid the insect in producing enough nutritional reserves to survive away from its host plant for prolonged periods whilst searching for a *Vitis* food source, which is also a significant consideration when developing suitable quarantine regulations. The storage capacity can also provide an energy source for copious egg production (Andrews et al., 2012; Kingston et al., 2009). Individual eggs produced by parthenogenetic apterous adults are 30–40% of the adult length and occupy 6–8% volume of the body cavity (Kingston et al., 2007a; Table 4.1). Adults develop multiple eggs, oviposit 3–6 eggs per day for 1–2 months and up to 360 eggs in a lifetime (Granett et al., 1983). Thus, there would be a relatively large space and energy requirement for egg production in the relatively small mature adult. However, fecundity is dependent on temperature, root and leaf gall size and genetic background of both the host and phylloxera (Herbert et al., 2010; Korosi et al., 2011; Makee, 2004).

Rilling and Steffan (1972) reared adult phylloxera on a ^{14}C-sucrose diet. The relative volume of diet ingested and the rate of respiration in radicole phylloxera were higher than in gallicole adults. Labelled ^{14}C compounds were primarily incorporated in the salivary glands and the ovaries and eggs laid within 4 h had detectable low ^{14}C levels. This highlights that the specialised digestive system of grape phylloxera can rapidly convert nutrients for egg production.

Table 4.1 Radicole grape phylloxera (a) external body dimensions and internal (digestive system and developing egg) measurements, (b) body cavity and internal (digestive system and developing egg) volume measurements from longitudinal sections of fourth instar and adult life stages and (c) adult grape phylloxera internal cold stage SEM dimension and volume measurements of the external body cavity and internal features

(a)

Life stage		External Body (μm)	Digestive system Midgut (μm)	Hindgut (μm)[a]	Developing Egg (μm)
Adult ($n=4$)	l	803 ± 46	488 ± 92	233 ± 117	233 ± 12
	w	308 ± 61	79 ± 35	28 ± 3.5	134 ± 14
Fourth instar ($n=2$)	l	680 ± 170	463 ± 124	180 ± 0	NA
	w	225 ± 78	98 ± 3.6	28 ± 3.5	NA

Mean values of maximum dimensions ± standard deviation, n value in parenthesis; developing egg measurements not applicable (NA) for fourth instar; length (l) width (w) of each feature indicated[a] adult n value = 2

(b)

Life stage	External Body (μm^3)	Digestive system Midgut (μm^3)	Hindgut (μm^3)[b]	Developing Egg (μm^3)
Adult ($n=4$)	$41 \times 10^6 \pm 16 \times 10^6$	$1.9 \times 10^6 \pm 1.5 \times 10^6$	$8.7 \times 10^4 \pm 2.3 \times 10^4$	$2.2 \times 10^6 \pm 0.43 \times 10^6$
Fourth instar ($n=1$)	$21 \times 10^6 \pm 17 \times 10^6$	$2.3 \times 10^6 \pm 0.78 \times 10^6$	$7.2 \times 10^4 \pm 1.8 \times 10^4$	NA

Mean values of maximum dimensions ± standard deviation, n value in parenthesis; developing egg measurements not applicable (NA) for fourth instar[b] adult n value for hindgut area measurement = 2

(c)

Measurement		External body ($n=2$)	Developing egg ($n=2$)	Midgut–hindgut junction ($n=1$)
Dimensions (μm)	l	718 ± 86	210 ± 7.8	26
	w	367 ± 28	126 ± 0.7	10
Volume (μm^3)		$51 \times 10^6 \pm 14 \times 10^6$	$1.7 \times 10^6 \pm 0.045 \times 10^6$	1.4×10^3

Value of maximum dimensions, n value in parenthesis; external body and developing egg mean ± standard deviation, absolute value for midgut–hindgut junction; length (l) and width (w) dimensions of each feature indicated

Source: Adapted from Kingston (2007).

Both gallicole and radicole phylloxera lack the characteristic looping structure (filter chamber) evident in aphids (Andrews et al., 2012; Ponsen, 1997), which indicates that the phylloxera does not require high fluid intake and feeds on a higher nutrient diet concentration, when compared to aphids, and therefore imbibes less diet volume.

4.2. Excretory and muscular system

The presence of a functional anal aperture for waste excretion in grape phylloxera remains unclear. Although studies have shown that an anus is indeed present (Andrews et al., 2012), honeydew production has not been reported in phylloxera and the anal region may be nonfunctional. Although alternate forms of waste secretion have been proposed, either through the use of modified salivary gland excretion (Anders, 1958a,b; Schäller, 1960; Sobetskiy and Derzhavina, 1973) or during the oviposition process (Andrews et al., 2012), no evidence to substantiate these hypotheses has been provided. Moreover, the high reproductive capacity (several hundred eggs per female—depending on food resources in the gall) means that energy requirements for egg production, the compartmentalised storage capacity of the digestive system and the lack of phloem feeding may reduce the need for copious waste excretion.

The skeletomuscular system of grape phylloxera has been described in a single light microscope study (Rilling, 1960) and could be examined further using electron microscopy. The presence of the hindgut and an anal opening with lateral muscles (but not dorsal muscles) indicates that waste excretion through the anus is physiologically possible and this may occur during oviposition (Andrews et al., 2012).

4.3. Salivary system

It has been speculated that the salivary system of grape phylloxera has features that differ from the Aphidoidea because it feeds on nonvascular tissue rather than the content of phloem cells. Morphological features of the salivary gland system of grape phylloxera were initially described in the early twentieth century (Grassi, 1912) and more detailed descriptions were subsequently made (Ponsen, 1997; Rilling, 1967).

Ponsen (2006) described the salivary gland system as consisting of a pair of accessory glands and transparent organ located in the pro- and mesothorax and a pair of principal glands in the mesothorax and metathorax and a transparent organ. The glands connect, via the salivary duct, to a relatively large salivary pump with a well-developed muscular system consisting of three pairs of muscles. This arrangement may be required to provide enough gustatory force to pump saliva into the parenchymal cells on which phylloxera feeds (Ponsen, 1997). Riley (1879) first speculated that the galls caused by phylloxera feeding are due to *more than mechanical action to some poisonous excretive substance.* Although the literature is relatively extensive regarding the role and composition of salivary excretions produced by phloem-feeding Aphidoidea (Miles, 1999), literature on salivary enzymes produced by parenchymal feeding Phylloxeridae is far less extensive.

Nodules, albeit superficially, appear similar to nodosities and have been induced on the grapevine roots by growing them in amino acid solutions (especially tryptophan, histidine and glutamic acid, either singly or in combination). Anders (1960a,b, 1961) suggested that host plant galls are induced as a response to saliva introduction into the parenchymal tissue and amino acids may be present in the saliva of grape phylloxera (Anders, 1957), although Miles (1968) disputed the validity of this claim. Anders (1958a,b, 1959, 1961) reported phylloxera saliva as pH 9 with 7% amino acids composed of glutamic acid, histidine, lysine, tryptophan and valine. In contrast, eleven amino acids and an unknown unidentified peptide were reported in phylloxera saliva with asparagine and glutamine having the highest concentrations (Schäller, 1960). The qualitative differences in amino acid profiles observed may be due to either different dietary source (Auclair, 1963) or poor detection sensitivity in these early studies. Schäller (1963a,b) also reported that phylloxera biotypes differed in their saliva composition.

Rilling (1967) described two types of salivary substance, one viscous (forming the protective salivary sheath around the stylet) and one watery in consistency. In aphids, viscous saliva contains free amino acids, to prevent premature gelling of sheath material in the salivary ducts and stylet canal. Watery saliva usually contains salivary enzymes including hydrolases and oxidases produced by the salivary glands (Miles, 1990, 1999). These enzymes may play a role in liquification of parenchymal cell contents (Ponsen, 2006). The possession of such a well-developed salivary system indicates that salivary enzymes are likely to play an important role in phylloxera–*Vitis* interactions.

Depending on the composition of its saliva when phylloxera inserts its stylet into the *Vitis* host, it may release saliva that could prevent a wound response or elicit a defence reaction in the roots. One way to determine role of salivary enzymes in phylloxera–*Vitis* interactions would be a proteomic analysis of saliva. This approach has provided evidence of a range of salivary enzymes produced by aphids (Harmel et al., 2008). Effectors (proteins that suppress host defence responses in order to facilitate infection and gall induction) (Haegeman, 2012) as constituents of the saliva are known to be involved in the suppression of plant defences or to interact with plant signalling or hormone pathways to promote the formation of the feeding site. To date, no effectors are defined for insect gall induction and formation and further studies are needed.

4.4. Reproductive system

The ovary structure of phylloxerids is largely undescribed. A recent study (Szklarzewicz et al., 2009) described the gross morphological characteristics of two closely related species *Phylloxera coccinea* and *P. glabra*. Szklarzewicz et al. (2009) concluded that one unusual feature of phylloxerids is that vitellogenesis (accumulation of cytoplasmic reserves) and choriogenesis (synthesis and secretion of precursors of egg envelopes) commence simultaneously, rather than when vitellogenesis is completed. The number of germ cells present in Phylloxeridae ovarioles ranges from 49 to 64 and is higher than that of Aphidoidea, which contain 32 germ cells. Phylloxeridae are exclusively oviparous and the predominantly viviparous Aphidoidea have evolved separately, which may explain the differences in germ cell number (Michalik et al., 2013).

Although there are some descriptions of the location of the oviduct, ovaries and gonophore in relation to the digestive system (Andrews et al., 2012), a detailed comparative study of gallicole and radicole grape phylloxera reproductive systems has not been conducted and represents an important knowledge gap.

4.5. External morphology

There are few recently published studies in which the external morphology of grape phylloxera has been described. Morphometric measurements of different life-stage features of radicole grape phylloxera have been determined using both light and scanning electron microscope (SEM) measurements including life-stage body dimensions and antennae and stylet length (De Klerk, 1979; Kingston et al., 2007a; Moleas et al., 1992; Schneider-Orelli and Leuzinger, 1924; Table 4.1; Fig. 4.2).

Anatomical features of the antennae were first described in light microscopy studies (Grassi, 1912) who described three primary components the flagellum, pedicel and scape. The flagellum has two placoid sensilla referred

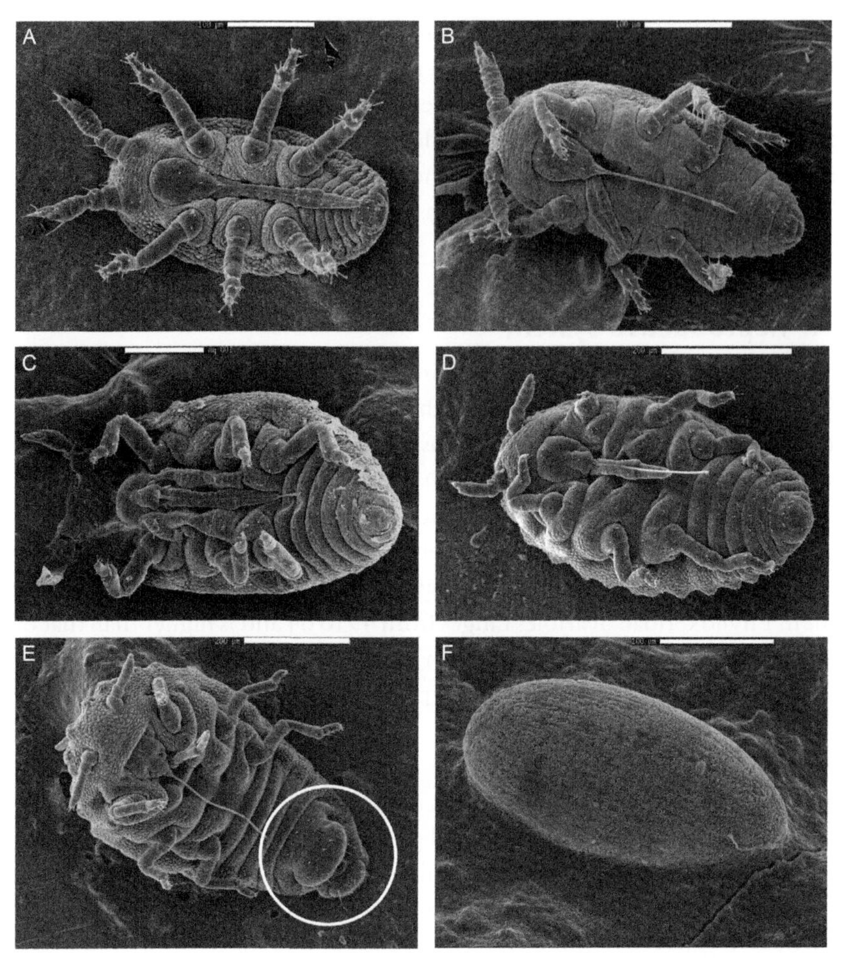

Figure 4.2 Cold stage SEM images of egg and apterous life stages of radicole grape phylloxera; (A) first instar, scale bar 100 μm; (B) second instar, scale bar 100 μm; (C) third instar, scale bar 100 μm; (D) fourth instar, scale bar 200 μm; (E) adult, the circle highlights the swelling of the posterior region, scale bar 200 μm; (F) egg, scale bar 100 μm. Source: *Kingston et al. (2007a).*

to as primary (found in all nymphal stages) and secondary rhinaria (only found in alate adults). The function of the rhinaria in grape phylloxera has not been characterised, but it has been proposed that they act as odour receptors to detect plant volatiles (Becker, 1952). A recent SEM study comparing nymph and alate antennae from different phylloxera genotypes indicates a morphological complexity in the alate antennae with additional sensory organs present (Huber et al., 2009a). The function of these antennal features for phylloxera and whether any morphological differences in antennal structure or function occur between different genetic strains of gallicole and radicole grape phylloxera are as yet undetermined, but they could play an important role in host plant location and subsequent population establishment.

4.6. Nutritional requirements

In the absence of data on the parenchymal cell contents of *V. vinifera* roots and leaves, there is limited information available regarding the specific nutritional requirements of either gallicole or radicole grape phylloxera. Ponsen (1997) suggested that parenchymal cell contents contain less fluid and carbohydrates and are higher in protein than the phloem sap diet of aphids. As phylloxera is a parenchymal feeder with no primary endosymbionts (Baumann, 2005), this makes determination of its essential nutritional feeding requirements and development of an artificial diet system more challenging than aphids. However, because of the lack of available data, the development of an artificial diet and *in vitro* artificial feeding system has been a challenge in grape phylloxera research. Most studies conducted to date have been based on adaptation of diets and feeding chambers used for phloem and foliar feeding Hemiptera, particularly aphids (Mittler, 1988).

Parenchymal cell contents are likely to have both a different pH and chemical composition compared to phloem sap. Phloem sap is characterised by a neutral pH (6.5–7.0), high osmotic pressure, high sugar content (primarily sucrose), essential and nonessential amino acids and a general absence of feeding deterrents or toxic secondary metabolites (Douglas, 2006). Most phloem feeders also have endosymbionts that can convert nonessential to essential amino acids. Consequently, there are few published studies where artificial diets have been tested to determine their suitability for grape phylloxera.

Studies on the chemistry of phylloxera leaf galls indicate that in general higher levels of N compounds and reduced levels of carbohydrates are present when compared with the diet of phloem-feeding aphids (Sobetskiy and Derzhavina, 1973). Up to 19 amino acids and two prominent polypeptides have been detected in leaf galls (Warick and Hildebrant, 1966). Using gallicole phylloxera, five diet formulations differing in sucrose and amino acid composition have been compared (Wöhrle, 1999). Artificial diets containing 5% sucrose were more successful than lower concentrations (<2.5%) in maintaining fourth instar survival for up to 15 days (Forneck and Wöhrle, 2003; Wöhrle, 1999) using a modified bioassay chamber design (Fig. 4.3A).

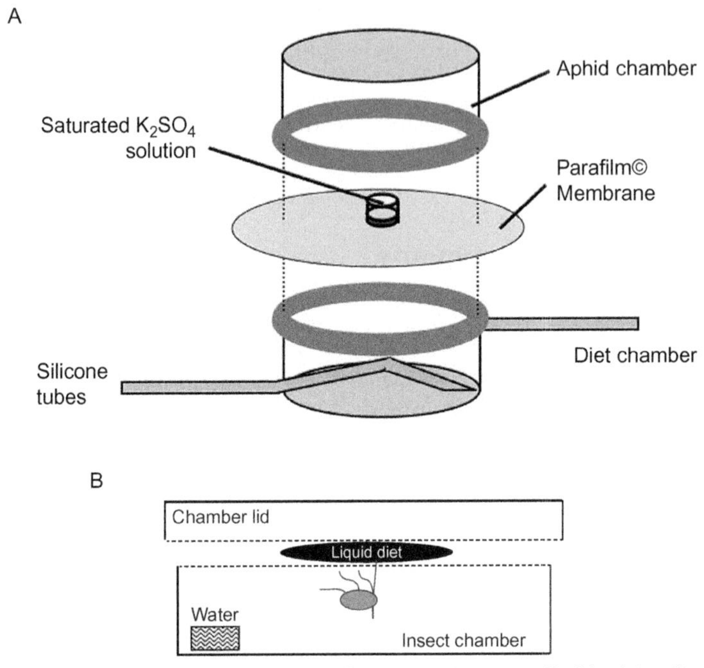

Figure 4.3 Diagrammatic representation of chamber design trialled for (A) gallicole and (B) radicole grape phylloxera artificial diet experiments. In (B) the solid line represents the 35 mm diameter Petri dish used for the diet chamber, the dashed line represents a layer of Parafilm. The insect is depicted on the diet membrane, inserting its stylet into the liquid diet. Humidity is controlled with inclusion of a water container. Source: *Kingston (2007) and Forneck and Wöhrle (2003).* (For colour version of this figure, the reader is referred to the online version of this chapter.)

Diet formulations for gallicole phylloxera are unlikely to be suitable for radicole phylloxera because free amino acid profiles differ between phylloxera-induced leaf galls and root galls (Kellow et al., 2004; Rilling et al., 1975). Rilling et al. (1974) compared gallicole and radicole grape phylloxera feeding on *Vitis* and found the essential amino acids glutamic acid, aspartic acid, tyrosine, alanine, glutamine and asparagine were acquired by the insect. However, the amounts accumulated differed between the two forms, with radicole accumulating higher amino acid concentrations. Radicole phylloxera may also require more sucrose in their diet than gallicole phylloxera (Rilling and Steffan, 1972).

Kingston (2007) screened 17 diet formulations for radicole phylloxera and observed improved survival when using a modified bioassay chamber

(Fig. 4.3B) and strongly acidic diet solution pH 4.5 compared with a weakly acidic diet pH 6.5. This is in contrast to most aphid diets that use a weakly acidic pH (6.0–6.5) to improve palatability and microbial control, even though optimal growth and reproduction of aphids is likely to be obtained with a neutral to alkaline pH to reflect the pH of phloem sap of preferred host plants (Mittler, 1988). Although the specific pH of *V. vinifera* parenchymal cell contents has not been characterised, *V. vinifera* leaf and root extracts have been shown to have pH values of 4.0 and 5.0, respectively (Kellow, 2000; Kingston, 2007). This also corresponds to other studies that have shown that phylloxera thrives in an acidic environment (Powell et al., 2013a). Using an acidic diet formulation, the maximum survival time recorded by radicole phylloxera on a diet consisting of sucrose (5%, w/v) and ten essential amino acids was 11 days, significantly extending survival by 3 days compared with a diet-only control (Kingston et al., 2007a). During this period, although ecdysis occurred, instar development was delayed and the insects did not develop to adulthood. This suggests a suboptimal diet and further studies are required to determine the chemical composition of the parenchymal cell contents in order to optimise dietary requirements. Kingston et al. (2007a) noted that the amino acid concentrations used in their study were higher than those recorded in *V. vinifera* root extracts (Kellow, 2000) and basing a diet on root extract profiles could potentially lead to a more suitable diet formulation. In future, once a near-optimal diet is developed, in addition to comparing gallicole and radicole phylloxera feeding, it would be important to consider the dietary preferences that may occur between different phylloxera genetic strains adapted to different *Vitis* hosts. In a recent study, differences in survival and development have been observed when feeding selected woolly apple aphid biotypes on artificial diets (Andrews and Powell, 2009), suggesting different biotype-specific nutritional requirements. Phylloxera artificial diet studies conducted to date also highlight the capacity of both leaf- and root-feeding phylloxera to survive for longer, in the absence of dietary nutrients, than observed in other studies on aphids. In most previously published studies, Aphidoidea generally have limited capacity to survive in the absence of diet (usually 3–4 days maximum). In contrast, phylloxera survived 6–7 days with no diet (Forneck and Wöhrle, 2003; Kingston et al., 2009). This prolonged survival ability is most likely related to the insects specialised digestive system and its capacity to store food (Andrews et al., 2012) with minimal waste excretion.

4.7. Endosymbiotic associations

Bacterial endosymbionts play key functional roles in insect nutritional physiology through provision of nutrient lacking in the diet or host utilisation (Feldhaar, 2011). They can be divided into two distinct groups: primary or obligate endosymbionts that are usually maternally transmitted and secondary endosymbionts that are facultative. Although the Aphididae have primary endosymbionts in the form of *Buchnera* spp. that are involved in provision of essential amino acids (Douglas, 2003), the literature indicates that no primary endosymbionts are present in Phylloxeridae (Baumann, 2005; Buchner, 1965; Fraenkel, 1952).

Both primary and secondary endosymbionts have been reported in the closely related clade Adelgidae although their role is unknown (Havill and Foottit, 2007). Evolutionary divergence of the Aphidoidea and Phylloxeridae occurred prior to infection of the Aphidoidea with *Buchnera* spp. around 100–250 Mya (Martinez-Torres et al., 2001). The lack of primary endosymbionts in grape phylloxera may in part reflect the different dietary source of the two groups with aphids primarily being phloem feeders and phylloxera feeding on parenchymal cells (Buchner, 1965).

Although no primary endosymbionts have been characterised from grape phylloxera, a relationship has been postulated between gallicole grape phylloxera and the bacteria *Pantoea agglomerans* (Vorwerk et al., 2007), yet the nature of this association is undefined. Kolberg et al. (2011) suggested the bacterium colonises the insect's integumental surface but is not maternally transmitted to its offspring. *P. agglomerans* appears transient in its association with gallicole phylloxera and has only been detected on the surface of eggs and adults. A possible protagonistic role may be that the bacteria solubilise substances or detoxify host plant secondary metabolites (Lawo and Forneck, 2011). Potentially, *P. agglomerans* may play a role in host-associated differentiation (HAD) between grape phylloxera genetic strains and *Vitis* species. *Phylloxera notabilis* Pergande has two bacterial endosymbionts present *P. agglomerans* and *Serratia marcescens* and their presence is dependent on the preferential host plant either pecan or water hickory (Medina et al., 2011). A potential symbiotic association between *P. agglomerans* and gallicole grape phylloxera needs further experimental evidence. A transient bacterial association was recently reported with radicole phylloxera. The bacterium isolated differed morphologically from that of both *Buchnera* and *P. agglomerans*; however, the species' identity and its symbiotic association remain unresolved (Kingston, 2007). New genomic sequencing techniques may assist in determining the presence or absence of bacterial symbionts.

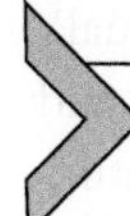

5. HOST–PLANT INTERACTIONS

5.1. *Vitis* rootstocks

Resistant rootstocks are the primary form of grape phylloxera management worldwide (Granett et al., 2001b). Most of the rootstocks in use today are based on hybrids of North American *Vitis* species. There are relatively few reported instances of rootstock failure and these occurred mainly where the parentage of a rootstock hybrid includes partial *V. vinifera* genetic background. Microsatellite markers for genetic resistance to phylloxera, which would assist in rootstock breeding for phylloxera resistance, have been examined (Powell et al., 2013b; Zhang et al., 2009).

It is important to have clear definitions of resistance, tolerance, immunity and susceptibility (Kellow, 2000). In the literature, there is no standard system or descriptors for rating 'resistance' of rootstocks to grape phylloxera. In addition, there is no standardised use of genetic markers (SSRs) to genotype phylloxera strains. However, it is apparent that it is important to determine biotype or genotype to ensure accuracy in rootstock recommendations. Screening to determine the level of resistance is important in view of the fact that the host–phylloxera interaction is affected by phylloxera 'biotypes' or genetic strains and consequently leads to differing degrees of resistance. Improved techniques for characterising phylloxera genotypes will assist further investigation of this apparent host–pest specificity.

Many early investigations of the phylloxera–grapevine interaction have used relatively limited techniques (Denisova, 1965; Niklowitz, 1954; Sobetskiy and Derzhavina, 1973) and have been conducted using grape phylloxera of undefined biotype or genetic background. Börner (1914) first highlighted the possibility of different biological 'races' of phylloxera. However, only relatively recently has the interaction between selected biotypes and genetic strains of grape phylloxera with several *Vitis* hybrid rootstocks, predominantly referred to as 'resistant' rootstocks, been studied using a variety of bioassay techniques.

Screening rootstocks for phylloxera resistance can be conducted by either field, glasshouse, or laboratory trials or ideally a combination of these. Studies of host plant resistance must be conducted using biotypes or clonal genetic strains. In this respect, our definition of a biotype would be a 'phylloxera population whose phenotype is determined by the interaction between host plants having different genes for resistance and the insects' ability to survive on the host plant', whereas a clonal genetic strain would be defined as 'a phylloxera population arising from a single clonal individual'. Different phylloxera strains may vary in their biological performance and preference to different *Vitis* hosts, and in addition, the resistance rating can be influenced by the screening method used (Table 4.2).

A range of rearing methods have been used to study interactions between grape phylloxera and *Vitis* hosts; these include excised roots (De Benedictis and Granett, 1993; De Benedictis et al., 1996; Granett et al., 1985, 1987; Makee et al., 2004), *in vitro* propagation (Askani and Beiderbeck, 1991; Forneck et al., 1996; Grzegorczyk and Walker, 1998; Kellow et al., 2002; Pelet et al., 1960), whole plants in plastic tubes (Forneck et al., 2001b; Yvon and Leclant, 2000), whole potted plants (Boubals, 1966b; Herbert et al., 2010; Pavloušek, 2012; Ramming, 2010) and field-grown grapevines (Boubals, 1966b; Trethowan and Powell, 2007). Each technique has advantages and disadvantages (Table 4.3).

5.1.1 Excised root bioassays

The excised root system can be used for individual genotype interaction studies but only allows an examination of radicole grape–phylloxera interactions and allows assessment of tuberosity formation, but not nodosity formation. It has also been shown that excised root bioassays over estimate grape phylloxera biotype or strain virulence and underestimate rootstock resistance, when compared with bioassays using field-grown (Granett et al., 2001a) or potted grapevines (Fergusson-Kolmes and Dennehy, 1993; Powell and Korosi, 2013).

5.1.2 *In planta and* in vitro *bioassays*

The *in vitro* system is relatively rapid, can be used for both radicole and gallicole grape phylloxera assessments and prevents cross contamination of strains due to their containment. However, *in vitro* grown plantlets allow only a realistic assessment of nodosity formation at the root tips and not tuberosity formation. Leaf galling can be monitored in aseptic dual cultures using intact plantlets. Both *in vitro* and excised root systems may over emphasise the susceptibility of the host (Grzegorczyk and Walker, 1998). This bioassay system operates in a soilless environment, which allows real-time observation and precise sampling of tissues affected by the host–pest interaction. All environmental conditions may be precisely controlled as are treatments (e.g. nutrition and plant signalling effects) of the host plant. However, any interaction on the rhizoplane between the soil, roots and radicole phylloxera cannot be evaluated.

Table 4.2 A summary of overall rankings for rootstock resistance, partial resistance and susceptibility to radicole grape phylloxera based on either (a) *in vitro* excised root screening under laboratory conditions or (b) *in planta* screening under glasshouse conditions using six clonal strains

Rootstock	Genotype Bioassay	G1 a	G1 b	G4 a	G4 b	G7 a	G7 b	G19 a	G19 b	G20 a	G20 b	G30 a	G30 b
Vitis vinifera	*V. vinifera*	S	S	S	S	S	S	S	S	S	S	S	S
Ramsey	*V. rupestris* × *V. candicans*	T	T	T	T	T	T	T	T	T	T	T	T
Schwarzmann	*V. rupestris* × *V. riparia*	R	R	R	R	T	T	T	T	T	T	T	T
Börner	*V. riparia* × *V. cinerea*	R	R	R	R	T	R	R	R	R	R	T	R
110 Richter	*V. berlandieri* × *V. rupestris*	T	T	T	T	T	R	T	R	T	T	T	R
140 Ruggeri	*V. berlandieri* × *V. rupestris*	T	R	T	T	R	R	R	R	R	R	R	R
420A	*V. berlandieri* × *V. riparia*	R	T	R	R	R	R	R	T	R	T	R	R
101-14	*V. riparia* × *V. rupestris*	R	R	R	R	T	R	R	R	R	T	R	R

a = excised grapevine root bioassay, b = potted grapevine bioassay, S = susceptible, T = tolerance or partial resistance, R = resistant
Adapted from Powell and Korosi (2013).

Table 4.3 A comparison of some techniques for assessing *Vitis* species interactions with *Daktulosphaira vitifoliae*

Bioassay technique	Strengths	Weaknesses
In vitro tissue culture	- Sterile, whole plant - Assessment of nodosity formation - Eliminates some variables, for example, soil - Relatively short time frame - Allows single clonal lineage interactions to be assessed - No strain cross contamination - Visible staged monitoring possible - Controlled environment, for example, temperature, day length - Suitable for gallicole and radicole - Accurate quantification of insect abundance and development in real time	- No assessment of tuberosity formation - No assessment of vine growth or vigour - No assessment of rootstock/scion compatibility - No assessment of soil interactions possible
Excised root bioassays	- Assessment of tuberosity formation effect - Detailed assessment of development, survival and reproduction, over time - Controlled environment, for example, temperature - Relatively short time frame - Allows single clonal lineage interactions to be assessed - Suitable for radicole - No strain cross contamination - Accurate quantification of insect abundance and development	- Roots detached so no whole plant defence response - Nonsterile environment allows fungal contamination - Gradual root decline - Does not assess vine growth or vigour, focuses on protective properties of the root material only - Unsuitable for gallicole - No assessment of nodosity formation Viability of excised root material depends on timing and site of collection
Whole potted plant	- Allows assessment of nodosity and tuberosity formation - Whole plant response	- Vines have a greater proportion of fibrous roots compared with field-grown vines

Table 4.3 A comparison of some techniques for assessing *Vitis* species interactions with *Daktulosphaira vitifoliae*—cont'd

Bioassay technique	Strengths	Weaknesses
	- Controlled environment, for example, temperature, water, lighting and soil - Elimination of soil pathogens (by heat sterilisation of soil) - Accurate quantification of insect abundance and development - Short time frame - Can determine which genotypes to assess (also soil types) - Rootstock/scion compatibility included - No strain cross contamination (providing mesh quarantine enclosures) are used - Suitable for gallicole and radicole	
Simple isolation chambers or root enclosures	- Allows visible observation of insect–root interactions - Allows assessment of nodosity and tuberosity formation - Whole plant response - Controlled environment, for example, temperature, water, lighting and soil - Elimination of soil pathogens (by heat sterilisation of soil) - Accurate quantification of insect abundance and development - Short time frame - Can determine which genotypes to assess (also soil types) - Rootstock/scion compatibility included	- Vines have a greater proportion of fibrous roots compared with field-grown vines - Vines are more prone to fungal infections

Continued

Table 4.3 A comparison of some techniques for assessing *Vitis* species interactions with *Daktulosphaira vitifoliae*—cont'd

Bioassay technique	Strengths	Weaknesses
Field	- Incorporates vine growth/vigour of rootstocks influence of soil type, vineyard management, (fruit composition and quality, compatibility of rootstock/scion) and seasonal differences - Can combine with trapping techniques to assess risk of aboveground dispersal - Extremely difficult, if not impossible, to prevent strain cross contamination especially in field sites where multiple phylloxera strains occur	- Cannot control as many variables as other techniques, for example, environmental, genotypes present - Cannot choose which genotypes to test unless at a site there is only one clonal strain - Cannot assess numbers on roots accurately - Relatively long time frame

In planta bioassays, where whole plants rather than tissue culture plantlets are used, for example, glasshouse-based trials, have the advantage when compared with excised root and *in vitro* systems that a whole plant response to phylloxera infestation can be monitored. Variables including temperature, day length, soil type and irrigation can also be controlled. Genetic strains of phylloxera can be prevented from cross contaminating control vines by using mesh enclosure both on the root itself and around the infested pot (Korosi et al., 2007). However, when conducting glasshouse trials using potted vines, the 'pot' environment itself may introduce artefacts that affect the moisture, oxygen, porosity and temperature status of the soil (Passioura, 2006). These variables if not controlled could affect phylloxera directly or indirectly by influencing the growth of the vine root system. Laboratory or glasshouse trials cannot replace field trials but they are a useful comparative predictor of resistance.

5.1.3 Field bioassays

Field trials for rootstock resistance allow for long-term monitoring of phylloxera under field conditions. Whole vine effects such as grape yield, leaf area index and root volume can also be monitored along with seasonal population dynamics of phylloxera and impacts of different vineyard management techniques. Using either whole potted plants or field-grown vines does allow both a whole plant response and an edaphic environment. However field trials do take longer to achieve results, cross contamination of phylloxera strains is more difficult to prevent and grapevine roots are less amenable to regular observation.

5.1.4 Radicole case studies

In conducting any study on grape–phylloxera interactions with *Vitis* hosts, ideally assessments of both host plant response to phylloxera infestation and the impact of the host plant on the insects' biological parameters should be assessed. By using this, combined approach insect life-table data could be used to determine 'biotype' status, and the host plant response can be used to develop rootstock resistance ratings.

Since the genetic characterisation of grape phylloxera in Australia, selected radicole strains have been specifically used for all rootstock screening (Korosi et al., 2007). Rootstock screening has involved comparing the phylloxerid's life-table parameters (namely, survival, development and fecundity) of these selected clonal phylloxera strains and the subsequent host plant response (root damage and galling) when feeding on different *Vitis* rootstocks. The bioassay systems utilised include excised root, tissue culture, potted vines and field vine bioassays. A number of studies in Australia have focused on root-galling phylloxera interactions using six (genotypes G1, G4, G7, G19, G20 and G30) of the 83 endemic genotypes as characterised using six mitochondrial markers (Umina et al., 2007). The studies highlight that development and fecundity of the strains differ on selected rootstocks and rootstock defensive response to different genetic strains differ markedly. In one early study using a relatively virulent genetic strain VWL-1 (syn. G4) phylloxera (Corrie et al., 1997) were unable to initiate nodosities on most of the vine types tested, one exception being the rootstock Ramsey (*V. rupestris* × *V. candicans*). G4 phylloxera was able to reproduce on Ramsey as it has in subsequent studies (Korosi et al., 2007; Powell and Korosi, 2013). Lower reproductive rates of phylloxera on Ramsey correlate with a lack of starch accumulation in root galls (Van Heeswijck et al., 2003). On resistant rootstocks (e.g. Börner (*V. riparia* × *V. cinerea*)), stylet probing resulted in a rapid hypersensitive-like response with root tissues developing a strong brown colour and necrosis, which predominantly led to death of first instars (El-Nady and Schroder, 2003). Two types of necrosis have been described:

type I, which was developed rapidly within 12–24 h after *D. vitifoliae* infestation, and type II, which developed in 2–5 days.

Although not fully characterised, a variety of antibiosis and antixenosis responses have been observed highlighting that rootstock 'resistance ratings' should be based purely on neither one screening technique nor screening of one phylloxera clonal lineage (Table 4.2).

5.2. Host-associated differentiation

HAD is the formation of genetically distinct host-associated populations. Evidence for this type of association has been observed with pecan leaf phylloxera, *P. notabilis* Pergande, that exhibits cyclic parthenogenesis on pecan and water hickory (Dickey and Medina, 2012). Evidence for host-associated clones of grape phylloxera with a preference for different *Vitis* genotypes has been observed under laboratory conditions when comparing endemic Californian, German, Austrian and Hungarian grape phylloxera strains (Forneck et al., 2001a, 2007; Kocsis et al., 1999; Omer et al., 1999b) and in field conditions in Australia (Corrie et al., 2003; Trethowan and Powell, 2007). Performance and development of some phylloxera strains increases on susceptible and partially resistant *Vitis* rootstocks when the inoculum levels increase (Omer et al., 1999c). These HAD relationships have implications for phylloxera management because although rootstock resistance may be stable when certain phylloxera strains are present at low population levels, the introduction of a different phylloxera lineage at higher levels may increase abundance more rapidly and lead to increased quarantine risk or potential breakdown in rootstock resistance.

5.3. Host plant defence response

When feeding radicole grape phylloxera pierce the root cortex with their stylet, following a single stylet pathway to the parenchymal cells on susceptible *V. vinifera* (Kellow et al., 2004). In contrast when feeding on two 'resistant' hybrids Ramsey (*V. rupestris* × *V. candicans*) and Schwarzmann (*V. rupestris* × *V. riparia*) or on Teleki 5C (*V. riparia* × *V. berlandieri*) multiple stylet tracks are observed suggesting repeated probing (Forneck et al., 2002; Kellow, 2000). Salivary excretions, which may be involved in induction of the plant response, have been observed when phylloxera feeds on *Vitis* roots (Rilling, 1967). However, the chemistry of salivary excretions in phylloxera is unknown. Once the parenchymal cell is located, the insect feeds on the parenchymal cell contents (Pollard, 1973; Rilling and Radler, 1960) and does not appear to penetrate vascular cells. Once penetrated, the parenchymal cells accumulate tannins (Niklowitz, 1954). However, there is limited information of the chemistry of the parenchymal cells.

5.3.1 Root response to root infestation

On susceptible *V. vinifera* hosts, the plant response is characterised by the development of nodosities on nonlignified roots and tuberosities on lignified roots with no observable necrotic response. The relative size of the nodosities produced is dependent on the phylloxera genetic strain, with smaller nodosities produced when less virulent strains are feeding (Powell, personal observation). On highly resistant *Vitis* hybrids such as Börner, a relatively rapid necrotic response has been observed and generally no gall induction (El-Nady and Schroder, 2003; Van Heeswijck et al., 2003). The necrotic zone induced on resistant *Vitis* species may isolate the phylloxera from healthy tissue and have a toxic effect (Karban and Baldwin, 1997). In contrast on partially resistant *Vitis* hybrids, a less marked necrotic root response is observed on lignified roots, sometimes accompanied by 'pseudo-tuberosity' formation (Korosi et al., 2007) and nodosities may develop on nonlignified roots.

Studies examining the host plant response to phylloxera infestation are limited and have used either *in vitro* or excised systems to examine the interactions on either susceptible or resistant hosts. Structural (Hofman, 1957; Niklowitz, 1954) and biochemical changes, including increased levels of carbohydrates, phenolics, starch, amino acids and amides in the host plant have been described during the formation of phylloxera-induced nodosities (Denisova, 1965; Forneck et al., 2002; Kellow, 2000; Schaefer, 1972, 1985; Sobetskiy and Derzhavina, 1973; Zotov, 1954). In some of these studies, comparison of root physiological response of susceptible *V. vinifera* has also been made using selected phylloxera 'biotypes' (Kellow et al., 2004).

Kellow et al. (2004) used a range of cDNA to screen for probes, to compare gene expression in uninfested roots and infested nodosities. As nodosities are formed, the root acts as a nutrient sink for sucrose, which is then converted to starch and starch accumulates in the cortex. There is little evidence of an enhanced defence-type response in susceptible roots. Enhanced levels of free amino acids particularly glutamine have been reported in nodosities (Kellow et al., 2004) and phylloxera-induced leaf galls (Sobetskiy and Derzhavina, 1973). Starch and amino acid accumulation in nodosities indicates a nutrient sink for phylloxera and enhanced transport of solutes through either plasmodesmata or apoplasticly from the phloem to the root cortex (Kellow et al., 2004). Ryan et al. (2000) using phylloxera-infested excised roots could not demonstrate any change in free amino acid levels in the roots, which indicates that the excised root system may not be the most appropriate system to study phylloxera–host–plant interactions because a whole plant response may differ.

Volatile production from host plants can be an important olfactory cue for insects in host plant location and selection and also may play a defensive role (Bruce et al., 2005; Heil, 2008). Insect herbivory also induces the synthesis of volatiles from plants, both locally and systemically (Ferry et al., 2004; Paré and Tumlinson, 1996). The mechanism through which radicole phylloxera locates its hosts and differentiates between susceptible and resistant hosts has not been described. Two recent studies have examined volatile production in grapevine roots infested with phylloxera. In both studies, *Vitis* hybrids were examined, with one study focusing on two resistant hybrids 5BB Kober (*V. berlandieri* × *V. riparia*) and 140 Ruggeri (*V. rupestris* × *V. berlandieri*) and one susceptible hybrid Kyoho (*V. vinifera* × *V. labrusca*) (Du et al., 2009) and one using 5C Teleki (*V. berlandieri* × *V. riparia*) (Lawo et al., 2011). In the rootstock 5C Teleki, the root tip response to phylloxera infestation is a change in metabolite expression. Several metabolites are increased, one decreases and another appears only in mature nodosities. These findings indicate that several defence-related pathways may be involved including the mevalonate and/or alternative isopentenyl pyrophosphate, the lipoxygenase and the phenylpropanoid pathways (Lawo et al., 2011). Polyphenol oxidases are anti-nutritive compounds and catalyse the oxidation of phenolics to quinones, which then cross react with proteins and amino acids (Chen, 2008). Aphid saliva also contains polyphenol oxidases, which can detoxify phenolics (Peng and Miles, 1988). The lipoxygenase pathway is involved in biosynthesis of jasmonic acid and methyl jasmonate and regulation of defence-related genes (Kessler and Baldwin, 2001). Some of these compounds have been implicated in aphid resistance (He et al., 2012). Methyl jasmonate induces a hypersensitive response accompanied by a defence gene response in grapevine leaves (Repka et al., 2001). The phenylpropanoid pathway includes synthesis of several plant secondary metabolites including phenolics, flavonoids and stilbenes that have been implicated in host plant defence systems.

In a study by Du et al. (2009), several volatile components were identified, some unique to resistant *Vitis* hybrids and some unique to susceptible *Vitis* hybrids. Other compounds such as linoleic acid were higher in resistant roots. Whether these responses are phylloxera-specific, rootstock-specific, genotype-specific or a more general defence or stress response is as yet unclear until further studies are conducted.

Manipulating host plant resistance through genetic modification of secondary metabolites that naturally occur in *Vitis* species, including *V. vinifera* (Franks et al., 2005), such as cyanogenic glucosides, may be a future management option for phylloxera. One preliminary study showed that genetically modified grapevine roots, expressing cyanogenic properties, reduced phylloxera fecundity under laboratory conditions (Franks et al., 2006).

Root cell wall chemistry could be involved in both physical and biochemical resistance towards radicole phylloxera. Lignin, cellulose and pectin have important structural attributes influencing Hemipteran resistance (Dreyer and Campbell, 1984, 2006; Park et al., 2006; Pontius et al., 2006). Plant cell walls can produce pectinases when cell walls are penetrated. In studies comparing physical and chemical properties of various rootstocks (Du et al., 2011; Qing et al., 2011) lignin, pectin and cellulose production levels differed among rootstocks, with higher levels being found in the roots of more resistant cultivars. Du et al. (2011) have therefore proposed that this type of assessment if developed further could potentially be used as a phylloxera-resistance index.

Enzyme activity is also altered in the roots of infested *Vitis* spp. upon phylloxera feeding. In *in vitro* studies comparing susceptible *V. vinifera* and resistant 5BB Kober (*V. berlandieri* × *V. riparia*), increased peroxidase, acid phosphatase and leucine aminopeptidase activity was observed for both rootstock types at the radicole phylloxera stylet insertion point and starch accumulation occurred in the nodosity (Forneck et al., 2000, 2002). It is worth noting however that although the salivary enzymes of phylloxera have not been characterised, aphids and whiteflies produce salivary enzymes including peroxidases, polyphenol oxidases, pectinases, phosphatases, cellulases and oxidases that aid penetration of plant tissues by the stylets (Funk, 2001; Miles, 1998). The level of host plant resistance could therefore be potentially influenced by the enzyme composition of the insect's saliva.

5.3.2 Foliar response to root infestation

With a view to developing early detection systems for radicole phylloxera presence on the roots, chemical fingerprinting techniques have been used to evaluate the induction of metabolic response of grapevine leaves to phylloxera root infestation (Benheim et al., 2012). These studies have used high-performance liquid chromatography (HPLC), mass spectrometry (MS) and nuclear magnetic resonance (NMR) techniques primarily using susceptible *V. vinifera* cultivars. In an NMR study conducted on both glasshouse and field-grown phylloxera-infested grapevines, infested with a single genetic strain, a transient change in α-linalool and a reduction in the ratio of linoleic acid to linolenic acid was observed (Tucker et al., 2007). In an MS study, several flavonol compounds including isorhamnetin glycoside, rutin, kaempferol glycoside and quercetin glycoside are markedly upregulated in the leaves when phylloxera is present on the root system of *V. vinifera* (Benheim et al., 2011).

In a field- and glasshouse-based study focusing on assessing photosynthetic pigment content, using HPLC, decreases in chlorophyll content and increases in photoprotective xanthophyll-cycle pigments and carotenoids (Düring, 1999) have also been shown (Blanchfield et al., 2006). These previsual pigment changes can be detected in <10 weeks postphylloxera infestation and may be due to stress imposed on the root system resulting in less water uptake and the need for reduced light-harvesting by the foliage and increased need to dissipate energy as heat. Although these changes may not be phylloxera-specific, the development of a photosynthetic pigment index could improve phylloxera detection that currently relies on visible symptoms of canopy stress appearing several years postinfestation (Herbert et al., 2008). Further validation under a range of environmental conditions using genetically diverse phylloxera and *Vitis* species is required to determine if these biochemical responses are actually phylloxera-specific and not just general disease indicators; only then will there be a potential opportunity to use chemical fingerprinting as a novel approach for phylloxera detection.

Remote sensing technologies, through either aerial or satellite observations, including infrared photography (Baldy et al., 1996; Frazier et al., 2004; Wildman et al., 1983), multispectral imagery (Edwards et al., 2004) and hyperspectral imagery (Renzullo et al., 2004) have been successfully used to identify canopy level weak spots in ungrafted *V. vinifera* canopies that are the result of phylloxera infestation. However, the weak spots cannot distinguish between canopy reduction induced by other abiotic and biotic stressors, unless a follow-up ground survey is conducted. If phylloxera-specific chemical responses in the foliage could be reliably quantified this could enhance area-wide remote sensing approaches to phylloxera detection and surveillance by characterisation of a unique spectral signature. Preliminary research has been conducted based on the use of hyperspectral reflectance spectrometry (Renzullo et al., 2006a,b), and some significant spectral bands for phylloxera prediction have been characterised.

In an anatomical study, using *in vitro* grown *Vitis* species, the leaf resistance response to gallicole phylloxera was examined on both resistant and susceptible varieties (Raman et al., 2009). In the susceptible variety 3309C (*V. riparia* × *V. rupestris*), mesophyll parenchyma cells were activated and metaplasied within in 3–6 h of feeding, which developed into nutritive tissue within 48 h and galls within 21 days. On resistant *V. vinifera* 'Portugieser Weißherbst' phenolics accumulated, indicative of a hypersensitive response, phylloxera died within 4 days and no galls developed. In the tolerant cultivar Weißer Amerikaner (*V. vinifera* × *V. labrusca*), an intermediate response occurred characterised by development of nutritive tissue and phenolic accumulation yet no gall development (Raman et al., 2009).

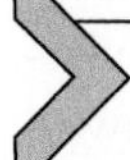

6. PHYLLOXERA INTERACTIONS

6.1. Feeding behaviour

Characterising the feeding behaviour of grape phylloxera on both susceptible and resistant *Vitis* species would increase our understanding of the fundamental interactions between phylloxera and its resistant and susceptible hosts. One method that could be further adapted for grape phylloxera is the EPG (electrical penetration graph) system (Tjallingii, 1978). Although EPG has been used widely for phloem-feeding Aphidoidea, it has been used for only a few parenchymal feeding aphids (Pointeau et al., 2012).

Despite the undoubted potential of utilising this technique to study phylloxera–host–plant interactions, there have only been two published EPG studies conducted on Phylloxeridae, one focused on the oak-feeding phylloxerid *P. coccinea* Heydon (Harrewijn et al., 1998) and one on root-galling *D. vitifoliae* (Kingston et al., 2007b). In the Harrewijn et al. (1998) study, four characteristic waveforms are described when *P. coccinea* feeds that may correspond to plant penetration (waveform C), cell penetration (potential drop), salivary injection (waveform G1) and cell sap ingestion (waveform G2). However, the biological meaning of each waveform can only really be speculative as no corresponding correlation studies were conducted. In a later EPG study, Kingston et al. (2007b) monitored probing behaviour of radicole *D. vitifoliae* on both excised root and whole plant tissue culture grapevines. Although successful recordings were obtained using both feeding systems, the success rate was higher using excised roots. A range of ten waveform types were described (Kingston, 2007). None of the waveforms described appear to have waveforms characteristic of phloem or xylem feeding, which provides evidence that phylloxera does not feed on vascular tissue. This contrasts with oak-feeding *P. coccinea* whose stylets have been observed in both parenchymal and xylem tissues (Kunkel, 1967).

There are several challenges in using the EPG system for feeding behavioural studies on radicole grape phylloxera including its relatively small size, its root-feeding habitat, its primary feeding location being the parenchymal cell (Petri, 1907) rather than phloem cells and the lack of characterised waveforms for the insect. Grape phylloxera first instars are relatively small (Kingston et al., 2007b) compared to most EPG studied aphids. A study conducted on lettuce root aphid *Pemphigus bursarius* L. highlighted that soil can cause the EPG signal to short circuit when in contact with the gold wire that is attached to the insect electrode (Cole et al., 1993). To reduce the limitations of both of these constraints, recent studies on smaller insects (such as whitefly and thrips) have used either 2.5 μm diameter platinum wire or 10 μm gold wire, instead of 12–25 μm gold wire (Harrewijn et al., 1996; Walker and Janssen, 2000), as is primarily used in aphid-feeding studies.

Phloem-feeding aphid waveforms have been well characterised in many studies, but not for individual life stages or to compare biotype probing behaviour. In phylloxera feeding behaviour studies, it is important to distinguish between the probing behaviours of the active dispersive first instars (which are important in colony establishment) compared to the more sedentary and reproductively mature adult stage. Some preliminary studies comparing feeding behaviour of first instar and adult radicole phylloxera of a single clonal lineage indicated that feeding behaviour does differ both between life stages and on different *Vitis* species (Kingston et al., 2009). Unfortunately, no corresponding correlation studies were conducted and therefore interpretative analysis can only be speculative. Nonetheless, future EPG studies could be conducted using different genetic strains of phylloxera to assist in understanding the mechanisms of resistance.

6.2. Field dynamics

The population dynamics of grapevine phylloxera on both grafted and ungrafted *V. vinifera* have been studied under commercial field conditions using a range of standard above- and belowground monitoring techniques. Monitoring techniques include (i) quantification of populations on roots by direct observation *in situ*, by taking core soil/root samples or by subsampling and excising roots (Granett et al., 2001a; Helm et al., 1991; Porten and Huber, 2003); (ii) pitfall traps (Powell et al., 2000) to monitor soil surface dispersal; (iii) soil emergence traps (Herbert et al., 2006; Powell et al., 2003); and (iv) sticky traps either wrapped around the trunk (Herbert et al., 2006; Powell et al., 2003) or placed within (King and Buchanan, 1986) or above the canopy (Powell et al., 2000).

Emergence and pitfall traps monitor only dispersive stages (first instar nymphs and alate adults) emerging on the soil surface in the spring and summer; trunk traps monitor dispersive stages moving up and down the grapevine trunk. Monitoring aboveground identifies key risk periods when phylloxera can be inadvertently transferred during normal vineyard operations. Root sampling monitors all life stages throughout the growing season on both lignified and nonlignified roots. However, this may require destructive sampling to assess life stages and this root removal process can be detrimental to vine health. Above and within canopy, sticky traps monitor wind-assisted dispersive life stages (King and Buchanan, 1986).

Field-based phylloxera monitoring has been conducted in North America (Omer et al., 1997), Europe (Porten and Huber, 2003), China (Zhang et al., 2010) and Australasia (Herbert et al., 2006; King and Buchanan, 1986). Each monitoring technique provided different aspects of the population dynamics.

In general terms, seasonal population dynamics are similar in all study regions, although there are some differences within and between seasons, due to changes in climatic variables, particularly temperature and rainfall. Populations overwinter on the root systems as first instar nymphs, which develop to adulthood and reproduce in the spring and reach peak abundance in the summer months; populations then decline in autumn. Two periods of peak abundance occur on the roots (Powell et al., 2000), which coincides with two periods of root growth in the spring and autumn (Freeman, 1983; Freeman and Smart, 1976). On *V. vinifera*, phylloxera initially establish on nonlignified roots and later in the season shift to lignified roots (Powell et al., 2000). Grape phylloxera numbers decrease as vines deteriorate, and a single or double peak of emergence aboveground occurs in summer. Size and timing of peak emergence vary between sites (Herbert et al., 2006). Phylloxera have a high reproductive capability, depending on genetic strain, and in a highly virulent strain, each asexual adult can produce several hundred viable eggs. The abundance and timing of active grape phylloxera is correlated with degree days (Herbert et al., 2006). The levels of peak abundance is also determined by a number of abiotic and biotic factors including chemical and physical characteristics of the soil (Powell et al., 2003), age and genetic background of the *Vitis* host (Omer et al., 2002) and the phylloxera

(Powell et al., 2003), soil temperature and depth within the soil profile (Gorkavenko and Gorkavenko, 1977; Herbert et al., 2006) and drought conditions (Helm et al., 1991).

7. BIOTIC AND ABIOTIC INTERACTIONS

As well as its genetic background and life cycle, a range of biotic and abiotic factors can influence the impact that grape phylloxera can have on its host plant. Some of these factors can be manipulated to improve the phylloxera management. However, there are still substantial knowledge gaps that need to be addressed.

7.1. Temperature

Very few life-table studies have been conducted to determine the impact of temperature on grape phylloxera. However, some studies to assess the impact of temperature on radicole phylloxera have been conducted under laboratory and field conditions. Prior to the 'discovery' of biotypes in California, life-table studies were conducted using the excised root system on phylloxera populations sourced from the Napa Valley and Davis regions. Reproductive rate, generation time and finite rate of increase were all determined and compared on susceptible *V. vinifera* and resistant *Vitis* hybrids. Survivorship to adulthood, establishment and daily egg production was higher on susceptible *V. vinifera* cultivars (Granett et al., 1983). A later study using excised roots demonstrated that for phylloxera 'biotype A', the optimal temperature range for nymphal survival was 21–28 °C, temperatures <16 °C and >36 °C reduced egg hatch, but eggs hatched at >7 °C, and at 32 °C, development to adulthood ceased (Granett and Timper, 1987). In a second laboratory bioassay, again using 'biotype A', the threshold temperature for establishment of a feeding site was 18 °C (Turley et al., 1996). Makee (2004) using phylloxera of unspecified 'biotype' determined the optimal range for nymph development was between 15 and 30 °C and that optimal egg production was at 25 °C. Connelly (1995) showed that temperatures >16 °C were required under laboratory conditions for development, but overall laboratory bioassay results were not representative of field conditions for development. Temperature thresholds have also been developed for foliar grape phylloxera reared on the resistant rootstocks 5BB Kober (Belcari and Antonelli, 1989) with optimal temperatures being between 25 and 30 °C.

From field observations in California, Davidson and Nougaret (1921) suggested that at <19 °C phylloxera hibernated, whereas Zhang et al., (2010) described phylloxera first instars becoming active as ground temperatures exceed 17 °C. In an excised root study comparing phylloxera populations collected at four geographically distinct regions in the Pacific Northwest, United States, although generation time varied under variable temperature regimes (Fisher and Albrecht, 2003), overall upper and lower temperature thresholds were <6 °C and >33 °C, respectively. Degree-day models have been developed for field populations of phylloxera in Italy (Belcari and Antonelli, 1989), the United States (Sleezer et al., 2011; Turley et al., 1996) and Australia (Herbert et al., 2010), yet none of these models takes into account the influence that temperature, soil properties and vine genetic background may have on different phylloxera strains.

An important observation made by Granett and Timper (1987) was that different 'biotypes' may have different optimal temperature ranges. This assumption has since been verified in a recent study comparing the influence of temperature and humidity on two selected radicole phylloxera clonal lineages (Korosi et al., 2012).

7.2. Water and nutrients

Much of the evidence for effective management of phylloxera through the application of additional water and nutrients is anecdotal and based on the premise that healthy grapevines may be able to cope with a phylloxera infestation better than unhealthy stressed grapevines.

Historically, water was first used as part of early eradication attempts for grape phylloxera by flooding of vineyards during the winter months (Riley, 1875). This submersion method caused a marked increase in grapevine vigour yet was only applicable to nonpermeable soils. Although still used in some parts of Europe (Campbell, 2004), flooding vineyards is generally uneconomical. Flooding requires access to large quantities of fresh water and additional fertiliser inputs, to replace water-soluble nutrients leached down the soil profile, and could cause damage to the grapevine root system.

In susceptible *V. vinifera* grapevines under potted and field conditions when roots became water stressed, this has been shown to favour phylloxera development (Helm et al., 1991). The use of irrigation water for improving the vigour of phylloxera-infested grapevines (Kopf, 2000) can only be regarded as a short-term option for phylloxera management as its efficacy would depend on the virulence of the phylloxera strains and other abiotic and biotic factors. In the case of low-virulence phylloxera strains, infested vines with healthy vigour may be better able to tolerate phylloxera in the short-term than low-vigour or stressed vines. However, other factors may preclude widespread use of this technique, including relative soil permeability (i.e. leaching) and availability and economic costs of irrigation water. In addition, because phylloxera has the ability to move up and down the soil profile and root system, it could avoid detrimental changes in soil moisture.

In laboratory trials, grape phylloxera first instars and eggs have been shown to survive for 8–21 days when submersed in water (Korosi et al., 2009; Powell et al., 2013a), indicating a high resilience to water immersion. The efficacy of the submersion treatment is also influenced by temperature and grape phylloxera life stage, with temperatures of less than 5 °C reducing survival of egg and crawler stages to 8 days (Korosi et al., 2009) and temperatures of 10 °C and above allowing survival for at least 21 days (Powell et al., 2013a). The survival of radicoles for this extended period, when submersed in water, is dependent upon the wettability of the tracheal system and whether the spiracles are open or closed. Insect spiracles are one of the few areas of the body from which high rates of evaporation of water can occur. Discontinuous gas exchange has been suggested to reduce water loss when insects have reduced activity and therefore need less oxygen (Schimpf et al., 2009; Williams et al., 2010) as the spiracles only open periodically to allow oxygen entry and water loss. The spiracular opening is supplied with muscles that could control opening or closure. During cold submersion, it is most likely that the metabolic rate is lowered with the reduced temperature, and therefore the storage of oxygen within the tracheal system will supply energy requirements for part of the time and as long as oxygen is dissolved in solution, oxygen will continue to be transferred into the tracheal system (similar to plastron in aquatic insects). If the trachea opens, water may enter the tracheal system and impair respiration, as water has a much lower diffusion coefficient for oxygen and lower oxygen carrying capacity than air (Forster and Woods, 2013). Morphological observations of the structural features of the tracheal system of phylloxera may help clarify the mechanism involved.

The influence of nutrient fertiliser treatment has been assessed on several aphid species and affects the insects' development and host–plant interactions by altering host plant nutritional quality. In a study using *Aphis gossypii* adult and nymph densities, r_m (intrinsic rate of increase) and fecundity were all positively correlated with nitrogen fertilisation (Nevo and Coll, 2001). Jahn et al. (2005) also showed that nitrogen fertiliser application resulted in higher survivorship, fecundity and r_m that were passed on to the second generation of rice aphid *Hysteroneura setariae* (Thomas).

The impact of nitrogen fertilisation on grapevine root development and phylloxera damage has received limited research attention. The application of a nitrogen source to grapevines could either directly affect root development or indirectly impact on grape phylloxera-feeding behaviour by influencing the N content and amino acid profile in the roots, which subsequently may affect the insects' population dynamics. In laboratory and field trials in Europe, using phylloxera-infested grafted rootstocks, reduced nodosity development has been observed (Kopf, 2000) following nitrogen fertiliser application, possibly through the effect of nitrate or ammonia, respectively.

7.3. Carbon dioxide

Atmospheric and soil CO_2 levels could potentially impact on phylloxera development as a range of responses to elevated CO_2, including modified fecundity, population density and development rates, have been observed in several aphid species (Holopainen, 2002). Elevated CO_2 levels alter plant chemistry and root phenology by modifying the C:N ratios in plant sap, which could then impact secondary metabolite, carbohydrate and amino acid profiles (Bezemer and Jones, 1998), and increasing cell production and cell expansion in plant roots (Lawler et al., 1997).

There is minimal information on the impact of atmospheric carbon dioxide on grape phylloxera. Granett et al. (1991) when testing carbon dioxide as a potential fumigant for radicole phylloxera reported CO_2 in laboratory tests *requires a high concentration and relatively long duration to be an effective control agent,* although the CO_2 concentrations were not specified.

7.4. Soil

Control of radicole grape phylloxera requires an understanding of how environmental factors impact on the insects' biology and interactions with the host plant belowground. Soil temperature, humidity, organic matter, salinity and CO_2 content can affect root-feeding insects' survival and host plant location (Gerson, 1996; Kühnelt, 1963). Soil conditions can also influence the rooting depth of different *Vitis* species (Hellman, 2003) and consequently the potential depth of radicole phylloxera within the soil profile. Radicole phylloxera have been observed on roots at soil depths of over 1 m (Buchanan, 1990; De Klerk, 1974). The ability of radicole grape phylloxera to move down the soil profile is likely to constrain the efficacy of some cultural, biological and chemical control options (Benheim et al., 2012) where soil interactions may occur.

Published literature on the interactions between soil and radicole phylloxera is relatively limited. Soil chemistry, texture, moisture, atmosphere and temperature are all likely to impact on phylloxera establishment, population development and dispersal and also the development of the host plant root system (Kühnelt, 1963). Most of the published and anecdotal evidence related to phylloxera distribution in soil focuses purely on textural properties particularly on the ratio of sand to clay content. In vineyard surveys conducted in South Africa phylloxera predominated in soils with less than 65% sand content (De Klerk, 1972). In Canadian and Californian vineyards, phylloxera are reportedly more common on clay soils rather than loam and sandy-loam soils (Bioletti, 1901; Nougaret and Lapham, 1928; Stevenson, 1964). Ermolaev (1990) suggested that the amount of silicon in sandy soil may be an important factor in reducing phylloxera abundance. The impact of soil on phylloxera is likely to be more complex than focusing on a single textural property. In a field study conducted on Austrian vineyards, Reisenzein et al. (2007) showed that a combination of pH, organic carbon and soil texture all influence phylloxera population abundance. In contrast, a glasshouse study showed that similar soil properties had minimal impact on establishment of phylloxera populations sourced from the Pacific Northwest region of North America (Chitikowski and Fisher, 2005).

Under Australian field conditions, some chemical factors have been highlighted, which may influence phylloxera abundance and dispersal. In this respect, Powell et al. (2003) reported that high levels of aluminium exchange capacity, toxic to grapevine roots, were related to higher phylloxera abundance. Aluminium is known to stress plant root systems inhibiting root growth and affecting root chemistry (Delhaize and Ryan, 1995; Dipierro et al., 2005). The spatial distribution of radicole phylloxera has also been linked to specific soil electrical conductivity ranges when using EM38 sensors (Bruce et al., 2011). Apparent electrical conductivity (EC_a) is related to soil textural and chemical properties including soil moisture, salinity and ionic content (Corwin and Lesch, 2005; Friedman, 2005). However, the specific property or properties, related to EC_a, that directly influence radicole phylloxera distribution in soil are yet to be determined. Other sensed data such as aerial or satellite acquired reflectance bioimagery, which focuses on stress symptoms in grapevine foliage, when combined with

EC_a data offer opportunities for integrated targeted phylloxera detection (Bruce et al., 2011).

Determining which soil properties influence phylloxera spatial distribution and emergence of dispersive life stages is also important for optimising sampling strategies for newly emerging detection tools including a phylloxera-specific molecular probe for soil samples (Herbert et al., 2008) and trapping techniques (Powell et al., 2009). Understanding the influence of soil properties on phylloxera could also lead to improved cultural management techniques.

Opportunities for controlling phylloxera through soil management such as tillage or through soil amendments, such as organic or physical mulches, have also received surprisingly limited research attention. Organic mulch applications can be used to modify physical and textural properties of the soil, making conditions more or less conducive to phylloxera establishment and survival on the root system, or they may directly affect the insects' ability to disperse through the soil to reach the soil surface or the grapevine roots. Contrasting results have been reported using this approach. In a Californian study, in which organically managed vineyards and conventionally managed phylloxera-infested vineyards were surveyed, a lower incidence of phylloxera-associated damage was reported in organically managed vineyards (Lotter et al., 1999). In two 3-year field studies conducted in Australia, composted green waste mulch applications increased the abundance and dispersal of phylloxera aboveground, whilst grape-pomace mulch formulations reduced phylloxera population levels under field conditions (Powell et al., 2006, 2007). In Germany, reduced phylloxera abundance was observed when spruce sawdust was applied as mulch, but whether this was due to changed physical or changed microbial properties of the soil was unclear (Huber et al., 2003). Although mulch formulation is likely to be an important factor in the efficacy of such treatments for phylloxera management, soil properties may also be influencing their efficacy. Studies in Europe have indicated that some vineyard soils may have suppressive properties (Huber et al., 2009b).

7.5. Root system

Nondestructive observations of root herbivory by grape phylloxera under edaphic conditions have received surprisingly limited research attention. Forneck et al. (2001b) developed a simple glasshouse-based isolation chamber that allowed continuous monitoring of phylloxera and root damage.

A system developed for monitoring of the root aphids *Eriosoma* sp. (Parker, 1974) could also be readily adapted for radicole phylloxera to study phylloxera–root interactions under contrasting soil conditions.

A single field study, using minirhizotrons, has been conducted (Bauerle et al., 2007), which examined the effect of phylloxera on root population dynamics, specifically focusing on root growth rate and tolerance to phylloxera attack. Two resistant rootstocks, 1103 Paulsen (*V. berlandieri* × *V. rupestris*) and 101-14 (*V. riparia* × *V. rupestris*), both grafted with *V. vinifera* 'Merlot' shoots, were used in the study and phylloxera favoured the faster-growing rootstock 1103 Paulsen that also had the largest proportion of growing root tips. Similar studies could be conducted comparing a wider range of rootstocks, phylloxera genetic strains and under different soil conditions. Alternatively, other experimental techniques such as X-ray tomography (Mankin et al., 2008) or synchrotron could potentially be adapted to study grape phylloxera behaviour. Understanding the fundamental interactions between phylloxera and soil properties could also lead to improved efficacy of biological control agents such as entomopathogenic nematodes and fungal pathogens that have been identified and screened for grape phylloxera control (English-Loeb et al., 1999; Kirchmair et al., 2004).

7.6. Root pathogens

An increased level of root damage and associated reduced grapevine vigour have been recorded when phylloxera infestations coincide with the presence of pathogenic fungi in the soil (Omer et al., 1995). Wounds in the root cortex caused by radicole grape phylloxera feeding create entry points for invasion by fungal pathogens. Experiments to verify this assumption have been conducted under laboratory and field conditions (Granett et al., 1998; Omer et al., 1995, 1999a). Under controlled conditions using inoculations of pathogenic fungi, root necrosis was recorded within 6 weeks (Edwards et al., 2007; Omer and Granett, 2000).

A range of fungal pathogens have been recorded in studies conducted in California and Australia including *Fusarium* spp., *Pythium ultimum*, *Cylindrocarpon* sp., *Cephalosporium* sp. and *Phaeoacremonium* sp. occurring in the roots of phylloxera-infested grapevines (Edwards et al., 2007; Granett et al., 1998; Omer and Granett, 2000). In Californian vineyards, peak fungal infection rates occur in spring and autumn (Granett et al., 1998).

Most of the reports implicating soil-borne fungal pathogens as a causative factor focus on the higher root damage levels in phylloxera-infested vineyards. There have been no published studies to determine if the pathogenic fungi induce physical and chemical effects on the grapevine root system that may then affect the insect indirectly through altering the host plant quality and quantity of available food source. Fungal infections of leaves have been shown to alter the amino acid, carbohydrate and secondary metabolite concentrations (e.g. Johnson et al., 2003) and have been shown to affect fecundity and development of a number of aphid species (see Hatcher, 1995).

8. QUARANTINE AND DISINFESTATION

Natural dispersal of grape phylloxera is limited to a few metres per season (King and Buchanan, 1986). However, human-assisted dispersal can occur by inadvertent transfer of active phylloxera life stages on infested viticulture machinery, clothing and footwear, planting material, soil, foliage, grapes, postharvest grape products and winery waste (Bishop et al., 2002; Deretic et al., 2003). Quarantine can therefore be used to minimise the spread of phylloxera. Phylloxera-specific quarantine protocols have been developed in a number of countries (EPPO, 1990; NVHSC 2009). The extent to which quarantine and associated disinfestation protocols are effective is however highly dependent on understanding the insects' biology and genetics and then scientifically validating disinfestation treatments. There are a number of disinfestation approaches used to ameliorate some risks of phylloxera transfer based on phylloxera's sensitivity to irradiation, fumigation, temperature extremes, humidity and liquid immersion.

Several treatments including gamma irradiation (Arya and Hildebrandt, 1969; Makee et al., 2008), methyl bromide fumigation (Litinov et al., 1985; Sakai et al., 1985), dry heat (Keen et al., 2002; Korosi et al., 2009), immersion in sodium hypochlorite (Dunstone et al., 2003; Grzegorczyk and Walker, 1997), ethanol (Grzegorczyk and Walker, 1997, 1998; Kellow et al., 1999), hot water (EPPO, 2009; Stonerod and Strik, 1996), cold water (Powell et al., 2013a) and composting (Bishop et al., 2002) have been tested against phylloxera.

In most instances, there is inconsistency in the biological parameters measured to test and compare treatment efficacy and a lack of data on both the biological and genetic status of the phylloxera tested. The only study published to date where a treatment has been tested against more than one genetic strain of phylloxera highlighted that lethal temperature and lethal humidity levels differ between strains (Korosi et al., 2012).

The effectiveness of any given disinfestation treatment may be impacted by phylloxera's ability to survive for several days in the absence of food, which is in part influenced by its digestive system and ability to store food reserves. Radicole grape phylloxera has been shown to survive for up to 8 days at 25 °C in the absence of a food source (Kingston et al., 2009) and when temperature is lowered its survival capacity is enhanced with the ability to survive up to 21 days at 10 °C when immersed in water (Powell et al., 2013a). Early researchers identified that hot water immersion treatment was a suitable treatment of rootstock cuttings with 50 °C for 1 min killing all phylloxera life stages (Balbiani, 1876). Sakai et al. (1985) showed that a hot water treatment (45 °C for 20 min) was equally effective. Minor modifications to this suggested treatment have been made but there appears to be no consistent recommendation internationally. The current recommendation for disinfestation of grapevine root material in Australia is either 54 ± 1 °C for 5 min or 50 ± 1 °C for 30 min (NVHSC, 2009), whilst the European recommendation is to preheat by dipping grape cuttings in water at 43 °C for 5 min, and then transferring to water at 52 °C for 5 min (EPPO, 1990); this treatment was later validated by Stonerod and Strik (1996). The efficacy of hot water disinfestation has not been tested against genetically diverse phylloxera strains.

The efficacy of disinfestation treatment can also be dependent on which life stage is tested. One study comparing the efficacy of sodium hypochlorite against two phylloxera life stages highlights that eggs are more resistant to treatment than first instar nymphs (Dunstone et al., 2003) presumably because the egg wall is less permeable than the insect's cuticle.

To ensure robustness of national and international quarantine protocols for grape phylloxera, a more comprehensive scientific validation of disinfestation protocols may be required, to take into consideration treatment efficacy against different genetic strains and different life stages.

9. CONCLUDING REMARKS AND FUTURE DIRECTIONS

Research into fundamental aspects of the physiology and biology of Phylloxeridae in general has received comparatively little attention when compared with Aphidoidea. Although grape phylloxera has received wider attention than other less economically important phylloxerids, there still remain significant knowledge gaps that need to be addressed. In part, the reason for this lack of research focus on grape phylloxera has been due to the considerable success in the use phylloxera-resistant rootstocks as an effective management option since the late nineteenth century. Other forms of phylloxera management, especially cultural and biological control, have received very limited attention except in a few countries where phylloxera is limited in its distribution, such as Australia and China. In these countries due to limited rootstock use much of the research focus has been on quarantine and detection.

The emergence of 'biotypes' that caused considerable economic loss for the Californian wine industry led to an increased research focus on understanding interactions between phylloxera strains and rootstocks and a better understanding of life cycle variations. This research was aided by improved DNA typing techniques over the last 20 years. Without doubt, the use of a genomics approach will further advance this knowledge more rapidly in the future as genomic sequences from both *Vitis* species and *D. vitifoliae* clones are being mapped (Delmotte et al., 2011; Jaillon et al., 2007). Phylloxeridae has a basal phylogenetic relationship to aphids (Ortiz-Rivas and Martínez-Torres, 2010) and therefore provides an interesting model for comparative genomics between aphids. Phylloxera genome sequencing will help fill the knowledge gaps between aphids and related Hemiptera.

The emerging field of metabolomics also opens up opportunities for the discovery and validation of biomarkers in both the root and leaves of *Vitis* species and also phylloxera life stages, which will improve our understanding of resistance mechanisms and aid in the development of chemical fingerprinting methods for novel grape phylloxera detection.

The use of a combined metabolomics and genomic approach could significantly advance our knowledge of fundamental biological characteristics of grape phylloxera and its interaction with *Vitis* species. For example, a proteomics approach could be used to characterise the, as yet unknown, salivary enzymes from grape phylloxera. This approach has recently been used for aphids (e.g. Carolan et al., 2009). A combined stylectomy and metabolomics approach may also allow characterisation of *Vitis* parenchymal cell contents. Stylectomy has been used to collect and subsequently analyse phloem sap from aphids (Varkonyi-Gasic et al., 2010). If used for phylloxera, the volume of fluid available for collection would be smaller than that produced by phloem-feeding aphids. However, methods for collection of small sap volumes are being improved (Palmer et al., 2013). Characterisation of parenchymal cell contents, using a metabolomics approach, would lead to an improved artificial diet formulation for grape phylloxera using existing or modified bioassay systems developed for both gallicole and radicole forms. It could also be used to compare nutritional requirements of different grape phylloxera strains and also for prescreening of antimetabolites for the development of genetically modified grapevines with phylloxera resistance and to investigate the roles of endogenous compounds in antibiosis or antixenosis.

Under future climate change scenarios, it is predicted that suitable climatic regions for growing *Vitis* species will change in geographic distribution (Jones, 2007); this could also lead to a change in the general distribution of grape phylloxera and impact on its development and population dynamics. Even though grape phylloxera is currently established in almost all grape-growing countries of the world, in some countries, it still does have limited distribution. The direct impact of elevated temperature, reduced water availability and elevated carbon dioxide on phylloxera genetic strains has not been examined in any depth and it is quite feasible that phylloxera strain distribution will change. Indirect changes could also potentially influence phylloxera through phenological changes in the host plant root biomass, root phenology, and nutritional quality, as has been suggested for other root-feeding aphids (Salt et al., 1996).

Our current knowledge of phylloxera biology is very limited and should be examined in more detail to ensure phylloxera does not become more of an issue to the viticulture industry in the future than it currently is.

ACKNOWLEDGEMENT

We are grateful to Philip Smith for proofreading this chapter.

REFERENCES

Anders, F., 1957. Über die gallenerregenden Agenzien der Reblaus (*Viteus* [Phylloxera] *vitifolii* Shimer). Vitis 1, 121–124.

Anders, F., 1958a. Aminosäuren als gallenerregende Stoffe der Reblaus (*Viteus* [Phylloxera] *vitifolii* Shimer). Experientia 14, 62–63.

Anders, F., 1958b. Das galleninduzierrende Prinzip der Reblaus (*Viteus vitifolii* Shimer). Verhandl. Deut. Zool. Ges. (Frankfurt), 353–363.

Anders, F., 1960a. Untersuchungen uber das cecidogene Prinzip der Reblaus (*Viteus vitifolii* Shimer), I Untersuchungen an der Reblausgalle. Biol. Zentralblatt 79, 47–58.

Anders, F., 1960b. Untersuchungen uber das cecidogene Prinzip der Reblaus (*Viteus vitifolii* Shimer), II Biologische Untersuchungen uber das galleninduzierende Sekret der Reblaus. Biol. Zentralblatt 79, 679–700.

Anders, F., 1961. Untersuchungen uber das cecidogene Prinzip der Reblaus (*Viteus vitifolii* Shimer), III Biochemische Untersuchungen uber das galleninduzierende Agens. Biol. Zentralblatt 80, 199–233.

Andrews, K.B., Powell, K.S., 2009. Understanding the fundamental interactions between woolly apple aphid and pome fruit: Final Report–APO6011. Horticulture Australia Limited, Sydney, p. 65.

Andrews, K.B., Kemper, D., Powell, K.S., Cooper, P.D., 2012. Spatial trade-offs in the digestive and reproductive systems of grape phylloxera. Austr. J. Zool. 59, 392–399.

Araya, J.E., 2001. Insects That Feed on Resistant Crops. University of Chile, Santiago.

Arya, H.C., Hildebrandt, A.C., 1969. Effect of gamma-radiation on callus growth of phylloxera gall and normal grape stem tissues in culture. Indian J. Exp. Biol. 7, 158–162.

Askani, A., Beierbeck, R., 1991. *In vitro* propagation of *Daktulosphaira vitifolii* Shimer (Homoptera, Phylloxeridae) on shoot and root cultures of a *Vitis* hybid. Vitis 30, 223–232.

Auclair, J.L., 1963. Aphid feeding and nutrition. Annu. Rev. Entomol. 8, 439–490.

Balbiani, M., 1876. Recherches sur la structure et sur la vitalité des oeufs du Phylloxeras. C. rendus hebdomadaires des séances de l'Académie des sciences 954–959.

Baldy, R., De Benedictis, J., Johnson, L., Weber, E., Baldy, M., Osborn, B., Burleigh, J., 1996. Leaf colour and vine size are related to yield in a phylloxera-infested vineyard. Vitis 35, 201–205.

Banerjee, A., Duflo, E., Postel–Vinay, G., Watts, T., 2007. Long run health impacts of income shocks: wine and phylloxera in 19th century France. Massachusetts Institute of Technology, Department of Economics, Working Paper Series, 07–03, 1–29.

Bauerle, T.L., Eissenstat, D.M., Granett, J., Gardner, D.M., Smart, D.R., 2007. Consequences of insect herbivory on grape fine root systems with different growth rates. Plant Cell Environ. 30, 786–795.

Baumann, P., 2005. Biology of bacteriocyte-associated endosymbionts of plant sap-sucking insects. Annu. Rev. Microbiol. 59, 155–189.

Becker, H., 1952. Beiträge zur Physiologies der Reblaus. PhD Thesis: University of Mainz, Germany.

Belcari, A., Antonelli, R., 1989. The influence of temperature on the development of pre-imaginal stages of *Viteus vitifoliae* (Fitch) (Rhynchota–Phylloxeridae). 3. Duration of larval development in epigeous generations at constant temperatures. In: Influence of environmental factors on the control of grape pests, diseases and weeds, Proc. Meet. European Community Experts Group, pp. 115–124.

Benheim, D., Rochfort, S., Ezernieks, V., Robertson, E., Potter, I.D., Korosi, G.A., Powell, K.S., 2011. Early detection of grape phylloxera (*Daktulosphaira vitifoliae* Fitch) infestation through identification of chemical biomarkers. Acta Horticult. 904, 17–24.

Benheim, D., Rochfort, S., Robertson, E., Potter, I.D., Powell, K.S., 2012. Grape phylloxera (*Daktulosphaira vitifoliae*)—a review of potential detection and alternative management options. Ann. Appl. Biol. 161, 91–115.

Bezemer, T.M., Jones, T.H., 1998. Plant–insect herbivore interactions in elevated atmospheric CO_2: quantitative analyses and guild effects. Oikos 82, 212–222.

Bioletti, F.T., 1901. The phylloxera of the vine. Univ. Calif. Coll. Agric. Bull. 131, 13–28.

Bishop, A.L., Powell, K.S., Gibson, T.S., Barchia, I.M., Wong, P.T.W., 2002. Mortality of grape phylloxera in composting organics. Aust. J. Grape Wine Res. 8, 48–55.

Blackman, R.L., Eastop, V.F., 1984. Aphids on the World's Crops: An Identification and Information Guide. John Wiley, Chichester, UK, p. 465.

Blanchfield, A.L., Robinson, S.A., Renzullo, L.J., Powell, K.S., 2006. Phylloxera-infested grapevines have reduced chlorophyll and increased photoprotective pigment content-can leaf pigment composition aid pest detection? Funct. Plant Biol. 33, 507–514.

Börner, C., 1914. Über reblaus-anfällige und immune Reben. Biol. Zentralblatt XXXIV, 1–8.

Botton, M., Walker, M.A., 2009. Grape Phylloxera in Brazil. Acta Hortic. 816, 39–40.

Boubals, D., 1966a. Heredite de la résistance au phylloxéra radicicole chez la vigne. Ann. Amelior. Plantes 16, 327–347.

Boubals, D., 1966b. Étude de la distribution et des causes de la résistance au phylloxéra radicicole chez les Vitacées. Ann. Amélior. Plantes 16, 145–184.

Boubals, D., 1994. Phylloxéra, Attention! En Allemagne, une race de phylloxéra radicicoliae plus aggressive attaquerait les raciness de SO4, 5BB, et 125AA (tous *riparia* x *Berlandieri*)! Prog. Agric. Vitic. 111, 374–375.

Bournier, A., 1976. Grape insects. Annu. Rev. Entomol. 22, 355–376.

Branas, J., 1963. Dépréssement du 41B en Charente, phylloxéra et court-noue. Prog. Agric. Vitic. 161, 178–186.

Breider, H., 1952. Beiträge zur Morphologie und Biologie der Reblaus *Dactylosphaera vitifolii* Shim. Z. Angew. Entomol. 33, 517–543.

Bruce, T.J.A., Wadhams, L.J., Woodcock, C.M., 2005. Insect host location: a volatile situation. Trends Plant Sci. 10, 269–274.

Bruce, R.J., Lamb, D.W., Hoffmann, A.A., Runting, J., Powell, K.S., 2011. Towards improved early detection of grapevine phylloxera (*Daktulosphaira vitifoliae* Fitch) using a risk-based assessment. Acta Horticult. 904, 123–131.

Buchanan, G. A., 1990. The distribution, biology and control of grape phylloxera, *Daktulosphaira vitifoliae* (Fitch), in Victoria. PhD Thesis, La Trobe University, Melbourne, Australia.

Buchner, P., 1965. Endosymbiosis of Animals with Plant Microorganisms. John Wiley and Sons, London, UK, p. 909.

Campbell, C., 2004. Phylloxera: How Wine Was Saved for the World. Harper Collins, London.

Carolan, J.C., Fitzroy, C.I., Ashton, P.D., Douglas, A.E., Wilkinson, T.L., 2009. The secreted salivary proteome of the pea aphid *Acyrthosiphon pisum* characterised by mass spectrometry. Proteomics 9, 2457–2467.

Carton, Y., Sorensen, C., Smith, J., Smith, E., 2007. Une coopération exemplaire entre entomologistes français et américains pendant la crise du Phylloxera en France (1868–1895), Annales de la Société entomologique de France (N.S.). Int. J. Entomol. 43, 103–125.

Chen, M.-S., 2008. Inducible direct plant defense against insect herbivores, a review. Insect Sci. 15, 101–114.

Chitikowski, R.L., Fisher, J.R., 2005. Effect of soil type on the establishment of grape phylloxera colonies in the Pacific Northwest. Am. J. Enol. Vitic. 56, 207–210.

Cole, R.A., Riggall, W., Morgan, A., 1993. Electronically monitored feeding behaviour of the lettuce root aphid (*Pemphigus bursarius*) on resistant and susceptible lettuce varieties. Entomol. Exp. Appl. 68, 179–185.

Compson, Z.G., Larson, K.C., Zinkgraf, M.S., Whitham, T.G., 2011. A genetic basis for the manipulation of sink-source relationships by the galling aphid *Pemphigus batae*. Oecologia 167, 711–721.

Connelly, A., 1995. Biology and demography of grape phylloxera, *Daktulosphaira vitifoliae* (Fitch) (Homoptera, Phylloxeridae), in Western Oregon. MSc Thesis. Oregon State University, Oregon.

Corrie, A.M., Hoffmann, A.A., 2004. Fine-scale genetic structure of grape phylloxera from the roots and leaves of *Vitis*. Heredity 92, 118–127.

Corrie, A.M., Buchanan, G., Van Heeswijck, R., 1997. DNA typing of populations of phylloxera (*Daktulosphaira vitifoliae* (Fitch)) from Australian vineyards. Aust. J. Grape Wine Res. 3, 50–56.

Corrie, A., Kellow, A.V., Buchanan, G.A., Van Heeswijck, R., 1998. Phylloxera biotypes in Australia. Aust. NZ. Grapegrower Winemaker 417, 32–36.

Corrie, A.M., Crozier, R.H., Van Heeswijck, R., Hoffmann, A.A., 2002. Clonal reproduction and population genetic structure of grape phylloxera, *Daktulosphaira vitifoliae*, in Australia. Heredity 88, 203–211.

Corrie, A.M., Van Heeswijck, R., Hoffman, A.A., 2003. Evidence for host-associated clones of grape phylloxera *Daktulosphaira vitifoliae* (Hemiptera, Phylloxeridae) in Australia. Bull. Entomol. Res. 93, 193–201.

Corwin, D.L., Lesch, S.M., 2005. Apparent soil electrical conductivity measurements in agriculture. Comput. Electronics Agric. 46, 11–43.

Crovetti, A., Rossi, E., 1987. Field and laboratory observations on some eco–ethological aspects of the grape phylloxera (*Viteus vitifoliae* (Fitch)). Influence of environmental factors on the control of grape pests, diseases and weeds. Proc. Meeting EC Experts' Group, Thessalonika, 6–8 October 1987, Commission of the European Communities, A. A. Balkema, Rotterdam, The Netherlands.

Davidson, W.M., Nougaret, R.L., 1921. The grape phylloxera in California. US Dept. Agric. Bull. 903, 1–127.

De Benedictis, J.A., Granett, J., 1992. Variability of responses of grape phylloxera (Homoptera, Phylloxeridae) to bioassays that discriminate between California biotypes. J. Econ. Entomol. 85, 1527–1534.

De Benedictis, J.A., Granett, J., 1993. Laboratory evaluation of grape roots as hosts of California grape phylloxera biotypes. Am. J. Enol. Vitic. 44, 285–291.

De Benedictis, J.A., Granett, J., Taormino, S.P., 1996. Differences in host utilisation by California strains of grape phylloxera. Am. J. Enol. Vitic. 47, 373–379.

De Klerk, C.A., 1972. Occurrence and distribution of the vine phylloxera *Phylloxera vitifoliae* (Fitch), in the Olifants River Irrigation area, Northwest Cape Province. Phytophylactica 4, 25–26.

De Klerk, C.A., 1974. Biology of *Phylloxera vitifoliae* (Fitch) (Homoptera: Phylloxeridae) in South Africa. Phytophylactica 6, 109–118.

De Klerk, C.A., 1979. An investigation of two morphometric methods to test for the possible occurrence of morphologically different races of *Daktulosphaira vitifoliae* (Fitch) in South Africa. Phytophylactica 11, 51–52.

Delhaize, E., Ryan, P.R., 1995. Aluminium toxicity and tolerance in plants. Plant Physiol. 107, 315–321.

Delmotte. F., Forneck, A., Powell, K.S., Rispe, C., Tagu, D., 2011. Proposal to sequence the genome of the grape phylloxera (*Daktulosphaira vitifoliae* Fitch). http://www.aphidbase.com/var/aphidbase/storage/htmlarea/2965/file/White-Paper-Phylloxera_25may2011-1.pdf.

Denisova, T.V., 1965. The phenolic complex of vine roots infested by phylloxera as a factor of resistance. J. Agric. Sci. (Russia) 10, 114–118.

Deretic, J., Powell, K.S., Hetherington, S., 2003. Assessing the risk of phylloxera transfer during post-harvest handling of wine grapes. Acta Horticult. 617, 61–66.

Dickey, A.M., Medina, R.F., 2012. Host-associated genetic differentiation in pecan leaf phylloxera. Entomol. Exp. Appl. 143, 127–137.

Dipierro, N., Mondelli, D., Paciolla, C., Brunetti, G., Dipierro, S., 2005. Changes in the ascorbate system in the response of pumpkin (*Cucurbita pepo* L.) roots to aluminium stress. J. Plant Physiol. 162, 529–536.

Douglas, A.E., 2003. The nutritional physiology of aphids. Adv. Insect Physiol. 31, 73–138.

Douglas, A.E., 2006. Phloem-sap feeding by animals, problems and solutions. J. Exp. Bot. 57, 747–754.

Downie, D.A., 2010. Baubles, bangles, and biotypes, a critical review of the use and abuse of the biotype concept. J. Insect Sci. 10, 1–18.

Dreyer, D.L., Campbell, B.C., 1984. Association of the degree of methylation of intercellular pectin with plant resistance to aphids and with induction of aphid biotypes. Cell. Mol. Life Sci. 40, 224–226.

Dreyer, D.L., Campbell, B.C., 2006. Chemical basis of host–plant resistance to aphids. Plant Cell Environ. 10, 353–361.

Dreyfus, L., 1889. Ueber Phylloxerinen. J.F. Bergmann, Wiesbaden, pp. 88.

Du, Y.P., Wang, Z.-S., Yang, Y., Zhao, Q., Zhai, H., Wang, Z.-Y., 2008. Nodosity formation and nutrition consumption in grape cultivars with different phylloxera resistance and infested by grape phylloxera. Acta Entomol. Sinica 51, 1050–1054.

Du, Y., Zheng, Q., Zhai, H., Jiang, E.S., Wang, Z.-Y., 2009. Selectivity of *Phylloxera viticola* Fitch (Homoptera, Phylloxeridae) to grape with different resistance and the identification of grape root volatiles. Acta Entomol. Sinica 52, 537–543.

Du, Y.P., Wang, Z.-S., Zhai, H., 2011. Grape root cell features related to phylloxera resistance and changes of anatomy and endogenous hormones during nodosity and tuberosity formation. Austr. J. Grape Wine Res. 17, 291–297.

Dunstone, R.J., Corrie, A.M., Powell, K.S., 2003. Effect of sodium hypochlorite on first instar phylloxera (*Daktulosphaira vitifoliae* Fitch) mortality. Austr. J. Grape Wine Res. 9, 107–109.

Düring, H., 1999. High light and water stress in grapevines: photoinhibition and photoprotection. Acta Horticult. 493, 45–55.

Edwards, J., Powell, K.S., Hackworth, P., Lamb, D., Lewis, M., 2004. Identification of phylloxera from high resolution infra-red imagery: a comparative study between airborne imagery types. Aust. NZ. Grapegrower Winemaker 488, 51–54.

Edwards, J., Norng, S., Granett, J., Powell, K.S., 2007. Relationships between grape phylloxera abundance, fungal interactions and grapevine decline. Acta Horticult. 733, 151–158.

El-Nady, M.F., Schroder, M.-B., 2003. Cytology of the hypersensitivity reaction in rootstocks-improvement of rootstock breeding. Acta Horticult. 617, 17–23.

English-Loeb, G., Villani, M., Martinson, T., Forsline, A., Consolie, N., 1999. Entomopathogenic nematodes for control of grape phylloxera (Homoptera: Phylloxeridae): a laboratory evaluation. Environ. Entomol. 28, 890–894.

EPPO, 1990. Data Sheets on Quarantine Pests, *Viteus vitifoliae.*

EPPO, 2009. PM 10/16, Hot water treatment of grapevine to control *Viteus vitifoliae.* EPPO Bull. 39, 484–485.

Ermolaev, A.A., 1990. Resistance of grape phylloxera on sandy soils. Agrokhimiya 2, 141–142.

Federov, S.M., 1959. The biological basis of phylloxera (*Dactylosphaera vitifolii* Schim., Homoptera, Phylloxeridae) control. Entomol. Rev. 38, 74–85.

Feldhaar, H., 2011. Bacterial symbionts as mediators of ecologically important traits of insect hosts. Ecol. Entomol. 36, 533–543.

Fergusson-Kolmes, L.A., Dennehy, T.J., 1991. Anything new under the sun: not phylloxera biotypes. Wines Vines 72, 51–56.

Fergusson-Kolmes, L.A., Dennehy, T.J., 1993. Differences in host utilisation by populations of North American grape phylloxera (Homoptera, Phylloxeridae). J. Econ. Entomol. 86, 1502–1511.

Ferry, N., Edwards, M.G., Gatehouse, J.A., Gatehouse, A.M.R., 2004. Plant–insect interactions: molecular approaches to insect resistance. Curr. Opin. Biotechnol. 15, 155–161.

Fisher, J.R., Albrecht, M.A., 2003. Constant temperature life table studies of populations of grape phylloxera from Washington and Oregon, USA. Acta Horticult. 617, 43–48.

Fong, G., Walker, M.A., Granett, J., 1995. RAPD assessment of California phylloxera diversity. Mol. Ecol. 4, 459–464.

Forneck, A., Huber, L., 2009. (A)sexual reproduction—a review of life cycles of grape phylloxera, *Daktulosphaira vitifoliae.* Entomol. Exp. Appl. 131, 1–10.

Forneck, A., Wöhrle, A., 2003. A synthetic diet for phylloxera (*Daktulosphaira vitifoliae* Fitch). Acta Horticult. 617, 129–134. http://www.actahort.org/books/617/617_18.htm.

Forneck, A., Walker, M.A., Merkt, N., 1996. Aseptic dual culture of grape (*Vitis* spp.) grape phylloxera (*Daktulosphaira vitifoliae* Fitch). Vitis 35, 95–97.

Forneck, A., Walker, M.A., Blaich, R., 2000. Genetic structure of an introduced pest, grape phylloxera (*Daktulosphaira vitifoliae* Fitch), in Europe. Genome 43, 669–678.

Forneck, A., Walker, M.A., Blaich, R., 2001a. Ecological and genetic aspects of grape phylloxera *Daktulosphaira vitifoliae* (Hemiptera: Phylloxeridae) performance on rootstock hosts. Bull. Entomol. Res. 91, 445–451.

Forneck, A., Walker, M.A., Blaich, R., Yvon, M., Leclant, F., 2001b. Interaction of phylloxera (*Daktulosphaira vitifoliae* Fitch) with grape (*Vitis* spp.) in simple isolation chambers. Am. J. Enol. Vitic. 52, 28–34.

Forneck, A., Kleinmann, S., Blaich, R., Anvari, S.F., 2002. Histochemistry and anatomy of phylloxera (*Daktulosphaira vitifoliae*) nodosites in young roots of grapevine (*Vitis* spp.). Vitis 41, 93–97.

Forster, T.D., Woods, H.A., 2013. Mechanisms of tracheal filling in insects. Biol. Rev. 88, 1–14.

Fraenkel, G., 1952. The role of symbionts as sources of vitamins and growth factors for their insect Hosts. Tijdschr. Entomol. 95, 183–195.

Franks, T.K., Hayasaka, Y., Choimes, S., Van Heeswijck, R., 2005. Cyanogenic glucosides in grapevine: polymorphism, identification and development patterns. Phytochemistry 66, 165–173.

Franks, T.K., Powell, K.S., Choimes, S., Marsh, E., Iocco, P., Sinclair, B.J., Ford, C.M., Van Heeswijck, R., 2006. Consequences of transferring three sorghum genes for secondary metabolite (cyanogenic glucoside) biosynthesis to grapevine hairy roots. Transgenic Res. 15, 181–195.

Frazier, P., Whiting, J., Powell, K.S., Lamb, D., 2004. Characterising the development of grape phylloxera infestation with multi-temporal near-infrared aerial photography. Aust. NZ. Grapegrower Winemaker 485a, 133–142.

Freeman, B.M., 1983. At the root of the vine. Aust. NZ. Grapegrower Winemaker 232, 58–64.

Freeman, B.M., Smart, R.E., 1976. Research note: a root observation laboratory for studies with grapevines. Am. J. Enol. Vitic. 27, 36–39.

Friedman, S.P., 2005. Soil properties influencing apparent electrical conductivity, a review. Comput. Electronics Agric. 46, 45–70.

Funk, C.J., 2001. Alkaline phosphatase activity in whitefly salivary glands and saliva. Arch. Biochem. Physiol. 46, 165–174.

Furkaliev, D.G., 1999. Wild grapevines used as phylloxera-resistant rootstocks. Aust. NZ. Grapegrower Winemaker. Annu. Tech. Iss. 28–31.

Gale, G., 2011. Dying on the Vine: How Phylloxera Transformed Wine. University of California Press, Berkeley, pp. 323.

Gerson, U., 1996. Arthropod root pests. In: Waisel, Y., Eshel, A., Kafkafi, U. (Eds.), Plant Roots: The Hidden Half, second ed. Marcel Dekker Inc., New York, pp. 797–809.

Gorkavenko, A.C., Gorkavenko, E.B., 1977. Particulars of the development of the root form of phylloxera. Zashchita Rastenii 3, 55–56.

Granett, J., 1990. Comparison of swellings caused by indoleacetic acid and tuberosities induced by grape phylloxera (Homoptera, Phylloxeridae). J. Econ. Entomol. 83, 494–499.

Granett, J., Timper, P., 1987. Demography of grape phylloxera, *Daktulosphaira vitifoliae* (Homoptera, Phylloxerideae), at different temperatures. J. Econ. Entomol. 80, 327–329.

Granett, J., Bisabri–Ershadi, B., Carey, J., 1983. Life tables of phylloxera on resistant and susceptible grape rootstocks. Entomol. Exp. Appl. 34, 13–19.

Granett, J., Timper, P., Lider, L.A., 1985. Grape phylloxera (*Daktulosphaira vitifoliae*) (Homoptera: Phylloxeridae) biotypes in California. J. Econ. Entomol. 78, 1463–1465.

Granett, J., Goheen, A.C., Lider, L.A., White, J.J., 1987. Evaluation of grape rootstocks for resistance to Type A and Type B grape phylloxera. Am. J. Enol. Vitic. 38, 298–300.

Granett, J., De Benedictis, J.D., Walker, M.A., Wolpert, J., Weber, E., 1991. Spread and Control of Biotype B Phylloxera. American Vineyard Foundation, California. http://www.avf.org/article.html?id=3600.

Granett, J., Omer, A.D., Pessereau, P., Walker, M.A., 1998. Fungal infections of grapevine roots in phylloxera-infested vineyards. Vitis 37, 39–42.

Granett, J., Omer, A.D., Walker, M.A., 2001a. Seasonal capacity of attached and detached vineyard roots to support grape phylloxera (Homoptera, Phylloxeridae). J. Econ. Entomol. 94, 138–144.

Granett, J., Walker, M.A., Kocsis, L., Omer, A.D., 2001b. Biology and management of grape phylloxera. Annu. Rev. Entomol. 46, 387–412.

Grassi, B., 1912. Contributo alla Conoscenza delle Fillosserini ed in Particolare della Fillosera della Vite. Tipografica Nazionale di G. Bertero EC, Rome, p. 456.

Grzegorczyk, W., Walker, M.A., 1997. Surface sterilization of grape phylloxera eggs in preparation for in vitro culture with *Vitis* species. Am. J. Enol. Vitic. 48, 157–159.

Grzegorczyk, W., Walker, M.A., 1998. Evaluating resistance to grape phylloxera in *Vitis* species with an *in vitro* dual assay culture. Am. J. Enol. Vitic. 49, 17–22.

Haegeman, A., Mantelin, S., Jones, J.T., Gheysen, G., 2012. Functional roles of effectors of plant-parasitic nematodes. Gene 492, 19–31.

Harmel, N., Letocart, E., Cherqui, A., Giordanengo, P., Mazzucchelli, G., Guillonneau, F., De Pauw, E., Haubruge, E., Francis, F., 2008. Identification of aphid salivary proteins: a proteomic investigation of *Myzus persicae*. Insect Mol. Biol. 17, 165–174.

Harrewijn, P., Tjallingii, W.F., Mollema, C., 1996. Electrical recording of plant penetration by western flower thrips. Entomol. Exp. Appl. 79, 345–353.

Harrewijn, P., Piron, P.G.M., Ponsen, M.B., 1998. Evolution of vascular feeding in aphids, an electrophysiological study. Proc. Exp. Appl. Entomol. N.E.V Amsterdam 9, 29–32.

Hatcher, P.E., 1995. Three-way interactions between plant pathogenic fungi, herbivorous insects and their host plants. Biol. Rev. 70, 639–694.

Havill, N.P., Foottit, R.G., 2007. Biology and evolution of Adelgidae. Annu. Rev. Entomol. 52, 325–349.

He, J., Chen, F., Chen, S., Guan, Z., Fang, W., Liu, Z., Lu, J., 2012. Expression profiles of defense-related genes in response to aphid inoculation in different *Chrysanthemum* cultivars. Acta Horicult. 929, 51–57.

Heil, M., 2008. Indirect defence via tritrophic interactions. New Phytol. 178, 41–61.

Hellman, E.W., 2003. Grapevine structure and function. In: Hellman, E.W. (Ed.), Oregon Viticulture. Oregon State University Press, Corvallis, Oregon, pp. 5–19.

Helm, K.F., Readshaw, J.L., Cambourne, B.C., 1991. The effect of drought on populations of phylloxera in Australian vineyards. Wine Ind. J. 6, 195–202.

Henke, O., 1961. Über die Bedeutung der Stickstoffverbindungen für die stoffwechselphysiologischen Beziehungen zwischen Parasit und Wirt am Biespeil Reblaus-Rebe. Phytopathol. Z. 41, 387–426.

Henke, O., 1963. Über den Stoffwechsel reblausanfälliger und -unanfälliger Reben. Phytopathol. Z. 47, 314–326.

Herbert, K.S., Hoffman, A.A., Powell, K.S., 2006. Changes in grape phylloxera abundance in ungrafted vineyards. J. Econ. Entomol. 99, 1774–1783.

Herbert, K.S., Powell, K.S., McKay, A., Hartley, D., Herdina, Ophel-Keller, K., Schiffer, M., Hoffmann, A.A., 2008. Developing and testing a diagnostic probe for grape phylloxera applicable to soil samples. J. Econ. Entomol. 101, 1934–1943.

Herbert, K.S., Umina, P.A., Mitrovski, P.J., Powell, K.S., Viduka, K., Hoffmann, A.A., 2010. Clone lineages of grape phylloxera differ in their performance on *Vitis vinifera*. Bull. Entomol. Res. 100, 671–678.

Herrmann, J.V., Forneck, A., 2001. Die natürlichen Feinde der Reblaus. Deutsche Weinmag. 14, 18–21.

Hofman, L., 1957. Die Histologie der Nodositäten verschiedener Rebensorten bei Reblausbefall. Vitis 1, 125–141.

Holopainen, J.K., 2002. Aphid response to elevated ozone and CO_2. Entomol. Exp. Appl. 104, 137–142.

Huber, L., Eisenbeis, G., Porten, M., Rühl, E.H., 2003. The influence of organically managed vineyard-soils on the phylloxera-populations and vigour of grapevines. Acta Hortic. 617, 55–59.

Huber, L., Michaelis, P., Eisenbeis, G., Forneck, A., 2009a. Antennal sensilla of grape phylloxera (*Daktulosphaira vitifoliae* Fitch). Acta Horticult. 816, 91–96.

Huber, L., Hoffmann, M., Ruhl, E.H., Kirchmair, M., 2009b. Disease suppressiveness of vineyard soils infested with grape phylloxera. Acta Horticult. 816, 41–51.

Jahn, G.C., Almazan, L.P., Pacia, J., 2005. Effect of nitrogen fertilizer on the intrinsic rate of increase of the rusty plum aphid, *Hysteroneura setariae* (Thomas) (Homoptera: Aphididae) on rice (*Oryza sativa* L.). Environ. Entomol. 34, 938–943.

Jaillon, O., Aury, J.M., Noel, B., Policriti, A., Clepet, C., Casagrande, A., et al., 2007. The grapevine genome sequence suggests ancestral hexaploidization in major angiosperm phyla. Nature 449, 463–467.

Johnson, S.N., Douglas, A.E., Woodward, S., Hartley, S.E., 2003. Microbial impacts on plant-herbivore interactions: the indirect effects of a birch pathogen on a birch aphid. Oecologia 134, 381–387.

Jones, G.V., 2007. Climate change: observations, projections, and general implications for viticulture and wine production. In: Essick, E., Griffin, P., Keefer, B., Miller, S., Storchmann, K. (Eds.), Economics Department Working Paper No. 7, Whitman College, Department of Geography, Southern Oregon University, Ashland, pp. 1–14.

Karban, R., Baldwin, I.T., 1997. Induced Responses to Herbivory. University of Chicago Press, Chicago, USA, p. 319.

Keen, B.P., Bishop, A.L., Gibson, T.S., Spohr, L.J., Wong, P.T.W., 2002. Phylloxera mortality and temperature profiles in compost. Aust. J. Grape Wine Res. 8, 56–61.

Kellow, A.V., 2000. A study of the interaction between susceptible and resistant grapevines and phylloxera. PhD Thesis. University of Adelaide, Adelaide.

Kellow, A.V., Corrie, A., Van Heeswijck, R., 1999. Surface sterilisation of phylloxera eggs for investigating grapevine-phylloxera interactions in tissue culture. Aust. J. Grape Wine Res. 5, 27–28.

Kellow, A.V., McDonald, G., Corrie, A., Van Heeswijck, R., 2002. *In vitro* assessment of grapevine resistance to two populations of phylloxera from Australian vineyards. Aust. J. Grape Wine Res. 8, 109–116.

Kellow, A.V., Sedgley, M., van Heeswijck, R., 2004. Interaction between *Vitis vinifera* and grape phylloxera, changes in root tissue during nodosity formation. Ann. Bot. 93, 581–590.

Kessler, A., Baldwin, I.T., 2001. Defense function of herbivore induced plant volatile emissions in nature. Science 291, 829–837.

King, P.D., Buchanan, G.A., 1986. The dispersal of phylloxera crawlers and spread of phylloxera infestations in New Zealand and Australian vineyards. Am. J. Enol. Vitic. 37, 26–33.

King, P.D., Rilling, G., 1985. Variations in the galling reaction of grapevines, evidence of different phylloxera biotypes and clonal reaction to phylloxera. Vitis 24, 32–42.

Kingston, K.B., 2007. Digestive and feeding physiology of grape phylloxera (*Daktulosphaira vitifoliae* Fitch). PhD Thesis. Australian National University, Canberra, Australia.

Kingston, K.B., Cooper, P.D., Powell, K.S., 2007a. Grape phylloxera external morphology observations under scanning electron microscopy. Acta Horticult. 733, 107–114. http://www.actahort.org/books/733/733_11.htm.

Kingston, K.B., Powell, K.S., Cooper, P.D., 2007b. Characterising the root-feeding habits of grape phylloxera using electrical penetration graph. Acta Horticult. 733, 33–45. http://www.actahort.org/books/733/733_3.htm.

Kingston, K.B., Powell, K.S., Cooper, P.D., 2009. Grape phylloxera, new investigations into the biology of an old grapevine pest. Acta Horticult. 816, 63–70. http://www.actahort.org/books/816/816_9.htm.

Kirchmair, M., Huber, L., Rianer, J., Strasser, H., 2004. *Metarhizium anisopliae*, a potential biological control agent against grape phylloxera. BioControl 49, 295–303.

Kislin, E.N., Nedov, P.N., Vilkova, N.A., 1990. Changes in endogenous content of abscisic acid in grape leaves damaged by phylloxera. Agric. Biol. (Russia) 5, 188–191.

Kocsis, L., Granett, J., Walker, M.A., Lin, H., Omer, A.D., 1999. Grape phylloxera populations adapted to *Vitis berlandieri* × *V. riparia* rootstocks. Am. J. Enol. Vitic. 50, 101–107.

Koennecke, T., Aigner, C., Specht, S., Lawo, N.C., Forneck, A., 2011. A stepwise assessment of *Daktulosphaira vitifoliae* infested grapevines in a Viennese vineyard site. Acta Horticult. 904, 59–62.

Koetting, O., Kossmann, J., Zeeman, S.C., Lloyd, J.R., 2010. Regulation of starch metabolism: the age of enlightenment?. Curr. Opin. Plant Biol. 13, 320–328.

Kolberg, R., Lawo, N.C., Forneck, A., 2011. Investigation on maternal transmission of the bacteria *Pantoea agglomerans* by grape phylloxera (*Daktulosphaira vitifoliae* Fitch). Acta Horticult. 904, 41–45.

Kopf, A., 2000. Untersuchungen zur Abundanz der Reblaus (*Dactylosphaera vitifolii* Shimer) und zur Nodositätenbildung in Abhängigkeit von Umweltfaktoren. Ph.D. Thesis, University of Hohenheim, Germany.

Korosi, G.A., Trethowan, C.J., Powell, K.S., 2007. Screening for rootstock resistance to grapevine phylloxera genotypes from Australian vineyards under controlled conditions. Acta Horticult. 733, 159–165.

Korosi, G.A., Trethowan, C.J., Powell, K.S., 2009. Reducing the risk of phylloxera transfer on viticulture waste and machinery. Acta Horticult. 816, 53–62.

Korosi, G.A., Carmody, B., Powell, K.S., 2011. Rootstock screening for phylloxera resistance under controlled conditions using selected phylloxera clonal lineages. Acta Horticult. 904, 33–39.

Korosi, G.A., Mee, P.T., Powell, K.S., 2012. Influence of temperature and humidity on mortality of grapevine phylloxera *Daktulosphaira vitifoliae* clonal lineages, a scientific validation of a disinfestation procedure for viticultural machinery. Aust. J. Grape Wine Res. 18, 43–47.

Kozhantschikov, J., 1928. New biometrical investigation of the phylloxera-races. Z. Indukt. Abstammungs Vererbungslehre. 47, 270–274.

Kühnelt, W., 1963. Soil-inhabiting Arthropoda. Annu. Rev. Entomol. 8, 115–136.

Kunkel, H., 1967. Zur Ernährungsweise zweir Phylloxera Arten (Phylloxeridae, Aphidina) an Eichenblättern. Z. Angew. Zool. 54, 517–525.

Lawler, I.R., Foley, W.J., Woodrow, I.E., Cork, S.J., 1997. The effects of elevated CO_2 atmospheres on the nutritional quality of Eucalyptus foliage and its interaction with soil nutrient and light availability. Oecologia 109, 59–68.

Lawo, N., Forneck, A., 2011. Presence of *Pantoea* sp. in leaf-galling phylloxera. Acta Horticult. 904, 63–66.

Lawo, N., Weingart, G., Schuhmacher, R., Forneck, A., 2011. The volatile metabolome of grapevine roots, first insights into the metabolic response upon phylloxera attack. Plant Physiol. Biochem. 49, 159–1063.

Lawo, N.C., Griesser, M., Forneck, A., 2013. Expression of putative expansin genes in phylloxera (*Daktulosphaira vitifoliae* Fitch) induced root galls of *Vitis* spp. Eur. J. Plant Pathol. 136, 383–391.

Lin, H., Walker, M.A., Hu, R., Granett, J., 2006. New simple sequence repeat loci for the study of grape phylloxera (*Daktulosphaira vitifoliae*) genetics and host adaptation. Am. J. Enol. Vitic. 57, 33–40.

Litinov, P.I., Glushkova, S.A., Chernei, L.B., 1985. Disinfestation of planting material containing root phylloxera. Zashchita Rastenii 7, 39–40.

Lotter, D.W., Granett, J., Omer, A.D., 1999. Differences in grape phylloxera-related grapevine root damage in organically and conventionally managed vineyards in California. Horticult. Sci. 34, 1108–1111.

Makee, H., 2004. Factors influencing mortality, fecundity and fertility of grape phylloxera (*Daktulosphaira vitifoliae* Fitch). Vitis 43, 49–50.

Makee, H., Charbaji, T., Ayyoubi, A., Idris, I., 2004. Evaluating resistance of some rootstocks to grape phylloxera with *in vitro and* excised root testing systems. In Vitro Cell. Dev. Biol. 40, 225–229.

Makee, H., Charbaji, T., Idris, I., Ayyoubi, Z., 2008. Effect of gamma irradiation on survival and reproduction of grape phylloxera *Daktulosphaira vitifoliae* (Fitch). Adv. Horicult. Sci. 22, 182–186.

Mankin, R.W., Johnson, S.N., Grinev, D.V., Gregory, P.J., 2008. New experimental techniques for studying root herbivores. In: Johnson, S.N., Murray, P.J. (Eds.), Root Feeders: An Ecosystem Perspective. CAB International, Oxon, pp. 30–43.

Marini-Filho, O.J., Fernandes, G.W., 2012. Stem galls drain nutrients and decrease shoot performance in *Diplusodon orbicularis* (Lythraceae). Arthro. Plant Interact. 6, 121–128.

Martinez-Torres, D., Baudes, C., Latorre, A., Moya, A., 2001. Molecular systematics of aphids and their primary endosymbionts. Mol. Phylogenet. Evol. 20, 437–449.

Mayet, V., 1890. Les Insectes de la Vigne. Masson, Montpellier, p. 528.

Medina, R.F., Nachappa, P., Tamborindeguy, C., 2011. Differences in bacterial diversity of host-associated populations of *Phylloxera notabilis* Pergande (Hemiptera, Phylloxeridae) in pecan and water hickory. J. Evol. Biol. 24, 761–771.

Michalik, A., Szklarzewicz, T., Wegierek, P., Wieczorek, K., 2013. The ovaries of aphids (Hemiptera, Sternorrhyncha, Aphidoidea): morphology and phylogenetic implications. Invert. Biol. 132, 226–240.

Miles, P.W., 1968. Insect secretions in plants. Annu. Rev. Phytopathol. 6, 137–164.

Miles, P.W., 1990. Aphid salivary secretions and their involvement in plant toxicoses. In: Campbell, R.K., Ekenbary, R.D. (Eds.), Aphid–Plant Genotype Interactions. Elsevier, Amsterdam, pp. 131–148.

Miles, P.W., 1998. Aphid salivary functions: the physiology of deception. In: Nieto, J.M., Dixon, A.F.G. (Eds.), Aphids in Natural and Managed Ecosystems. University of León, León, France, pp. 255–263.

Miles, P.W., 1999. Aphid saliva. Biol. Rev. 74, 41–85.

Mittler, T.E., 1988. Applications of artificial feeding techniques for aphids (Section 8.13). In: In: Minks, A.K., Harrewijn, P. (Eds.), Aphids: Their Biology, Natural Enemies and Control Vol. B. Elsevier, Amsterdam, pp. 145–169.

Moleas, T., Addante, R., Bari, G., 1992. Osservazioni sulla variabilità della lunghezza degli stiletti di neanidi di *Viteus vitifolii* (Fitch) (Rhynchota–Phylloxeridae) in rapporto alle generazioni. Boll. Lab. Ent. 49, 61–68.

Molnár, J.G., Neméth, Cs, Máyer, J., Jahnke, G.G., 2009. Assessment of leaf-galling incidence on European grapevines in Badascony, Hungary. Acta Horticult. 816, 97–104.

Nevo, E., Coll, M., 2001. Effect of nitrogen fertilization on *Aphis gossypii* Homoptera: Aphididae): variation in size, color, and reproduction. J. Econ. Entomol. 94, 27–32.

Niklowitz, W., 1954. Histologische studien an Reblausgallen und Reblausabwehrnekrosen. Phytopath. Z. 24, 299–340.

Nougaret, R.L., Lapham, M.H., 1928. A study of phylloxera infestation in California as related to types of soils. US Dept. Agric. Tech. Bull. 20, 1–39.

NVHSC, 2009. National Phylloxera Management Protocol. National Vine Health Steering Committee, Australia, 32pp.

Nysterakis, F., 1946. Nouvelle interprétation de la formation des cécidies. C.R. Seances Acad. Sci. Ser. III, Sci. Vie. 222, 133–1134.

Omer, A.D., Granett, J., 2000. Relationship between grape phylloxera and fungal infections in grapevine roots. J. Plant Dis. Protect. 107, 285–294.

Omer, A.D., Granett, J., De Benedictus, J.A., Walker, M.A., 1995. Effects of fungal root infections on the vigour of grapevines infested by root feeding grape phylloxera. Vitis 34, 165–170.

Omer, A.D., Granett, J., Downie, D., Walker, M.A., 1997. Population dynamics of grape phylloxera in Californian vineyards. Vitis 36, 199–205.
Omer, A.D., Granett, J., Wakeman, R.J., 1999a. Pathogenicity of *Fusarium oxysporum* on different rootstocks. J. Phytopathol. 147, 433–436.
Omer, A.D., Granett, J., Shebelut, C.W., 1999b. Effect of attack intensity on host utilization in grape phylloxera. Crop Protect. 18, 341–347.
Omer, A.D., Granett, J., Kocsis, L., Downie, D., 1999c. Preference and performance response of California grape phylloxera to different *Vitis* rootstocks. J. Appl. Ent. 123, 341–346.
Omer, A.D., Granett, J., Walker, M.A., 2002. Influence of plant growth stage on grape phylloxera (Homoptera, Phylloxeridae) populations. Environ. Entomol. 31, 120–126.
Ortiz-Rivas, B., Martínez-Torres, D., 2010. Combination of molecular data support the existence of three main lineages in the phylogeny of aphids (Hemiptera: Aphididae): the basal position of the subfamily Lachninae. Mol. Phylogenet. Evol. 55, 305–317.
Palmer, L.J., Palmer, L.T., Pritchard, J., Graham, R.D., Stangoulis, J.C., 2013. Improved techniques for measurement of nanolitre volumes of phloem exudate from aphid stylectomy. Plant Methods 9, 18.
Paré, P.W., Tumlinson, J.H., 1996. Plant volatile signals in response to herbivore feeding. Florida Entomol. 79, 92–103.
Park, S.J., Huang, Y.H., Ayoubi, P., 2006. Identification of expression profiles of sorghum genes in response to greenbug phloem-feeding using cDNA subtraction and microarray analysis. Planta 223, 932–947.
Parker, N.J.B., 1974. A method for culturing root aphids. Entomol. Monthly Magazine 110, 229–231.
Passioura, J.B., 2006. The perils of pot experiments. Funct. Plant Biol. 33, 1075–1079.
Pavloušek, P., 2012. Screening of rootstock hybrids with *Vitis cinerea* Arnold for phylloxera resistance. Central Eur. J. Biol. 7, 708–719.
Pelet, F., Hildebrandt, A.C., Riker, A.J., Skoog, F., 1960. Growth *in vitro* of tissues isolated from normal stems and insect galls. Am. J. Bot. 47, 186–195.
Peng, Z., Miles, P.W., 1988. Studies on the salivary physiology of plant bugs: function of the catechol oxidase of the rose aphid. J. Insect Physiol. 34, 1027–1033.
Petri, L., 1907. Studi sul marciume delle radici nelle viti fillosserate. Memoire R. Staz. Pathol. Veget. 7, 56–65.
Pointeau, S., Ameline, A., Laurans, F., Sallé, A., Rahbé, Y., Bankhead-Dronnet, S., Lieutier, F., 2012. Exceptional plant penetration and feeding upon cortical parenchyma cells by the woolly poplar aphid. J. Insect Physiol. 58, 857–866.
Pollard, D.G., 1973. Plant penetration by feeding aphids (Hemiptera, Aphidoidea): a review. Bull. Ent. Res. 62, 631–714.
Ponsen, M.B., 1997. A histological description of the alimentary tract and related organs of Phylloxeridae (Homoptera, Aphidoidea). Wageningen Agric. Univ. Papers 97–1, 1–77.
Ponsen, M.B., 2006. A histological description of the alimentary tract and related organs of Adelgidae (Homoptera, Aphidoidea). Wageningen Agric. Univ. Papers 06–1, 1–103.
Pontius, J.A., Hallett, R.A., Jenkins, J.C., 2006. Foliar chemistry linked to infestation and susceptibility to Hemlock Woolly Adelgid (Homoptera: Adelgidae). Environ. Entomol. 35, 112–120.
Porten, M., Huber, L., 2003. An assessment method for the quantification of *Daktulosphaira vitifoliae* (Fitch) (Hem., Phylloxeridae) populations in the field. J. Appl. Entomol. 127, 157–162.
Powell, K.S., 2008. Grape phylloxera: an overview. In: Johnson, S.N., Murray, P.J. (Eds.), Root Feeders: An Ecosystem Perspective. CAB International, Oxon, pp. 96–114.

Powell, K.S., 2012. A holistic approach to future management of grapevine phylloxera. In: Bostanian, N., Vincent, C., Isaacs, R. (Eds.), Arthropod Management in Vineyards: Pests, Approaches and Future Directions. Springer Science and Business Media BV, London, UK, pp. 239–251.

Powell, K.S., Korosi, G.A., 2013. Resistant rootstocks—making the right choice to protect against endemic strains of grapevine phylloxera. In: Proc 15th Australian Wine Industry Technical Conference, Sydney, Australia. Winetitles, Sydney, (in press).

Powell, K.S., Brown, D., Dunstone, R., Hetherington, S., Corrie, A., 2000. Population dynamics of phylloxera in Australian vineyards and implications for management. In: Powell, K.S., Whiting, J. (Eds.), Proceedings of the International Symposium on Grapevine Phylloxera Management. Department of Natural Resources and Environment, Melbourne, Australia, pp. 7–19.

Powell, K.S., Slattery, W.J., Deretic, J., Herbert, K.S., Hetherington, S., 2003. Influence of soil type and climate on the population dynamics of grapevine phylloxera in Australia. Acta Horticult. 617, 33–37.

Powell, K.S., Burns, A., Norng, S., Granett, J., McGourty, G., 2006. Influence of composted green waste on the population dynamics and dispersal of grapevine phylloxera *Daktulosphaira vitifoliae*. Agric. Ecosyst. Environ. 119, 33–38.

Powell, K.S., Trethowan, C.J., Blanchfield, A.L., Norng, S., 2007. Composted winery waste and its influence on grape phylloxera in ungrafted vineyards. Acta Horticult. 733, 143–150.

Powell, K.S., Korosi, G.A., Mackie, A.M., 2009. Monitoring grape phylloxera populations using simple non-destructive trapping systems. Acta Horticult. 816, 29–34.

Powell, K.S., Bruce, R.J., Korosi, G.A., 2013a. Assessing the risk of phylloxera survival during white grape processing. VIth International Phylloxera Symposium, Bordeaux-France, August 28-30, 2013. INRA, Bordeaux. p. 23.

Powell, K.S., Korosi, G.A., Clingeleffer, P., Walker, R., Thomas, M.R., Smith, H., Smith, B., 2013b. Genetic markers for phylloxera resistance: preliminary observations using two endemic phylloxera genetic strains. VIth International Phylloxera Symposium, Bordeaux-France, August 28-30, 2013. INRA, Bordeaux. p. 28.

Printz, Y.I., 1937. Contribution to the question of the changes in the virulence of phylloxera lignin of different biotypes. Plant Protect. Leningrad 12, 137–142.

Qing, Z., Yuanpeng, D., Yan, D., Heng, Z., 2011. Changes of secondary metabolites content and phenolics metabolism-associated enzymes in phylloxera infested nodosities. Acta Horticult. 904, 67–76.

Quaglia, F., Rossi, E., 1987. Susceptibility of some cultivars of ungrafted European grape–vines to the foliar attack of the grape phylloxera (*Viteus vitifoliae* (Fitch)), observations of the triennium 1982–1984. In: Integrated Pest Control in Viticulture. Proceedings of a Meeting of the EC Experts Group, Portoferraio. Commission of the European Communities. A. A. Balkema, Rotterdam, The Netherlands.

Raman, A., 2011. Insect–plant interactions: the gall factor. In: Seckbach, J., Dubinsky, Z. (Eds.), All Flesh Is Grass. Cellular Origin, Life in Extreme Habitats and Astrobiology, vol. 16. Springer, pp. 119–146.

Raman, A., Beiderbeck, R., Werner, H., 2009. Early subcellular responses of susceptible and resistant *Vitis* taxa to feeding by grape phylloxera *Daktulosphaira vitifoliae*. Bot. Helvet. 119, 31–39.

Ramming, D.W., 2010. Greenhouse screening of grape rootstock populations to determine inheritance of resistance to phylloxera. Am. J. Enol. Vitic. 61, 234–239.

Raspi, A., Antonelli, R., 1987. Grape phylloxera (*Viteus vitifoliae* (Fitch)) infestation on American vine leaves in a nursery at San Pietro a Grado (Pisa) during the years 1982–1983. In: Integrated Pest Control in Viticulture. Proceedings of a Meeting of the EC Experts Group, Portoferraio. Commission of the European Communities. A. A. Balkema, Rotterdam, The Netherlands.

Reisenzein, H., Baumgarten, A., Pfeffer, M., Aust, G., 2007. The influence of soil properties on the development of the grape phylloxera population in Austrian viticulture. Acta Horticult. 733, 13–23.

Renzullo, L.J., Held, A.A., Powell, K.S., Blanchfield, A.L., 2004. Remote sensing phylloxera infestation: current capabilities and future possibilities for early detection. Aust. NZ. Grapegrower Winemaker 485a, 126–130.

Renzullo, L.J., Blanchfield, A.L., Guillermin, R., Powell, K.S., Held, A.A., 2006a. Comparison of prospect and HPLC estimates of leaf chlorophyll contents in a grapevine stress study. Int. J. Remote Sens. 27, 817–823.

Renzullo, L.J., Blanchfield, A.L., Powell, K.S., 2006b. A method of wavelength selection and spectral discrimination of hyperspectral reflectance spectrometry. IEEE Trans. GeoSci. Remote Sens. 44, 1986–1994.

Repka, V., Fischerova, I., Silharova, K., 2001. Methyl jasmonate induces a hypersensitive–like response of grapevine in the absence of a virulent pathogens. Vitis 40, 5–10.

Riley, C.V., 1875. The grape phylloxera. In: Seventh Annual Report of the State Entomologist, St Louis, Missouri, pp. 90–121.

Riley, C.V., 1879. The nervous system and salivary glands of phylloxera. Psyche 2, 225–228.

Rilling, G., 1960. Das Skelettmuskelsystem der ungeflügelten Reblaus (*Dactylosphaera vitifolii* Shimer). Vitis 2, 222–240.

Rilling, G., 1967. Die Speicheldrusen der Reblaus (*Dactylosphaera vitifolii* Shimer). Vitis 6, 136–150.

Rilling, G., Radler, F., 1960. Die kontrollierbare Aufzucht der Reblaus auf Gewebekulturen von Reben. Naturwissenschaften 23, 547–548.

Rilling, G., Steffan, H., 1972. Untersuchungen zur Physiologie der Reblaus (*Dactylosphaera vitifolii* Shimer): Stoffwechsel von Saccharose–^{14}C(U) in Beziehung zu Reblaustyp und Umweltfaktoren [Investigation into the physiology of the vine aphid (*Dactylosphaera vitifolii* Shimer), metabolism of sucrose–^{14}C(U) in relation to aphid form and environmental factors.]. Z. Angew. Entomol. 72, 43–58.

Rilling, G., Rapp, A., Steffan, H., Reuther, K.H., 1974. Freie und gebundene Aminosauren de Reblaus (*Dactylosphaera vitifolii* Shimer) und Moglichkeiten ihrer Biosynthese aus Saccharose-^{14}C (U). Z. Angew. Entomol. 77, 195–210.

Rilling, G., Rapp, A., Reuther, K.-H., 1975. Veränderungen des Aminosäurengehaltes von Rebenorganen bei Befall durch die Reblaus (*Dactylosphaera vitifolii* Shimer). Vitis 14, 198–219.

Ritter, A., Vorwerk, S., Blaich, R., Forneck, A., 2007. Adaptational potential of grape phylloxera (*Daktulosphaira vitifoliae*) clonal lineages. Mitt. Klosterneuburg 57, 116–122.

Rosen, H.R., 1916. Development of the *Phylloxera vastatrix* leaf gall. Am. J. Bot. 3, 337–360.

Rossi, E., 1988. La filossera della vite. Info. Fitipat. 9, 17–22.

Roush, T.L., Granett, J., Walker, M.A., 2007. Inheritance of gall formation relative to phylloxera resistance levels in hybrid grapevines. Am. J. Enol. Vitic. 58, 234–241.

Ryan, F.J., Omer, A.D., Aung, L.H., Granett, J., 2000. Effects of infestation by grape phylloxera on sugars, free amino acids, and starch of grapevine roots. Vitis 39, 175–176.

Sakai, H., Tsutsumi, Y., Kawai, A., Sato, S., Takano, T., Takahashi, T., 1985. Methyl bromide fumigation and hot water treatment of grapevine stocks against the grape phylloxera, *Viteus vitifoliae* Fitch. Res. Bull. Plant Protect. Serv. Jpn. 21, 67–69.

Salt, D.T., Fenwick, P., Whittaker, J.B., 1996. Interspecific herbivore interactions in a high CO_2 environment: root and shoot aphids feeding on Cardamine. Oikos 77, 326–330.

Schaefer, H., 1972. Über Unterschiede im Stoffwechsel von reblausvergallten und gesunden Rebenblättern. J. Phytopathol. 75, 285–314.

Schaefer, H., 1985. Metabolic differences between phylloxera root galls and healthy grapevine roots. Wein–Wissenschaft 40, 219–227.

Schäller, G., 1960. Untersuchungen über den Aminosäuregehalt des Speicheldrüsensekretes der Reblaus (*Viteus* [Phylloxera] *vitifolii* Shimer), Homoptera. [The amino acid content of the salivary gland secretion of the vine phylloxera, *Viteus vitifolii* Shimer, Homoptera]. Entomol. Exp. Appl. 3, 128–136.

Schäller, G., 1963a. Untersuchungen über die Abhängigkeit der Gallenbildung und Nekrosereaktion vom physiologischen Zustand der Wirtspflanzen am Beispiel Rebe—Reblaus. J. Phytopathol. 46, 269–275.

Schäller, G., 1963b. Biochemische Rassentrennung bei der Reblaus (*Viteus vitifolii* Shimer) durch Speichelanalysen. Zool. Jahrb. Physiol. 70, 278–283.

Schimpf, N., Matthews, P.G.D., Wilson, R.S., White, C.R., 2009. Cockroaches breathe discontinuously to reduce respiratory water loss. J. Exp. Biol. 212, 2773–2780.

Schneider-Orelli, O., Leuzinger, H., 1924. Vergleichende Untersuchungen zur Reblausfrage. Beibl. zur Viertjahrsschrift. 39, 1–51.

Sleezer, S., Johnson, D.T., Lewis, B., Goggin, F., Rothrock, C., Savin, M., 2011. Foliar grape phylloxera, *Daktulosphaira vitifoliae* (Fitch) seasonal biology, predictive model, and management in the Ozarks region of the United States. Acta Horticult. 904, 151–156.

Sobetskiy, L.A., Derzhavina, M.A., 1973. A contribution to the study of the physiology of the feeding of the vine phylloxera, *Viteus vitifolii* FITCH (Homoptera, Phylloxeridae). Entomol. Rev. 52, 357–361.

Song, G.-C., Granett, J., 1990. Grape phylloxera (Homoptera, Phylloxerideae) biotypes in France. J. Econ. Entomol. 83, 489–493.

Stellwaag, F., 1928. Die Weinbauinsekten der Kultureländer. Paul Parey, Berlin.

Sterling, C., 1952. Ontogeny of the phylloxera gall of grape leaf. Am. J. Bot. 39, 6–15.

Stevenson, A.B., 1964. Seasonal history of root-infesting *Phylloxera vitifoliae* (Fitch) (Homoptera: Phylloxeridae) in Ontario. Can. Entomol. 96, 979–987.

Stevenson, A.B., 1970. Strains of the grape phylloxera in Ontario with different effects on the foliage of certain grape cultivars. J. Econ. Entomol. 63, 135–138.

Stonerod, P., Strik, B., 1996. Hot-water dipping eradicates phylloxera from grape nursery stock. HortTechnology 6, 381–383.

Strapazzon, A., Girolami, V., 1983. Foliar phylloxera infestations (*Viteus vitifoliae* (Fitch)) with completion dell'olociclo on *Vitis vinifera* (L.) engaged. Redia 66, 179–194.

Sullivan, V., 1996. New rootstocks stop vineyard pest for now. Calif. Agric. 50, 7–8.

Sun, Q., Chen, Y., Wang, H.B., Downie, D.A., Zhai, H., 2009. Origin and genetic diversity of grape phylloxera in China. Acta Entomol. Sinica 52, 885–894.

Szklarzewicz, T., Jankowska, W., Wieczorek, K., Węgierek, P., 2009. Structure of the ovaries of the primitive aphids *Phylloxera coccinea* and *Phylloxera glabra* (Hemiptera, Aphidinea, Phylloxeridae). Acta Zool. 90, 123–131.

Tjallingii, W.F., 1978. Electronic recording of penetration behaviour by aphids. Entomol. Exp. Appl. 24, 521–530.

Trethowan, C.J., Powell, K.S., 2007. Rootstock-phylloxera interactions under Australian field conditions. Acta Horticult. 733, 115–122.

Tucker, D., Lamb, D., Powell, K.S., Blanchfield, A.L., Brereton, I.M., 2007. Detection of phylloxera infestation in grapevines by NMR methods. Acta Horticult. 733, 173–181.

Turley, M., Granett, J., Omer, A.D., De Benedictis, J., 1996. Grape phylloxera (Homoptera, Phylloxerideae) temperature threshold for establishment of feeding sites and degree-day calculations. Environ. Entomol. 25, 842–847.

Umina, P.A., Corrie, A.M., Herbert, K.S., White, V.L., Powell, K.S., Hoffmann, A.A., 2007. The use of DNA markers for pest management-clonal lineages and population biology of grape phylloxera. Acta Horticult. 733, 183–195.

Van Heeswijck, R., Bondar, A., Croser, L., Franks, T., Kellow, A., Powell, K.S., 2003. Molecular and cellular events during the interaction of phylloxera with grapevine roots. Acta Horticult. 617, 3–16.

Varkonyi–Gasic, E., Gould, N., Sandanayaka, M., Sutherland, P., MacDiarmid, R.M., 2010. Characterisation of microRNAs from apple (*Malus domestica* 'Royal Gala') vascular tissue and phloem sap. BMC Plant Biol. 10, 159.

Velasco, R., Zharkikh, A., Troggio, M., Cartwright, D.A., Cestaro, A., et al., 2007. A high quality draft consensus sequence of the genome of a heterozygous grapevine variety. PLoS ONE 2 (12), e1326. http://dx.doi.org/10.1371/journal.pone.0001326.

Vidart, M.V., Mujica, M.V., Bao, L., Duarte, F., Bentancourt, C.M., Franco, J., Scatoni, I.B., 2013. Life history and assessment of grapevine phylloxera leaf galling incidence on *Vitis* species in Uruguay. SpringerPlus. 2, 181.

Vorwerk, S., Forneck, A., 2006. Reproductive mode of grape phylloxera (*Daktulosphaira vitifoliae*, Homoptera, Phylloxeridae) in Europe, molecular evidence for predominantly asexual populations and a lack of gene flow between them. Genome 49, 678–687.

Vorwerk, S., Forneck, A., 2007. Analysis of genetic variation within clonal lineages of grape phylloxera (*Daktulosphaira vitifoliae* Fitch) using AFLP fingerprinting and DNA sequencing. Genome 50, 660–667.

Vorwerk, S., Martinez-Torres, D., Forneck, A., 2007. *Pantoea agglomerans*-associated bacteria in grape phylloxera (*Daktulosphaira vitifoliae*, Fitch). Agric. Forest Entomol. 9, 57–64.

Walker, G.P., Janssen, J.A.M., 2000. Electronic recording of whitefly (Homoptera: Aleyrodidae) feeding and oviposition behavior. In: Walker, G.P., Backus, E.A. (Eds.), Principles and Applications of Electronic Monitoring and Other Techniques in the Study of Homopteran Feeding Behaviour. Thomas Say Publications in Entomology, Lanham, USA, pp. 172–200.

Walker, A., Granett, J., Omer, A., Lin, H., Kocsis, L., Forneck, A., Porten, M., 1998. Are phylloxera feeding on 5C rootstock in Europe? Pract. Winery Vineyard 3, 24–28.

Wapshere, A.J., Helm, K.F., 1987. Phylloxera and *Vitis*: an experimentally testable coevolutionary hypothesis. Am. J. Enol. Vitic. 38, 216–222.

Warick, R., Hildebrant, A., 1966. Free amino acid contents of stem and phylloxera gall tissue cultures of grape. Plant Physiol. 41, 573–578.

Wegner-Kiss, G., 2007. Der reblaus einhalt gebieten. Badische Winzer. May, 24–26.

Wildman, W.E., Nagaoka, R.T., Lider, L.A., 1983. Monitoring spread of grape phylloxera by colour infrared aerial photography and ground investigation. Am. J. Enol. Vitic. 34, 83–94.

Williams, R.N., Shambaugh, G.F., 1987. Grape phylloxera (Homoptera: Phylloxeridae) biotypes confirmed by electrophoresis and host susceptibility. Ann. Entomol. Soc. Am. 81, 1–5.

Williams, C.M., Pelini, S.L., Hellmann, J.J., Sinclair, B.J., 2010. Intra-individual variation allows and explicit test of the hygric hypothesis for discontinuous gas exchange in insects. Biol. Lett. 6, 274–277.

Wöhrle, A., 1999. Versuche zur Kunstlichen Ernahrung von Reblausen (*Dactulosphaira vitifoliae* Fitch). MSc Thesis. Universitat Hohenheim, Stuttgart-Hohenheim, Germany.

Yvon, M., Leclant, F., 2000. Intraspecific variability in grape phylloxera (*Daktulosphaira vitifoliae* Fitch) in France: development of rearing techniques. Acta Horticult. 528, 581–586.

Yvon, M., Peros, J.P., 2003. Variation in aggressiveness and genetic diversity of grape phylloxera in southern France. J. Int. Sci. Vigne Vin. 37, 77–84.

Zhang, J., Hausmann, L., Eibach, R., Welter, L.J., Töpfer, R., Zyprian, E.M., 2009. A framework map from grapevine V3125 (*Vitis vinifera* 'Schiava grossa' × 'Riesling') × rootstock cultivar 'Börner' (*Vitis riparia* × *Vitis cinerea*) to localize genetic determinants of phylloxera root resistance. Theor. Appl. Genet. 119, 1039–1051.

Zhang, H.-G., Liu, C-H., Zhong, X-H., Wang, Z-Y., Sun, H-S., Fan, X-C., Liu, X-M., Shen, H-B., 2010. Investigation on population dynamics of grape phylloxera (*Daktulosphaira vitifoliae* Fitch) in vineyards of Xi'an and Shanghai. Acta Entomol. Sinica 37, 291–296.

Zotov, V.V., 1954. Characteristics of carbohydrate metabolism in the roots of grapes infected with phylloxera. Biokhimiya (Moscow) 19, 649–651.

Zotov, V.V., Gadiyev, R.S., 1975. Stimulators of growth of gall tissues in phylloxera-infected grape plants. Sel'sko-khozyaistvennaya Biol. 10, 241–244.

Zotov, V.V., Svetlyakova, R.I., Sokolovskaya, T.I., Storozhuk, E.M., Kucher, A.A., 1966. Physiology of the resistance of grape to phylloxera. Sel'sko-khozyaistvennaya Biol. 1, 410–420.

Molecular Approaches for Studying Root Herbivores

Carly M. Benefer, Rod P. Blackshaw
Centre for Agricultural and Rural Sustainability, Plymouth University, Plymouth, United Kingdom

Contents

Abstract

The use of molecular techniques in insect ecology has expanded rapidly, allowing ever more challenging questions to be addressed. Compared to their aboveground equivalents, root herbivore molecular ecology has received less attention, despite essentially the same ecological questions being of importance in both the above- and belowground ecosystems. Studies so far have concentrated on economically important taxa, using mitochondrial and nuclear DNA sequencing and a variety of markers to investigate the species identity and relationships, population dynamics and dispersal, distribution, feeding behaviour and interactions with other organisms. Although this has proved useful for elucidating these aspects of their ecology, there remains a need to focus on the functioning of root

Advances in Insect Physiology, Volume 45
ISSN 0065-2806
http://dx.doi.org/10.1016/B978-0-12-417165-7.00005-2

herbivores in the soil ecosystem. Application of new and emerging technologies developed for aboveground systems will increasingly be applied to those belowground, allowing a focus on root herbivore biology and ecology in the context of ecosystem processes and systems ecology. For the foreseeable future, however, the use of molecular techniques is likely to remain dominated by the need to address pragmatic research questions about specific taxa, notably pests.

1. INTRODUCTION

A rise in the number and accessibility of DNA techniques and their application to entomology has led to investigation of, among other aspects, the genetic diversity, phylogeography and species identity and taxonomic relationships of a wide range of invertebrates. Numerous published studies have used molecular markers in tandem with data obtained in the field or laboratory to answer ecological questions, focusing on both applied (e.g. pest management and conservation biology) and theoretical aspects of the ecology of various species, but the vast majority of these are on aboveground insect herbivores, which are generally conspicuous, accessible and easy to manipulate both in the field and the laboratory using well-described methods. For example, only 10 of the 249 articles on insect herbivores published in the journals *Molecular Ecology* and *Molecular Ecology Resources* (formerly *Molecular Ecology Notes*) are on root-feeding taxa. This bias towards the study of aboveground herbivores has been noted (e.g. Blossey and Hunt-Joshi, 2003; Hunter, 2001) and the distribution, abundance, activities and effects of many important belowground species remain underresearched. This reflects a general situation where we know far more about above- than belowground herbivory. A large proportion of the studies that have been carried out on root antagonists focus on nematodes, for which researchers have adopted molecular approaches similar to those used in soil microbiology (e.g. Powers et al., 2011; Waite et al., 2003) to investigate their biology and ecology partially because their microscopic size and morphologically cryptic nature makes this particularly difficult. Many other taxa have been overlooked, yet molecular genetic methods have the potential to overcome some of the difficulties associated with the study of root herbivores and allow the investigation of areas of their ecology for which we currently lack a detailed understanding. This has become increasingly recognised by researchers in this field over the last decade and some important advances have been made, but further opportunities remain, particularly as new methods are developed.

Most molecular genetic techniques revolve around the extraction of DNA from the organism of interest followed by polymerase chain reaction (PCR) to amplify DNA, allowing it to be subsequently manipulated and characterised (Freeland, 2005). Using primers, either universal or species-specific, to target different regions of DNA allows a large degree of flexibility in the questions that can be asked. Mitochondrial DNA (mtDNA) is often used because of its high mutation rate, ease of manipulation and lack of recombination and is therefore useful for investigating diversity within and between populations, for example, for indirectly inferring dispersal from gene flow and determining evolutionary relationships. Molecular markers are used extensively in ecology to quantify levels of genetic variation. Allozymes (enzyme polymorphisms) determined using isoelectric focusing/isoenzyme electrophoresis (separation of proteins using gel electrophoresis) and monoclonal antibodies are two of the first protein-based techniques used in ecological research. Restriction fragment length polymorphisms (RFLPs), based on variation in DNA sequences rather than proteins, and PCR–RFLP, a variation on this targeting specific regions of DNA, are now less common but can still be useful for certain applications. Microsatellites are often used in population genetic studies. These are stretches of DNA that consist of tandem repeats of 1–6 bp, and their use is based on allele size differences between individuals, caused by mutations arising during slippage in replication (Freeland, 2005). Other markers include randomly amplified polymorphic DNA (RAPD; Welsh and McClelland, 1990; Williams et al., 1990), for which PCR is used to generate fragments using a randomly selected 10-bp primer (Ritland and Ritland, 2000) and amplified fragment length polymorphisms (AFLPs), which use restriction digest of genomic DNA followed by amplification of a subset of fragments (Vos et al., 1995). These techniques do not require prior sequence knowledge and so development time can be short (though see Section 3.2), but RAPDs are no longer commonly used for population genetics because of lack of reproducibility and therefore reliability (Freeland, 2005). For many of the previously mentioned methods, the DNA sequence that is being targeted is not known, but DNA sequences obtained through Sanger sequencing (a widely used DNA sequencing method) can be used directly to quantify genetic variation via base-pair differences between individuals. Related to this, single-nucleotide polymorphisms are single-base-pair positions in a sequence that are variable

Table 5.1 Summary of molecular techniques, including target DNA region (where applicable) and their application to the study of root herbivore taxa

Application	Molecular technique	Target DNA region	Taxa	Review section
Predator–prey interactions	Species-specific PCR fragments	*COI*	Scarab larvae (*Melolontha melolontha*)	2.1
			Scarab larvae (*Phyllopertha horticola*)	
	qPCR	*COI/tRNA-Leu*	Corn rootworm (*Diabrotica* spp.)	
		COI	Scarab larvae (*M. melolontha/Amphimallon solstitiale*)	
Plant–herbivore interactions	Species-specific PCR fragments	*rbcL/trnL*plastid/*trn*T-F chloroplast (plants)	Wireworms (*Agriotes* spp.)	2.2
Endosymbionts	Mitochondrial (mtDNA) sequencing	*12S-N4* and *CB2H-C2R*	Corn rootworm (*Diabrotica barberi*)	2.3
		16S rRNA, *ftsZ* and *wsp* (*Wolbachia*)		
		16S rRNA	Wireworms (*Limonius canus/Agriotes lineatus*)	
		Metagenome sequencing	Cutworm (*Agrotis ipsilon*)	
Genetic diversity	Allozymes	Protein-based	Corn rootworm (*Diabrotica* spp.)	3
	RFLPs (PCR–RFLPs, T-RFLP)	*12S-N4* and *CB2H-C2R/ITS1*	Corn rootworm (*Diabrotica* spp.)	

	Microsatellites	–	Corn rootworm (*Diabrotica* spp.)	
			Gall-forming root aphids (*Pemphigus* spp.)	
			Ant-associated root aphids (multiple Aphididae)	
			Root weevil (*Diaprepes abbreviatus*)	
	AFLPs	–	Gall-forming root aphids (*Pemphigus* spp.)	
	RAPDs	–	Root weevil (*Diaprepes abbreviatus*)	
	Mitochondrial (mtDNA) sequencing	*16S rRNA* and *COI*	Root weevil (*Diaprepes abbreviatus*)	
		16S rRNA	Wireworms (multiple Elateridae)	
		COII	Scarab larvae (*Antitrogus parvulus*)	
	Nuclear (nuDNA) sequencing	*ITS1*	Corn rootworm (*Diabrotica barberi*)	
Species identification	Mitochondrial (mtDNA) sequencing	*Cytb*	Leatherjackets (*Tipula* spp.)	4.1
		16S rRNA/COI	Wireworms (multiple Elateridae)	
	RAPDs	–	Gall-forming root aphids (*Pemphigus* spp.)	
	RFLPs (PCR–RFLPs, T-RFLP)	*COII*	Scarab larvae (multiple Scarabidae)	
		16S rRNA	Wireworms (*Agriotes* spp.)	
	Isoelectric focusing	Protein-based	Leatherjackets (*Tipula* spp.)	
	Multiplex PCR	*COI*	Wireworms (*Agriotes* spp.)	

Continued

Table 5.1 Summary of molecular techniques, including target DNA region (where applicable) and their application to the study of root herbivore taxa—cont'd

Application	Molecular technique	Target DNA region	Taxa	Review section
Association of life stages	Mitochondrial (mtDNA) sequencing	*COI*	Gall-forming root aphids (multiple Eriosomatini)	4.2
		COI and *16S rRNA*	Scarab larvae (multiple Scarabidae)	
	Nuclear (nuDNA) sequencing	*28S rRNA*	Scarab larvae (multiple Scarabidae)	
Phylogenetic relationships	Allozymes	Protein-based	Corn rootworm (*Diabrotica* spp.)	4.3
	RFLPs (PCR–RFLPs, T-RFLP)	*NADH 4*	Corn rootworm (*Diabrotica* spp.)	
	Mitochondrial (mtDNA) sequencing	*COI, COII*	Corn rootworm (*Diabrotica* spp.)	
		COI	Gall-forming root aphids (multiple Fordini)	
	Nuclear (nuDNA) sequencing	*ITS1*	Corn rootworm (*Diabrotica* spp.)	
		EF-1 α/LWO	Gall-forming root aphids (multiple Fordini)	
Species distribution	RFLPs (PCR–RFLPs, T-RFLP)	*16S rRNA*	Wireworms, leatherjackets, Bibionid and Sciarid fly larvae	4.4
	Multiplex PCR	*COI*	Wireworms (*Agriotes* spp.)	
Quantification	Monoclonal antibodies	Protein-based	Slugs (*Deroceras reticulatum*)	4.5

These studies have been examined in further detail in the specified review section.

between individuals and provide another way of determining sequence variation at multiple loci. Quantitative polymerase chain reaction (qPCR) can be used to determine the amount of DNA present in a sample, whereby a molecular tag is incorporated into the PCR. When the tag binds to DNA, it fluoresces and this is recorded at each cycle as a fluorometric reading, which is proportional to the amount of DNA present. More recently, next-generation sequencing (NGS) approaches, which allow massively parallel, or deep, sequencing of DNA fragments, have allowed whole genomes or communities to be sequenced in a time- and cost-effective manner, paving the way for larger-scale studies and opening up new opportunities for the study of certain aspects of insect ecology.

The use of these markers depends upon both the question under study and practicality (time and resources) and as such a variety of methods have been used in different disciplines within root herbivore ecology (Table 5.1). This chapter aims to describe and summarise the applications of these techniques, drawing on studies already undertaken and providing suggestions for the focus of future work.

2. SPECIES INTERACTIONS

The nature of the soil environment, and the size of the organisms within it, makes it difficult to study the behavioural ecology of belowground insects without manipulation of conditions in usually less than realistic laboratory experiments. One area in which molecular ecological techniques have helped to overcome these practical issues is feeding ecology. Aboveground insect herbivores have been extensively studied, for example, in terms of host–plant interactions (e.g. Pieterse and Dicke, 2007; Whiteman and Jander, 2010) and predator–prey food webs (Hereward and Walter, 2012), but there are relatively few studies on soil insect herbivores specifically.

Soil organisms have been recognised to play a vital role in nutrient cycling and decomposition and as such have been studied in relation to their role in the soil food web. This has mainly been limited to the mesofauna, bacteria and fungi (through the use of stable isotopes, e.g. Crotty et al., 2011, 2012); however, researchers have started to consider species-specific interactions of the macrofauna in the soil food web (Eitzinger et al., 2013). This has been achieved by gut content analysis, using molecular probes that target specific regions of plant or herbivore DNA present in whole-organism

DNA extracts using PCR, without the need for extraction and dissection of the gut. These PCR products can then be subjected to gel electrophoresis and identified based on the size of the product produced or fluorescently labelled and analysed using fragment analysis using an automated sequencer. Traditionally, monoclonal antibodies (reviewed in Sheppard and Harwood, 2005) and isoenzyme electrophoresis (e.g. Amalin et al., 2000; Traugott, 2003) have been used for gut content analysis, but diagnostic PCR protocols are now more commonly adopted since they are cost- and time-efficient and use widely available equipment, and once the tool is developed, it can be applied by other researchers and/or modified to detect specific DNA targets, depending upon the question under study (Juen and Traugott, 2006).

2.1. Predator–prey interactions

As in studies of aboveground insects (Sheppard and Harwood, 2005), sections of the mtDNA *COI* gene have been targeted in the present studies. This region of DNA is commonly used for species identification through barcoding (see Section 4.1) and contains both conserved and variable regions, meaning universal primers (Folmer et al., 1994) and species-specific primers can be employed. As such, this makes it an easily manipulated target for use in gut content analysis. Juen and Traugott (2005) developed the first PCR-based approach for studying soil insect predator–prey dynamics using white grubs, also known collectively as canegrubs and chafers (Scarabidae larvae; *Melolontha melolontha*), in feeding experiments with larvae of the ground beetle *Poecilus versicolor*, a common above- (as adults) and belowground (as larvae) predator. They tested the effect of digestion time on detection of target DNA of different lengths, using primers designed specifically to amplify *M. melolontha* DNA. Importantly, for future studies, they found that DNA could be detected from even small amounts of predation (i.e. when larvae were fed a single egg) and that the primers were able to amplify DNA from feeding on both carcasses and fresh prey at similar levels. This was not correlated with rate of digestion and did not differ significantly between lengths of fragment targeted (ranging from 175 to 585 bp). It was therefore not possible to distinguish between scavenging and active predation—a possible limitation to this type of study. This protocol was applied in an investigation of the predator guild of another pest scarab, the garden chafer *Phyllopertha horticola,* but using newly designed primers specifically targeting this species (Juen and Traugott, 2007). *P. horticola* DNA was successfully detected in a variety of predators (mostly in Geophilidae and predatory beetle larvae), showing that the larvae are a potentially important prey item in the soil food web and that these taxa could therefore be important natural enemies of this pest. This is especially relevant since earthworm DNA was only found in 2.7% of tested individuals, yet this taxon is known to be a widespread and highly abundant and important food source for soil predators.

Using a slightly different approach, Lundgren et al. (2009) considered predation of *Diabrotica virgifera virgifera* (Coleoptera: Chrysomelidae) larvae using qPCR to detect *D. virgifera* DNA sequences in arthropod predator guts. Primers that amplified a segment of the *COI* and tRNA-*Leu* genes of *D. virgifera* were designed and used together with feeding experiments with various predators. This semiquantitative method not only detects prey DNA but gives an idea of the amount present. It was found that more predators were positive for *D. virgifera* DNA in the egg stage than the larval stage and predators differed in their propensity to consume *D. virgifera*. This suggests that timing might be an important factor in predation (i.e. longevity of egg and larval stages) and that a range of arthropod species (ground beetles, harvestmen, wolf spiders and predaceous mites) are important in regulating this pest, which has implications for agricultural management. However, the study mainly concerned surface-active predators caught in pitfall traps; focusing on predators within the soil column may provide a different picture. Using the same approach, a further study (Lundgren and Fergen, 2011) evaluated whether winter vegetation increased larval predation by predators aboveground and in the soil column. Twenty predator taxa from the soil column and 33 taxa from the soil surface were found to have consumed the larvae, with Carabidae being one of the most abundant in both, showing this pest has a diverse predator community, which suggests conservation efforts should be targeted at this level rather than at specific taxa. qPCR correlated well with pest damage only when a predation index was used taking into account DNA quantity, predator abundance and relative frequency of detection but, combined with predation intensity observations using restrained larvae, gave an overall view of these predator–prey interactions.

Eitzinger et al. (2013) recently developed a PCR assay for the detection of prey DNA in soil-dwelling centipedes (*Lithobius* spp.) from forest soils, targeting a range of invertebrate taxa at the family level, and others have developed assays for soil-dwelling predacious beetle larvae-consuming taxa at other trophic levels (e.g. detritivores: earthworms and Collembola; Eitzinger and Traugott, 2011), which could potentially be adapted for assessing interactions of root herbivores specifically.

As shown in aboveground predator–prey systems, there are various parameters that can affect the ability to detect and quantify prey within predators, such as predator identity and size of the target DNA molecule (Sheppard and Harwood, 2005). These effects are little known for root herbivores, but some aspects have recently been studied by Waldner et al. (2013), who tested prey detection intervals for DNA fragments of different sizes, meal size and relative biomass increase in four different soil-dwelling predators of white grub. Whereas meal size and predator biomass had no significant effect (as for similar aboveground studies), prey detection intervals were affected by the size of DNA fragments, for which smaller intraspecific differences were found for the medium-sized fragment in beetles, and differed between, but were similar within, taxa. These results imply that primer sensitivity and predator identity may affect observed outcomes. In terms of quantification, there are many interacting factors that lead to the final amount of food present in the gut, and as such, it is not possible to make inferences about the number of prey eaten, rather an indication of which species are involved and their relative predation capabilities (Lundgren et al., 2009).

2.2. Plant–herbivore interactions

Studies on the dietary choices of root-feeding insects have also only recently started to be explored. Staudacher et al. (2011b) developed a method for detecting the presence of wheat and maize DNA in wireworms (Coleoptera: Elateridae; click beetle larvae) using a similar protocol as the predation studies previously mentioned but targeting genes that are more relevant for plant identification (*COI* is not thought to be suitable since it evolves too slowly). General primers were developed to amplify fragments of *rbcL* and *trnL* plastid DNA, often used for plant DNA barcoding and also specifically for maize and wheat. The plants could be detected for up to 72 h postfeeding (the maximum tested), and the specific primers made it possible to determine which plant DNA was present because of differentially sized products, though further testing would be needed to determine the postfeeding limits of detection. Following on from this, Wallinger et al. (2013), using a multiplex PCR approach (Wallinger et al., 2012), determined that plant identity, but not wireworm species identity, and level of decay were important in the detection of plant DNA, with decayed material having a lower detection rate than fresh material and wheat detection being significantly decreased over time as compared to maize.

Only one study has applied a developed molecular method, in combination with stable isotopes, to specifically investigate aspects of the ecology of root herbivores, again using *Agriotes* wireworms as a model species (Staudacher et al., 2013b). Using plants that are more or as attractive to pests than the cultivated crop has been proposed to prevent wireworm damage (Landl and Glauninger, 2013; Vernon et al., 2000), and this study investigated wireworm feeding behaviour in the presence of increased plant diversity and whether this could protect the maize crop. The protocol described by Wallinger et al. (2012) was used for gut content analysis, while stable isotopes were used to assess the long-term diet. The presence of a range of plant species did in fact lure the wireworms away from the main crop, increasing its yield, while it was shown that the larvae were actively feeding on the other plant species, switching to these from the maize crop previously consumed. This not only confirms previous observations but also shows that using a mixture of noncrop plants planted at the same time as the crop potentially provides protection from damage. However, some plants were consumed more often than others and this differed between seasons, so further work using these techniques is needed to determine the most appropriate plant mix for optimum results.

2.3. Endosymbionts

The importance of symbionts in insect–plant interactions is becoming increasingly recognised (Frago et al., 2012), partly due to the relative ease in which this can now be studied due to advances in molecular genetic and genomic techniques. Gut symbionts allow adaptation to different food plants, protection from predators or stress and may also promote speciation by causing reproductive and ecological isolation of host populations (Moran et al., 2008). *Wolbachia* species are intracellular, maternally inherited bacterial parasites that can cause cytoplasmic incompatibility (preventing infected males fertilising the eggs of uninfected females), feminisation of males and killing of male embryos. Using primers targeting *Wolbachia* genes, the presence of this endosymbiont was investigated in northern corn rootworm, *Diabrotica barberi* (Coleoptera: Chrysomelidae) populations in north central United States (Roehrdanz and Levine, 2007). Previous studies found two distinct mtDNA clades east and west of this boundary suggesting little mixing of populations (Roehrdanz et al., 2003) and since *Wolbachia* had been detected in populations from the east of this area, this was postulated as a possible cause. mtDNA (*12S-N4* and *CB2H-C2R* genes) was amplified for the corn rootworm to gain mitochondrial haplotypes and primers targeting *Wolbachia* genes (*16S rDNA*, *ftsZ* and *wsp* genes) used to determine its presence in rootworm DNA extracts. Two different strains were detected on either side of the boundary, each associated with a different mtDNA clade, suggesting the strains are not compatible, but testing with nuclear DNA may help to determine whether *Wolbachia* is reducing gene flow and increasing population differentiation or other factors are also involved. Practically, if these delineated populations have different traits affecting their control (e.g. insecticide resistance), this could have an impact on pest management programmes.

Some endosymbionts have been considered for genetic manipulation as biocontrol agents, especially for root-feeding pests such as wireworms for which pesticide application is difficult and broad-spectrum pesticides are now limited. Microbial agents have been trialled for wireworms in the past with limited success, but a modified form of known symbionts in certain species may increase their efficacy. Lacey et al. (2007) carried out a survey, using *16S rRNA*, of gut bacteria associated with *Limonius canus* wireworms, which are often associated with potato damage, for this purpose. A number of potential control candidates were found, including species that have already been manipulated to produce insecticidal toxins and antagonistic rhizosphere-associated species, which could be added to seed dressings or used in a trap cropping approach (*Rahnella aquatilis* is common in the wheat rhizosphere). Other bacterial species have also been isolated and tested for their pathogenicity using the same approach for *Agriotes lineatus* wireworms, though testing in the field is needed to confirm their efficacy in control (Danismazoglu et al., 2012).

Metagenomic analysis of gut symbionts in three different herbivorous insects, grasshopper (*Acrida cinerea*; Orthoptera), cutworm (*Agrotis ipsilon*; Lepidoptera) and termite (*Nasutitermes* sp.; Isoptera: Termitidae), recently revealed that gut symbiont composition was related to their function in biomass degradation and nutrient biosynthesis (Shi et al., 2013). Furthermore, gut symbionts and insects coevolved based on insect food preferences in favour of optimal biomass degradation, nutrient utilisation and other species-specific factors related to lifestyle. For example, cutworm contained a higher diversity of symbionts than the other herbivores, possibly related to its generalist feeding habit. This deep-sequencing approach revealed many more symbionts than traditional techniques such as denaturing gradient gel electrophoresis followed by *16S rRNA* sequencing, including many unculturable (but taxonomically described) species.

3. GENETIC DIVERSITY

Investigating the genetic diversity within and between populations provides much insight into the biology and ecology of species and is an important contributor to the long-term conservation of populations, particularly under environmental change. Though it may be relatively straightforward to distinguish populations of aboveground insects, for which boundaries can be more obvious, the soil represents a continuous habitat, which makes it difficult to determine where one population ends and another begins or indeed whether organisms are spatially structured in the same way as they are aboveground. Dispersal is a key mechanism for maintaining genetic diversity and viable populations, and for pest species, knowledge of this is particularly relevant for devising management strategies, taking into account spatial and temporal aspects of species' biology.

At present, little is known of invertebrate movement within the soil. Though some sophisticated techniques have been employed to look at the *in situ* movement of soil-dwelling invertebrates, for example, X-ray microtomography (Johnson et al., 2004) and controlled laboratory and field experiments (Murray et al., 2010; Williams, 2012), this fails to provide information on the extent of movement within and between populations in the field. Using stable isotopes, Schallhart et al. (2011) attempted to track the movement of *Agriotes obscurus* wireworms (click beetle larvae) between fields, though none was detected (probably because there was a food supply present). However, since many root herbivores are insect larvae with aboveground adult stages, which can be assessed more easily, the majority of studies carried out to date have focused on adults. Few root-feeding taxa have been studied but, where they have, several techniques have been used to determine the population genetics of economically or ecologically important species.

3.1. Corn rootworm

Molecular markers have been utilised to investigate the population genetics of corn rootworm, which comprise a complex of *Diabrotica* species that are of particular economic importance in maize systems; larvae feed on crop roots belowground, while adults inhabit the aboveground plant. Early studies using allozymes (Krysan et al., 1989; McDonald et al., 1985) and PCR–RFLP (based on the nuDNA *ITS1* region; Szalanski et al., 1999) found low levels of differentiation in geographically isolated populations of western corn rootworm (*D. v. virgifera;* WCR), Mexican corn rootworm (*D. v. zeae*; MCR) and northern corn rootworm (*D. barberi*; NCR), suggesting high dispersal ability and limited barriers to gene flow. In the PCR–RFLP study, it was also found that for some mitochondrial regions (*18S* and *5.8S rRNA*), there was no variation between subspecies, suggesting a recent common evolutionary history. Because of the lack of variation at these mitochondrial regions, microsatellite markers have since been developed for all three species (Kim and Sappington, 2004; Kim et al., 2008; Waits and Stolz, 2008). Kim and Sappington (2005) used these microsatellites for the WCR and found that there were high levels of genetic diversity but little differentiation between populations (though there was a significant correlation between geographic and genetic distances overall). The authors concluded that although it is possible these beetles have a high migration rate, it is more probable that populations have had insufficient time to become significantly genetically different since their expansion eastwards across the United States around 50 years previously. More recently, Chen et al. (2012) used microsatellites for WCR to genotype populations with high, intermediate and low levels of resistance to methyl parathion and aldrin, finding significant differentiation between populations with high and low levels of resistance and between geographically separated populations. As well as showing how genetic and biological species information can be combined to increase understanding of species' ecology, the long-term study of corn rootworm also shows how marker choice has evolved and is important in quantifying genetic variation at different organisational levels.

3.2. Root-associated aphids

Due to their status as major agricultural pests, much work has been carried out on the genetic diversity and evolutionary history of aphids and their association with crops (Hales et al., 1997). There are, however, root-associated species, which have attracted less attention, but are interesting because of their life cycle. The lettuce root aphid (*Pemphigus bursarius*), for example, sexually reproduces on black poplar and winged forms then migrate to members of the Compositae (including lettuce) as secondary hosts where they parthenogenetically reproduce on the roots. After several generations, winged forms emerge and they migrate back to the primary host (Miller et al., 2003). Microsatellites have been developed (Miller et al., 2000) and applied (Miller et al., 2003) to investigate the genetic structure of populations on the primary and secondary hosts. The population structure of this species was found to be irregular. Several sample locations showed significantly different allele frequencies when compared, but there was no significant relationship between genetic and geographic distances. Excess homozygosity was observed at a number of sites (i.e. populations were not in Hardy–Weinberg equilibrium), and inbreeding and local bottlenecks, based on observational data of galls on poplar, which contained no aphids but sometimes anthocorid (Heteroptera: Anthocoridae) predators, were put forward as reasons for this spatial structure. In addition, there were allele frequency differences between populations on poplar and lettuce, which may be accounted for by secondary host races and occurrence of anholocyclic (viviparous females give birth to only viviparous females, clonally reproducing) and holocyclic (viviparous females produce oviparous females or males, which go on to sexually reproduce). In this case, the use of microsatellite markers increased the understanding of the population dynamics of this species, though further work is needed to determine the mechanisms, which account for the structure that was found. Since there are differences in the dispersal abilities of above- and belowground organisms, particularly those such as aphids that are wind-dispersed aboveground, the spatial scales of study and dispersal may differ and as such this should be taken into account in future studies.

Understanding the population dynamics of species with this type of life cycle becomes more complicated when two morphologically similar species are involved. In this case, Chen et al. (2009) used both RAPDs and AFLPs to differentiate species and determine the population dynamics and occurrence of *Pemphigus populitransversus* and *Pemphigus obesinymphae*, which have a primary poplar host and secondary cruciferous host, namely, cabbage, on which they are important pests in Texas. The occurrence of two separate species was confirmed (previously only determined using morphology) and they were found at different times of the year but co-occurred for extended periods, having implications for previous studies that assumed one, continuously present species. Most aboveground alate adults, determined morphologically, were *P. obesinymphae* and clearer RAPD banding patterns were found for this species in the soil, which the authors suggest could be down to their greater presence. However, many factors can affect the efficiency of the PCR and the increased band intensity may therefore be attributable to causes other than simply more individuals of this species being present. Methodologically, RAPDs proved to be superior to AFLPs; this technique was more efficient, cheaper and with higher numbers of polymorphic bands (many were missing from the AFLPs in tests with DNA of known species). These molecular methods were particularly useful for determining the apterous soil stage of the species, which is more difficult to identify morphologically, but required fresh samples for the best results (those preserved in 100% ethanol contained less and lower-quality DNA).

Other root-associated aphid species also have interesting life histories, which have consequences for their genetic diversity and population structure that may be in contrast to aboveground species. The yellow meadow ant *Lasius flavus* found in grasslands in Europe tends a number of aphid species in exchange for honeydew (and also eats them). Ivens et al. (2011) developed 26 microsatellite markers for studying the effects of these mutualistic relationships on the population genetic structure of four species of ant-associated aphids and then applied them (Ivens et al., 2012b) to determine the extent of dispersal, mode of reproduction and potential for dispersal between nests by winged forms. All species were found to be anholocyclic with few alate individuals and there was a strong correlation between genetic and geographic distances for two species, which confirms previous observations and suggests that there is limited dispersal between nests. However, some dispersal was observed since some clonal lineages were spread along the sampling transect. As with *Pemphigus* species, sexual reproduction usually occurs on the primary host, but these were not present nearby, which could have had an effect on the results. Their presence in the underground nests all year may have had an impact too and is in contrast to their aboveground equivalents, which often maintain a holocyclic lifestyle, highlighting how it is important to study these belowground populations independently. A further study (Ivens et al., 2012a) looked into the spatial structure in more detail and found that more than half of the mounds contained a single species with a single clone, but where multiple clones did exist, they were spatially separated. As well as giving information on aphid biology, it gives much insight into the behaviour of these ants, which seem to eat the younger instars and maintain lower levels of honeydew-producing adults in a similar way that humans farm livestock.

3.3. Root weevil

The genetic diversity of other beetle species has also been investigated. The root weevil (*Diaprepes abbreviatus*) is a serious invasive pest introduced from the Caribbean to the United States in the 1960s, since spreading and becoming associated with more than 300 plant species (Ascunce et al., 2008). Larvae feed on roots and adults on leaves. Studies on populations in Florida using esterase polymorphisms and RAPD–PCR markers (Bas et al., 2000) produced different results dependent on the marker used; esterase and protein can be changed by physiological analysis (they depend upon active genes) and optimum PCR conditions are critical to the success of RAPD markers, reducing their reliability for population genetic studies. Since then, eight microsatellite markers have been developed (though published studies using them are not yet available; Ernst et al., 2006).

Finally, *16S rRNA* and *COI* mitochondrial markers have been used to characterise populations and determine dispersal in Florida and Dominican populations, their supposed origin (Ascunce et al., 2008). There was low genetic diversity and one haplotype in each of six sampled populations, with low numbers of substitutions between the three haplotypes found in total, suggesting a recent introduction. That the same haplotype was found in the east and west, despite populations being discontinuous, suggests some form of human transportation on infected plants (movement has been shown to be slow and localised previously). The authors suggest that using fine-scale neutral markers such as microsatellites may help to resolve the differences seen using between mtDNA and esterase polymorphism and RAPDs, which suggest more widely differentiated populations than *COI* or *16S rRNA* sequences. None of the *COI* sequences for the Dominican samples matched those from Florida, suggesting these may have been imported from another Caribbean island. However, there was some uncertainty in the morphological identification and a high genetic distance between species varieties found in Dominica and Florida within this genus, which requires further analysis.

3.4. Wireworms

A similar study to that of Ascunce et al. (2008) has been carried out for wireworms in Canada (Benefer et al., 2013). However, unlike all previous studies on the genetic diversity of root herbivores, this focused on the damaging, soil-inhabiting larval stage rather than the aboveground adults. Wireworms are click beetle larvae that feed on crop roots, causing high levels of economic damage in some areas. There are several invasive species in Canada (including those in the genus *Agriotes*, which are the main UK pest species) for which information on the distribution and species identity is lacking. Phylogenetic analyses were conducted using mtDNA *16S rRNA* to assess genetic variation in populations across Canada. While there was low intraspecific variation for most of the species (low sample size prevented further analysis or strong conclusions to be drawn for these), there was relatively high variation between *Hypnoidus bicolor* individuals and a significant relationship between genetic and geographic distances, with genetic separation of populations from the east to west of Canada. Further investigation revealed that there may be two different, cryptic *Hypnoidus* species present. There was also some evidence for genetic variation related to geographic location for *Limonius californicus*, for which samples from British Columbia in the west were separated from those in Saskatchewan and Alberta in the mid to midwest areas. Since the distance between these locations is large (up to 1460 km), the opportunity for populations to mix is probably limited. High cryptic diversity and possible nuclear mtDNA (numts) were limitations to the study, but it provided preliminary data as a starting point for further investigation into pest species identity and distribution in Canada.

Like for root weevil, it is recognised that it would be advantageous to have fine-scale data on wireworm population genetics. Previous attempts had been made to develop microsatellite markers at Plymouth University (Benefer, 2011; J. Ellis, personal communication) for *A. lineatus*, *A. obscurus* and *A. sputator*, common European pests, without success; none were found to be polymorphic on initial screening. Benefer (2011) then developed AFLP markers to investigate population structure and dispersal of these species in the United Kingdom, but due to numerous methodological problems (including lack of reproducibility between runs), an optimised set of markers could not be produced. Work is now under way to develop microsatellite markers using NGS data for the same species (C. Benefer, personal communication), whereby fragment libraries were generated from known adult genomic DNA and all three sequenced on a 454 GS FLX platform, the data mined for microsatellite loci and primers designed using freely available software. This protocol cuts out the lengthy development time associated with traditional cloning techniques and produces hundreds of loci, which can be selected depending on their likely ability to amplify the target loci (taking into account primer and loci attributes). So far, 20–25 microsatellite loci per species have been developed using this protocol and are being screened across individuals.

3.5. Canegrubs

In a follow-up study to that of Miller et al. (1999) where pest scarab larvae were identified using PCR–RFLP and genetic variation was found to correspond with species distributions (see Section 4.1), Miller and Allsopp (2005) investigated the phylogeography of the scarab beetle *Antitrogus parvulus* in southeastern Queensland. The females of this species are thought to be poor dispersers, mating and laying eggs close to the emergence site, while males are strong fliers. This type of mating behaviour and other ecological factors play a role in population structure, having implications for pest management, and as such, the effect of habitat fragmentation and maternal dispersal ability were focused on here, using cytochrome oxidase II (*COII*) sequences analysed by analysis of molecular variance (AMOVA), nucleotide divergence (as a measure of sequence divergence among individuals between populations), Mantel test for isolation by distance and neighbour-joining (NJ) phylogenetic analysis. Most of the genetic variation was accounted for at the regional or population level, while intrapopulation variation was low, and there was a highly significant relationship between geographic distance and gene flow, with migration limited by geographic distance. This was reflected in the NJ tree in which populations from the same regions were clustered together, sharing more than one haplotype. These results suggest that females do not move far, but when they do, they move to the next closest soil patch. Two populations were anomalous, with high gene flow compared to others, possibly due to other factors, for example, wind flow, playing a role in their dispersal. Other populations were found to be somewhat genetically distinct despite their close proximity due to geographic barriers. Soil-type preference and rainfall/moisture could also be important, as seen in similar species, which suggests knowledge of dispersal ability and ecological preferences may help in targeting management practices for scarab pests in this region.

4. SPECIES IDENTIFICATION AND PHYLOGENY

Some of the studies already mentioned have shown how important it is to be able to identify the species concerned correctly (see Section 3). This has been made possible for many insect taxa through the sequencing of diagnostic regions of DNA that are variable between species. Though other DNA regions can and are used, this DNA barcoding approach commonly refers to the use of a 648 bp region of the mtDNA *COI* gene. This has been somewhat controversial, mainly where it is applied to the identification of new species because of the unpredictable range of intraspecific variation (Blaxter et al., 2005; Hebert et al., 2003; Janzen, 2004; Taylor and Harris, 2012). Nevertheless, it has been extensively applied to a range of invertebrate taxa, again mainly aboveground, providing information on species ecology that was previously inaccessible. For soil-dwelling species, the use of such barcoding methods provides a relatively quick, easy and affordable way of identifying morphologically similar or cryptic species that usually require a great deal of skill and time using microscopy and that for many immature stages is not possible at all, such as for many Coleopteran and Dipteran larvae, or microscopic and mesofaunal organisms. Species identification through the use of sequence data has been adopted for root-feeding taxa such as nematodes (from the early 1990s, e.g. Okimoto et al., 1991, and more recently, e.g. Floyd et al., 2002) including using metagenomic, NGS approaches (Porazinska et al., 2009). Others include springtails (Collembola; Porco et al., 2012; Yu et al., 2012) and earthworms (Huang et al., 2007; Porco et al., 2012). To date, root-feeding taxa have been identified by a combination of molecular markers and sequence data.

4.1. Identification of larval pests

There are 19 pest species of Melolonthini (Coleoptera: Scarabaeidae) in Australia whose larvae cause economic levels of damage to sugarcane by feeding on their roots, but of these, five are indistinguishable using morphological techniques and have similar behaviours and overlapping distributions. Therefore, cytochrome *c* oxidase II subunit gene (*COII*) sequences were used to develop a diagnostic tool for species identification (Miller et al., 1999). Firstly, sequences from known adults, which are more easily determined, were used to produce a phylogenetic tree. This information on species relationships and mtDNA variation (base-pair differences) was used to select restriction enzymes producing diagnostic RFLPs for use in PCR–RFLP for analysis using gel electrophoresis, negating the need to sequence larvae. The phylogenetic analysis also revealed genetic differentiation of populations of one species (*Lepidiota negatoria*), which could be related to biogeographic boundaries to gene flow, and no differentiation for another species (*Antitrogus consanguineus*), suggesting uninterrupted gene flow. As well as providing a tool to identify important pest species that could be useful in pest management, this study also suggests that the sequence data may be more broadly useful for assessing species' biology, ecology and behaviour.

Crane fly larvae, or leatherjackets (Diptera: Nematocera), were some of the first root herbivore species to be subjected to molecular identification using isoelectric focusing, based on the position and number of protein bands (Humphreys et al., 1993). *Tipula oleracea* and *T. paludosa* are pests of cereal crops and grassland and this method was used to assess the species responsible for the recent (at the time) destruction of winter cereals in

Northern Britain, since larvae of these species are difficult to differentiate using morphological characteristics and time-consuming to rear to adults. A survey of Northern Ireland and Scotland revealed that *T. oleracea* was rare in grassland, possibly due to different moisture and vegetation preferences to *T. paludosa*, and that it might only become damaging when adult dispersal is restricted (eggs are usually more widely dispersed due to better female flight ability). This technique had not been used since this survey, for any published study at least, and has now been replaced by the use of sequence data. Rao et al. (2006) used mitochondrial cytochrome B (*cytB*) to distinguish these same two leatherjacket species in Oregon, the United States. Although there are other *Tipula* species present in this state, these particular species are invasive, with *T. paludosa* becoming a pest of pastures, lawns and golf courses but with less information on the host plants of *T. oleracea*. In contrast, the native species, which look very similar, are thought to feed on organic matter rather than the grass itself. Indeed, of all the samples identified using the technique, only the invasive species were found in urban landscapes (excluding one *T. tristis*), while only native species were found in cultivated grass. The invasive species were also found in cultivated peppermint with very few native species. The authors suggest further work would be needed to ascertain whether the native species are actively feeding on the crop, but the tool could help to identify the invasive species that are likely to cause damage, preventing indiscriminate pesticide application on finding leatherjackets. This method (Rao et al. 2006) is relatively laborious and costly when compared with other PCR-based techniques and *Tipula* research would benefit from the development of a different method.

The use of molecular techniques for the study of wireworm identity, distribution and phylogeography has increased rapidly in recent years. Ellis et al. (2009) developed a terminal restriction fragment length polymorphism (T-RFLP) approach to identify three UK pest species (*A. obscurus*, *A. sputator* and *A. lineatus*). As with the other larval pests mentioned so far, these species are almost impossible to identify without a great deal of time and expertise. Firstly, primers were developed for the mitochondrial *16S rRNA* region and adults of each species from locations in the United Kingdom, Europe and Canada sequenced. Based on this sequence data, restriction enzymes were chosen for their ability to produce fragments of different sizes dependent on species identity. A PCR in which the reverse primer is fluorescently labelled was carried out, then digested with the chosen enzymes and subjected to fragment analysis on an automated sequencer, differentiating it from the similar PCR–RFLP method, which uses gel electrophoresis to discriminate DNA fragments. This method was subsequently used by Benefer et al. (2010, 2012) to look further into the distribution of these species (see Section 4.3) and the *16S rRNA* sequencing protocol used to identify Canadian wireworm species (Benefer et al., 2013; phylogenetic analyses described in Section 4.4). While there are thought to be approximately 30 economically important species in Canada, lack of a robust identification technique has limited studies on the distribution and occurrence of pest species or aspects of their biology and ecology, which could provide useful information for control. Importantly, this study showed that identification to morphospecies was not always reliable due to identification errors (morphological characters were difficult to differentiate) and cryptic species. In a similar scenario to that of chafer communities (Ahrens et al., 2007; Section 4.2), although adults are fairly easy to distinguish, the larvae for some species have not been described, and due to their long life cycle (3–5 years depending on climate), it is not practical to rear these to adults for the majority of biological and ecological projects. The use of *16S rRNA* sequences may therefore be useful as a first step in linking adult and larval life stages (including eggs) and initiating morphological description of larvae.

Using a PCR-based approach, Staudacher et al. (2011a) identified further pest species within the genus *Agriotes* that are of agricultural importance. This technique focused on the *COI* region, first amplified using universal primers or newly designed primers in some cases (for degraded DNA) using adult samples from across Europe. Once general sequences were obtained for 20 *Agriotes* species, a set of primers were developed targeting nine of the most important species for use in a multiplex PCR assay and tested using the original 20 species plus other soil invertebrates found together with *Agriotes* populations. It was then applied to over 900 *Agriotes* larvae, 83% of which produced diagnostic bands when using gel electrophoresis. Of those that did not produce bands using the multiplex PCR, several were found to be related *Adrastus* spp., which are morphologically very similar. The study also found that there was a high genetic similarity between *A. proximus* and *A. lineatus*, which also have only minor morphological differences as adults and are attracted to the same sex pheromone in the field. Further analysis of both the pheromone composition and *COI* sequences of these two species has been carried out using a different section of the *COI* region (Vuts et al., 2012), finding again very high sequence similarity between the two species (>99%) and also a highly similar pheromone profile suggesting a very close relationship between the species. Since sex pheromones are used for monitoring adult presence and activity in the field as a surrogate for wireworm presence in the soil, it is important to understand the relationship of these species, and further work is currently underway on overlapping populations.

The *COI* region has also been used to identify wireworm species in the Midwestern United States, where a complex of species causes damage to maize (Lindroth and Clark, 2009). Larvae were collected from the Midwestern and Eastern United States and identified using dichotomous keys and comparison to museum specimens. Sequences comprising the majority of the *COI* gene were obtained for eleven of the 15 known economically important species. For the *Melanotus* species, museum specimen sequence data were compared to unknown larvae from four species that could not be distinguished and also compared with the full wireworm dataset, allowing identification based on their grouping with known species in maximum likelihood analysis. The ability to identify these species provides insight into the environments in which they are found (e.g. soil temperature and moisture preferences), allowing targeting of these areas for pest control.

4.2. Association of life stages

Aphids with morphologically different life stages that exploit taxonomically different host plants are typically difficult to assign to the correct species. Zhang et al. (2008) used *COI* sequences from Eriosomatinae aphids found on *Gramineae* (grass) roots, as the secondary host, and those found on *Ulmus* (elm) species, as the primary host, to identify species and associate the different life stages, thereby using them as a diagnostic tool. Using sequence data already available from adult specimens, a phylogeny was produced and the unknown morph sequences clustered with *Tetraneura chinensis*, confirming the species identity. The authors then described in detail the morphology of the secondary morph, allowing inclusion in future identification keys.

Ahrens et al. (2007) sequenced both mitochondrial (cytochrome oxidase subunit 1; *cox1* and 16S ribosomal RNA; *rrnL*) and nuclear (*28S rRNA*) DNA from chafer (see Sections 2.1 and 3.5) communities in order to associate adult and larval life stages from tropical lowlands of Nepal. Although adult taxonomy is well developed and identification is possible (albeit by taxonomic experts), new species are being discovered and the relation of the root-feeding larvae, which have limited morphologically characterising features, to the adults aboveground is unknown. These authors used more in-depth phylogenetic analyses to align sequences to species since intraspecific variation can make this difficult when using absolute methods of sequence divergence and especially for unknown, large assemblages of species. Combined mitochondrial and nuclear gene trees (using maximum likelihood) and statistical parsimony, whereby populations are subdivided into subgroups, and population aggregation analysis, which combines populations that are uniform for a particular character state, percentage differences (*p*-distance) and AMOVA and species delimitation methods for estimating species boundaries from the tree were used to assign species membership. This resulted in 24 species of which 19 could be associated with adults and identified using Linnaean names (nearly 93% of unknown larvae were identified to species). The *rrnL* and *cox1* networks produced using the statistical parsimony analyses were congruent, and there was much lower divergence for the *28S* sequences, which has implications for gene marker choice in such studies. As for the aphid study, the use of these techniques for determining species and then searching for diagnostic morphological characters of both larvae and adults is advocated. In terms of relating this to species biology, the species assemblages found above- and belowground can depend upon larval mortality, which varies both spatially and temporally; in particular, larvae and adults were often not found in the same season, which can confound attempts to associate larvae with adults. With a sampling scheme that takes these factors into account, it may be possible to use such DNA-based taxonomy to further investigate spatiotemporal patterns and geographic variation.

4.3. Phylogenetic relationships

As well as simply for species identification, a number of studies have used this DNA sequence data to produce molecular phylogenies for some taxa for further information on their evolutionary relationships. This can provide information on the rates and pattern of evolution and species diversification, providing insight into species ecology. Aboveground, this has been much used to investigate plant–herbivore interactions (e.g. Jurado-Rivera et al., 2009; Pinzon-Navarro et al., 2010) and the taxonomy of specific genera (e.g. Bell et al., 2004; Ortiz-Rivas et al., 2009). However, fewer studies have focused on community genetics, coevolution and plant–herbivore interactions in the belowground context (though see Hiltpold et al., 2013; Rasmann and Agrawal, 2011).

Diabrotica (corn rootworm) species have already been discussed in terms of predator–prey interactions (Section 2.1) and genetic diversity (Section 3.1), but studies have also considered their phylogenetic relationships. Little was known of phylogenetic relationships within this genus due to their morphological similarity. Allozymes have been used to determine that these species are represented in two distinct groups (*virgifera* and *fucata*) (Krysan et al., 1989) and molecular markers (mtDNA *NADH 4* with PCR–RFLP) were developed for differentiation of southern corn rootworm, WCR and NCR (Szalanski and Powers, 1996). Nuclear (*ITS1*) and mtDNA (*COI* and *COII*) genes have since been used to construct the phylogeny of *Diabrotica* species (Clark et al., 2001; Szalanski et al., 2000), which were in support of allozyme and morphological data with the same *virgifera* and *fucata* groupings observed. As adults are found in ecosystems with permanent perennial grasses, the *virgifera* group has been inferred to have an ancestor that might have evolved as a grass roots specialist (Branson and Krysan, 1981; Krysan and Smith, 1987), and the authors point out an interesting route to go down would be to assess phylogenetic relationships of larval associations of the *virgifera* group with their preferred hosts and female host–plant oviposition preferences, taking the focus to plant–herbivore interactions and evolutionary relationships.

Using a combination of nuDNA (translation elongation factor 1a; *EF-1* α) and mtDNA (*COI*), morphological and biological characters, Zhang and Qiao (2007) examined the evolution of gall morphology in Fordini aphids from China and Israel. Using samples collected on different primary hosts (*Pistacia* or *Rhus* species; the secondary hosts are grass roots or mosses), phylogenetic relationships of species were assessed, and it was found that host–plant affinity was an important character in subtribe division, with separate clades in the phylogenetic tree for *Pistacia*- and *Rhus*-feeding species. Gall traits also differed between species and over time seem to have become more adapted to manipulating the host plant, also reflected in the phylogenetic tree. This study focused on only the aboveground primary host. In the past, failure to take into account the secondary host morphs has led to taxonomic difficulties, with researchers creating synonymies. Further investigation (Ortiz-Rivas et al., 2009) using transfer experiments (of larvae from winged primary host morphs to wheat shoots), morphometric data of the wingless aphids reproducing on the roots and sequences from nuclear long-wavelength opsin, *EF-1a* and mtDNA, found no correlation of the phylogenetic relationships with primary host specialisation, but some aspects of gall morphology (gall capacity) were related to species position in the phylogenetic tree, as found previously. As in other studies of gall-forming aphids (see Section 3.2), it was difficult to distinguish secondary host morph species using morphology, but this was possible using nuDNA and mtDNA sequences. The data brought into question the characterisation of some species and suggested further ecological and biological data are needed to inform gall-forming aphid taxonomy.

4.4. Distribution

Studies on the spatial distribution of root-feeding organisms have been somewhat limited. While it is often found that organisms have patchy distributions within the soil, related to a range of biotic, chemical and physical factors, interactions between taxa and the scale of sampling are largely unknown. Benefer et al. (2010) investigated the spatial distribution of four root-feeding insect larvae pests (wireworms, leatherjackets, Sciarid (Diptera: Sciaridae) and Bibionid (Diptera: Bibionidae) flies) and their possible interactions over different sampling scales. Wireworms were identified to three species using T-RFLP (Ellis et al., 2009; see Section 4.1), while the other insects were defined using morphological characteristics, where possible. By separating the wireworms to species, differences in their abundance, composition and spatial distribution became apparent; for example, more variance in *A. obscurus* distribution was accounted for at the field scale than the site scale (locations sampled within fields) and their associations with other wireworm species changed from the field (largest) to the core scale (smallest). This has implications for grouping the three *Agriotes* wireworm UK pest species together in ecological studies as has traditionally been the case, assuming they are similar in biology and ecology (mainly because of a lack of reliable identification method; see Section 4.1). The same pattern was true of Bibionid larvae, separated to two species morphologically, but Sciarid larvae and leatherjackets could not be identified morphologically and so a similar result may be apparent when species are considered individually.

Other wireworm studies have considered their occurrence and above–belowground distribution in agricultural land, using molecular identification techniques. Staudacher et al. (2013a) assessed *Agriotes* distribution in relation to climatic and soil parameters in Austria using multiplex PCR (Staudacher et al., 2011a; see Section 4.1). This also included newly developed primers for *Adrastus* spp., which are morphologically very similar to *Agriotes*. Six out of 14 known species were identified from 85 sites, with more than 50% of sites containing more than one species with species-specific differences in their distribution and abundance apparent across the country. Altitude, annual precipitation, pH and water permeability was correlated with occurrence, with differences being found between species suggesting different environmental preferences, accounting for their distribution on a national scale. Ninety-two of the 1242 individuals were found to be *Adrastus* species. That not all known *Agriotes* species were found and some morphologically similar samples were from different genera suggests it is important to have knowledge of the species present in the soil, since the presence of wireworms with this general morphology does not necessarily mean damage will occur. The relation to climatic and soil data allowed distinction of two ecological groups, confirming some previous observations and providing information that could be used together with other species-specific factors to predict their presence and in different agricultural regions.

Few studies have considered how populations of above- and belowground stages of root herbivores are distributed, despite there being many ecologically and economically important species that spend their larval stage in the soil. As already discussed, UK *Agriotes* wireworms were historically considered together as a pest complex, without differentiation of individual species. Benefer et al. (2012) again used T-RFLP (Ellis et al., 2009) to identify three species of wireworm in the soil to their conspecific adult male click beetles trapped aboveground using sex pheromone traps. Sex pheromone traps are used to monitor adult male presence and activity and assume that this reflects wireworm distribution in the soil. However, no *A. lineatus* wireworms were identified from soil samples despite adults being the most numerous species captured and present in all fields. Twelve of 72 wireworm samples subjected to T-RFLP failed to produce the expected fragment sizes, and subsequent sequencing at the *16S rRNA* region revealed them to be 'non-*Agriotes*'—there was no conclusive match to the three UK *Agriotes* species or any other wireworm species sequenced at *16S rRNA*. Phylogenetic analyses (Benefer, 2011) suggested that one of these unknown species may be related to *Athous haemorrhoidalis*, an occasional pest, and *Denticollis linearis*. This result shows that the relationship between aboveground adults and belowground larvae is not necessarily straightforward and that sex pheromone traps may therefore be misleading in terms of the proportion and distribution of species encountered, confirming the deductions of Blackshaw et al. (2009) and Blackshaw and Hicks (2013). In addition, other species may be involved in damage if found in large enough populations. Using multivariate analysis, different associations were found between wireworms and environmental variables when considered as a group or separated to species, emphasising, as per Staudacher et al. (2011a), that although these species are closely related, they may have different ecological and biological preferences that shape their distributions and should be considered separately.

4.5. Quantification

Soil sampling and extraction techniques (e.g. soil coring followed by heat extraction) are widely used to obtain samples for further analysis using a range of sampling designs. Although there are inherent biases and practical drawbacks to their use (as with any sampling method, whether above- or belowground), this remains the most practical way to estimate root herbivore abundance in the field, though acoustic monitoring techniques have been used to map soil insect populations (e.g. Brandhorst-Hubbard et al., 2001; Zhang et al., 2003). One molecular ecological technique that has been used is monoclonal antibodies and enzyme-linked immunosorbent assays (ELISA), for quantifying slug density in the soil (McKemey et al., 2006). Slugs are particularly labour-intensive to sample, the traditional method involving collecting blocks of soil from the field and slowly flooding to drive them to the surface. Monoclonal antibodies were tested and selected based on their ability to detect slug proteins, but not those of other invertebrates, from soil samples, and then experiments were carried out using ELISA to calibrate slug biomass, comparing slug protein equivalents against known biomass, and field experiments comparing flooding and the ELISA-based technique. The two techniques produced similar estimates of slug density in the field, proving its utility in studies aiming to determine slug densities in general, though not species-specific estimates. One complication was that its sensitivity varied by soil type; though it worked well in natural soils, it failed in commercially available humus-rich soil, possibly due to high levels of organic matter. Despite this, the technique has the potential to overcome some of the difficulties associated with sampling these pests and could be applied to other root herbivores.

Another molecular approach to quantification involves the use of the number of DNA sequence reads per individual derived from NGS data. However, there have been mixed results from studies so far, with some authors finding little correlation between the number of sequence reads and the number of individuals of a species (Binladen et al., 2007; Deagle et al., 2013; Porazinska et al., 2009), while others have shown that read abundance generally reflected the number of individuals per species (Porazinska et al., 2010). This could be associated with taxon-specific variation in copy number per cell, tissue cell density or environmental persistence or due to technical factors during amplification, for example, using tagged primers, stringent bioinformatics filtering methods or sequencing platform (Deagle et al., 2013), and would need to be properly evaluated using controlled experiments and data analysis for application in root herbivore studies. Since this technology and associated quantification issues are only recently being evaluated, our understanding of the factors involved should improve as further studies on a wider range of taxa are carried out.

5. SUMMARY AND FOCUS FOR FUTURE WORK

To date, a variety of molecular techniques, including DNA sequencing of mitochondrial and nuclear genes, markers such as microsatellites, AFLPs, RFLPs and RAPDs and qPCR or a combination of these, have been used to study root herbivore ecology. Particularly when combined with other observational and experimental methods, the data have proved useful in elucidating species identity and relationships, population dynamics and dispersal, distribution, feeding behaviour and interactions with other root herbivores, predators, symbionts and plants. The majority of research has been carried out on belowground larval stages of economically important species, particularly Coleoptera (Scarabidae, Elateridae and Chrysomelidae), with most studies focusing on the aboveground adult stage, which is involved in dispersal, ultimately responsible for the distribution of the root-feeding stage in the soil and more practical to obtain. In common with their aboveground counterparts, root-feeding aphids have been the subject of a number of molecular studies due to their interesting, from an evolutionary point of view, and complicated two-host life cycle. Other less extensively covered taxa include Lepidoptera (*Agrotis* cutworms), Curculionidae (*Diaprepes* root weevils), Tipulidae (*Tipula* crane fly larvae—leatherjackets), Bibinoid and Sciarid fly larvae and a mollusc (*Deroceras reticulatum* slugs, not an insect but included as an important non-root herbivore and as explanation of a potentially useful method).

In general, the literature reviewed here shows that root herbivory research lags behind that for aboveground herbivores in the application of molecular approaches. Although the full range of available techniques has been used and in similar ways to that of aboveground herbivores, it is clear that the taxa coverage is much smaller (only studies on 10 taxa, though comprising several species, were identified during the literature search; Table 5.1) and that in some cases this is still in a method development or optimisation phase. For example, protein-based gut content analysis techniques have been used to study predator–prey interactions in aboveground herbivores since the late 1980s and DNA approaches since the 1990s, while they have only been applied to that of root-feeding taxa relatively recently (Juen and Traugott, 2005). In addition, we found few published studies that had applied the developed methods to test specific hypotheses. Despite the relatively recent adoption of molecular techniques for some taxa (e.g. Elaterid wireworms), there has been a surge in studies using these types of methods to answer questions on species' ecology, which was not possible using traditional techniques, for example, species-specific relationships between adult and larval distributions in *Agriotes* wireworms, the interaction between *Wolbachia* parasites and corn rootworm population dynamics and differences in distribution of native and invasive Tipulid pest species. While this has opened up new areas of research in many cases, extension of these studies and application of new and emerging technologies are likely to further increase the practicality and value of this research.

The dominance of (relatively few) pest studies has led to a level of pragmatism in the questions that have been addressed; the focus has been on pest management-related issues rather than research to understand the functioning of root herbivores within the soil ecosystem (but see Hiltpold et al., 2013), linking back to the research lag mentioned earlier. This means that some clearly important questions have not yet been addressed. For example, most root herbivores have patchy distributions, but the mechanisms underlying this are poorly understood. The development of molecular approaches to identify siblings is an essential precursor to understanding the role of adult females in oviposition site selection as a biotic factor, hence allowing the effect of abiotic influences to be better addressed. It is also desirable to be able to distinguish the sex of root herbivores in spatiotemporal studies, and here, there is a challenge to develop a molecular approach to substitute for laborious dissections. Dispersal, scaling and metapopulation dynamics are also topics that have received scant attention in root herbivore research. Direct observation is rarely possible in the soil and our current knowledge is largely based on inferences derived from statistical analyses but, as the work of Benefer et al. (2012) has shown, such conclusions can be influenced by the sampling method. Genetic studies are likely to be less susceptible to this effect and, we suggest, will drive these topics forward in the future.

A further current limitation to root herbivore research is that sampling and extraction methods are shaped by old technologies (e.g. Tullgren funnels) and have little relevance to the spatial scale at which the herbivores interact with the soil system. It is also desirable to be able to sample at the spatial scales relevant to specific taxa. Here, lessons can be learned from soil microbiologists and the recovery of whole-soil DNA for NGS. For this to be effective for the study of root herbivores necessitates robust sequence data being available for target species and the ability to scale extraction methods. For some herbivores, such as nematodes, samples are inevitably bigger than the space occupied by the animal and extension of the semiquantitative methods used (Porazinska et al., 2010) to larger herbivores and much bigger soil samples would be desirable.

At a broader level, there are potential applications in community ecology and above–belowground interactions that extend current studies and put root herbivore biology and ecology in the context of ecosystem processes and systems ecology. NGS techniques now make it possible to investigate the genetics of whole communities of organisms (metagenetics). Gut content analyses have a much bigger contribution to make to understanding the role(s) of root herbivores in food web and system functioning, and such methods could be applied to the analysis of root–herbivore interactions in soil food webs (extension of feeding ecology studies, Pompanon et al., 2012), species-specific surveys of endosymbiont diversity and links with root–herbivore functional diversity, large-scale phylogenetic analysis using several genetic markers across many related species for use in root–herbivore host–plant coevolutionary studies and fast and cost-effective isolation of genetic markers for use in phylogeographic and population genetic studies (e.g. Bai et al., 2010; Perry and Rowe, 2011). Other applications such as transcriptome characterisation, allowing genes expressed and their functions in different life stages or species to be assessed (Ekblom and Galindo, 2011), could be useful in understanding the genetics behind variation in traits between individuals and species. Similarly, gene expression profiling, often used to assess responses of plants to herbivory, could provide the other side of the picture and enable understanding of responses of root herbivores to environmental cues including parasites and temperature. The use of such newly emerging methods for nonmodel organisms would allow a view of the functional aspect of root herbivore ecology.

The application of molecular methods has spread across the biological sciences and enabled researchers to address ever more challenging questions. These are early days in their adoption for the study of root herbivores, but we can anticipate their increasing use as awareness of their potential and the development of molecular skills expands. It is, however, likely that this expanded use will continue to be dominated by the need to address pragmatic research questions about specific taxa, notably pests.

ACKNOWLEDGEMENT

We would like to thank Philip Smith for proofreading this chapter.

REFERENCES

Ahrens, D., Monaghan, M.T., Vogler, A.P., 2007. DNA-based taxonomy for associating adults and larvae in multi-species assemblages of chafers (Coleoptera: Scarabaeidae). Mol. Phylogenet. Evol. 44, 436–449.

Amalin, D.M., Peña, J.E., McSorley, R., 2000. Gut content analysis of three species of sac spiders by electrophoresis. Fla. Entomol. 83, 489–492.

Ascunce, M.S., Ernst, J.A., Clark, A., Nigg, H.N., 2008. Mitochondrial nucleotide variability in invasive populations of the root weevil *Diaprepes abbreviatus* (Coleoptera: Curculionidae) of Florida and preliminary assessment of *Diaprepes* sp. from Dominica. J. Econ. Entomol. 101, 1443–1454.

Bai, X., Zhang, W., Orantes, L., Jun, T.-H., Mittapalli, O., Rouf Mian, M., Michel, A.P., 2010. Combining next-generation sequencing strategies for rapid molecular resource development from an invasive Aphid species, *Aphis glycines*. PLoS One 5, e11370.

Bas, B., Dalkilic, Z., Peever, T.L., Nigg, H.N., Simpson, S.E., Gmitter, F.G., Adair, R.C., 2000. Genetic relationships among Florida *Diaprepes abbreviatus* (Coleoptera: Curculionidae) populations. Ann. Entomol. Soc. Am. 93, 459–467.

Bell, K.L., Yeates, D.K., Moritz, C., Monteith, G.B., 2004. Molecular phylogeny and biogeography of the dung beetle genus *Temnoplectron* Westwood (Scarabaeidae: Scarabaeinae) from Australia's wet tropics. Mol. Phylogenet. Evol. 31, 741–753.

Benefer, C.M., 2011. The molecular and behavioural ecology of click beetles (Coleoptera: Elateridae) in agricultural land (Ph.D.). Plymouth University, Plymouth.

Benefer, C., Andrew, P., Blackshaw, R.P., Ellis, J.S., Knight, M., 2010. The spatial distribution of phytophagous insect larvae in grassland soils. Appl. Soil Ecol. 45, 269–274.

Benefer, C.M., Knight, M.E., Ellis, J.S., Hicks, H., Blackshaw, R.P., 2012. Understanding the relationship between adult and larval *Agriotes* distributions: the effect of sampling method, species identification and abiotic variables. Appl. Soil Ecol. 53, 39–48.

Benefer, C., Herk, W.G., Ellis, J.S., Blackshaw, R.P., Vernon, R.S., Knight, M.E., 2013. The molecular identification and genetic diversity of economically important wireworm species (Coleoptera: Elateridae) in Canada. J. Pest. Sci. 86, 19–27.

Binladen, J., Gilbert, M.T., Bollback, J.P., Panitz, F., Bendixen, C., Nielsen, R., Willerslev, E., 2007. The use of coded PCR primers enables high-throughput sequencing of multiple homolog amplification products by 454 parallel sequencing. PLoS One 2, e197.

Blackshaw, R.P., Hicks, H., 2013. Distribution of adult stages of soil insect pests across an agricultural landscape. J. Pest. Sci. 86, 53–62.

Blackshaw, R.P., Hicks, H., Vernon, R.S., 2009. Sex pheromone traps for predicting wireworm populations: limitations to interpretation. In: Collier, R. (Ed.), IOBC/WPRS Bulletin Working Group "Integrated Protection of Field Vegetables", Proceedings of the Meeting at Porto, Portugal, 23–29 September, 2007, pp. 17–21.

Blaxter, M., Mann, J., Chapman, T., Thomas, F., Whitton, C., Floyd, R., Abebe, E., 2005. Defining operational taxonomic units using DNA barcode data. Phil. Trans. R. Soc. Lond. B Biol. Sci. 360, 1935–1943.

Blossey, B., Hunt-Joshi, T.R., 2003. Belowground herbivory by insects: influence on plants and aboveground herbivores. Annu. Rev. Entomol. 48, 521–547.

Brandhorst-Hubbard, J.L., Flanders, K.L., Mankin, R.W., Guertal, E.A., Crocker, R.L., 2001. Mapping of soil insect infestations sampled by excavation and acoustic methods. J. Econ. Entomol. 94, 1452–1458.

Branson, T.F., Krysan, J.L., 1981. Feeding and oviposition behavior and life cycle strategies of *Diabrotica*: an evolutionary view with implications for pest management. Environ. Entomol. 10, 826–831.

Chen, N., Liu, T.-X., Sétamou, M., French, J.V., Louzada, E.S., 2009. Molecular identification and population dynamics of two species of *Pemphigus* (Homoptera: Pemphidae) on cabbage. Insect Sci. 16, 115–124.

Chen, H., Wang, H., Siegfried, B.D., 2012. Genetic differentiation of western corn rootworm populations (Coleoptera: Chrysomelidae) relative to insecticide resistance. Ann. Entomol. Soc. Am. 105, 232–240.

Clark, T.L., Meinke, L.J., Foster, J.E., 2001. Molecular phylogeny of *Diabrotica* beetles (Coleoptera: Chrysomelidae) inferred from analysis of combined mitochondrial and nuclear DNA sequences. Insect Mol. Biol. 10, 303–314.

Crotty, F.V., Blackshaw, R.P., Murray, P.J., 2011. Tracking the flow of bacterially derived ^{13}C and ^{15}N through soil faunal feeding channels. Rapid Commun. Mass Spectrom. 25, 1503–1513.

Crotty, F.V., Adl, S.M., Blackshaw, R.P., Murray, P.J., 2012. Using stable isotopes to differentiate trophic feeding channels within soil food webs. J. Eukaryot. Microbiol. 59, 520–526.

Danismazoglu, M., Demir, İ., Sevim, A., Demirbag, Z., Nalcacioglu, R., 2012. An investigation on the bacterial flora of *Agriotes lineatus* (Coleoptera: Elateridae) and pathogenicity of the flora members. Crop. Prot. 40, 1–7.

Deagle, B.E., Thomas, A.C., Shaffer, A.K., Trites, A.W., Jarman, S.N., 2013. Quantifying sequence proportions in a DNA-based diet study using Ion Torrent amplicon sequencing: which counts count? Mol. Ecol. Resour. 13, 620–633.

Eitzinger, B., Traugott, M., 2011. Which prey sustains cold-adapted invertebrate generalist predators in arable land? Examining prey choices by molecular gut-content analysis. J. Appl. Ecol. 48, 591–599.

Eitzinger, B., Micic, A., Körner, M., Traugott, M., Scheu, S., 2013. Unveiling soil food web links: new PCR assays for detection of prey DNA in the gut of soil arthropod predators. Soil Biol. Biochem. 57, 943–945.

Ekblom, R., Galindo, J., 2011. Applications of next generation sequencing in molecular ecology of non-model organisms. Heredity 107, 1–15.

Ellis, J.S., Blackshaw, R.P., Parker, W., Hicks, H., Knight, M.E., 2009. Genetic identification of morphologically cryptic agricultural pests. Agric. For. Entomol. 11, 115–121.

Ernst, J.A., Ascunce, M.S., Clark, A.M., Nigg, H.N., 2006. Polymorphic microsatellite loci for *Diaprepes* root weevil (*Diaprepes abbreviatus* L.). Mol. Ecol. Notes 6, 1–3.

Floyd, R., Abebe, E., Papert, A., Blaxter, M., 2002. Molecular barcodes for soil nematode identification. Mol. Ecol. 11, 839–850.

Folmer, O., Black, M., Hoeh, W., Lutz, R., Vrijenhoek, R., 1994. DNA primers for amplification of mitochondrial cytochrome *c* oxidase subunit I from diverse metazoan invertebrates. Mol. Mar. Biol. Biotechnol. 3, 294–299.

Frago, E., Dicke, M., Godfray, H.C.J., 2012. Insect symbionts as hidden players in insect–plant interactions. Trends Ecol. Evol. 27, 705–711.

Freeland, J., 2005. Molecular Ecology. John Wiley and Sons, Chichester, UK.

Hales, D.F., Tomiuk, J., Wohrmann, K., Sunnucks, P., 1997. Evolutionary and genetic aspects of aphid biology: a review. Eur. J. Entomol. 94, 1–55.

Hebert, P.D.N., Cywinska, A., Ball, S.L., deWaard, J.R., 2003. Biological identifications through DNA barcodes. Proc. R. Soc. B Biol. Sci. 270, 313–321.

Hereward, J.P., Walter, G.H., 2012. Molecular interrogation of the feeding behaviour of field captured individual insects for interpretation of multiple host plant use. PLoS One 7, e44435.

Hiltpold, I., Bernklau, E., Bjostad, L.B., Alvarez, N., Miller-Struttmann, N.E., Lundgren, J.G., Hibbard, B.E., 2013. Nature, evolution and characterisation of rhizospheric chemical exudates affecting root herbivores. Adv. Insect Physiol. 45, 97–157.

Huang, J., Xu, Q., Sun, Z.J., Tang, G.L., Su, Z.Y., 2007. Identifying earthworms through DNA barcodes. Pedobiologia 51, 301–309.

Humphreys, I.C., Blackshaw, R.P., Stewart, R.M., Coll, C., 1993. Differentiation between larvae of *Tipula paludosa* and *Tipula oleracea* (Diptera: Tipulidae) using isoelectric focusing, and their occurrence in grassland in northern Britain. Ann. Appl. Biol. 122, 1–8.

Hunter, M.D., 2001. Out of sight, out of mind: the impacts of root-feeding insects in natural and managed systems. Agric. For. Entomol. 3, 3–9.

Ivens, A.B.F., Kronauer, D.J.C., Boomsma, J.J., 2011. Characterisation and cross-amplification of polymorphic microsatellite loci in ant-associated root-aphids. Conserv. Genet. Resour. 3, 73–77.

Ivens, A., Kronauer, D., Pen, I., Weissing, F., Boomsma, J., 2012a. Ants farm subterranean aphids mostly in single clone groups—an example of prudent husbandry for carbohydrates and proteins? BMC Evol. Biogeosci. 12, 1–12.

Ivens, A.B., Kronauer, D.J., Pen, I., Weissing, F.J., Boomsma, J.J., 2012b. Reproduction and dispersal in an ant-associated root aphid community. Mol. Ecol. 21, 4257–4269.

Janzen, D.H., 2004. Now is the time. Philos. Trans. R. Soc. B Biol. Sci. 359, 731–732.

Johnson, S.N., Read, D.B., Gregory, P.J., 2004. Tracking larval insect movement within soil using high resolution X-ray microtomography. Ecol. Entomol. 29, 117–122.

Juen, A., Traugott, M., 2005. Detecting predation and scavenging by DNA gut-content analysis: a case study using a soil insect predator–prey system. Oecologia 142, 344–352.

Juen, A., Traugott, M., 2006. Amplification facilitators and multiplex PCR: tools to overcome PCR-inhibition in DNA-gut-content analysis of soil-living invertebrates. Soil Biol. Biochem. 38, 1872–1879.

Juen, A., Traugott, M., 2007. Revealing species-specific trophic links in soil food webs: molecular identification of scarab predators. Mol. Ecol. 16, 1545–1557.

Jurado-Rivera, J.A., Vogler, A.P., Reid, C.A., Petitpierre, E., Gomez-Zurita, J., 2009. DNA barcoding insect–host plant associations. Proc. R. Soc. B Biol. Sci. 276, 639–648.

Kim, K.S., Sappington, T.W., 2004. Isolation and characterization of polymorphic microsatellite loci in the boll weevil, *Anthonomus grandis* Boheman (Coleoptera: Curculionidae). Mol. Ecol. Notes 4, 701–703.

Kim, K.S., Sappington, T.W., 2005. Polymorphic microsatellite loci from the western corn rootworm (Insecta: Coleoptera: Chrysomelidae) and cross-amplification with other *Diabrotica* spp. Mol. Ecol. Notes 5, 115–117.

Kim, K.S., Coates, B.S., Hellmich, R.L., Sumerford, D.V., Sappington, T.W., 2008. Isolation and characterization of microsatellite loci from the European corn borer, *Ostrinia nubilalis* (Hübner) (Insecta: Lepidoptera: Crambidae). Mol. Ecol. Resour. 8, 409–411.

Krysan, J.L., Smith, R.F., 1987. Systematics of the *virgifera* species group of *Diabrotica* (Coleoptera: Chrysomelidae: Galerucinae). Entomography 5, 375–484.

Krysan, J.L., McDonald, I.C., Tumlinson, J.H., 1989. Phenogram based on allozymes and its relationship to classical biosystematics and pheromone structure among eleven Diabroticites (Coleoptera: Chrysomelidae). Ann. Entomol. Soc. Am. 82, 574–581.

Lacey, L.A., Unruh, T.R., Simkins, H., Thomsen-Archer, K., 2007. Gut bacteria associated with the Pacific Coast wireworm, *Limonius canus*, inferred from 16s rDNA sequences and their implications for control. Phytoparasitica 35, 479–489.

Landl, M., Glauninger, J., 2013. Preliminary investigations into the use of trap crops to control *Agriotes* spp. (Coleoptera: Elateridae) in potato crops. J. Pest. Sci. 86, 85–90.
Lindroth, E., Clark, T.L., 2009. Phylogenetic analysis of an economically important species complex of wireworms (Coleoptera: Elateridae) in the midwest. J. Econ. Entomol. 102, 743–749.
Lundgren, J.G., Fergen, J.K., 2011. Enhancing predation of a subterranean insect pest: a conservation benefit of winter vegetation in agroecosystems. Appl. Soil Ecol. 51, 9–16.
Lundgren, J.G., Ellsbury, M.E., Prischmann, D.A., 2009. Analysis of the predator community of a subterranean herbivorous insect based on polymerase chain reaction. Ecol. Appl. 19, 2157–2166.
McDonald, I.C., Krysan, J.L., Johnson, O.A., 1985. Genetic variation within and among geographic populations of *Diabrotica barberi* (Coleoptera: Chrysomelidae). Ann. Entomol. Soc. Am. 78, 271–278.
McKemey, A.R., Glen, D.M., Wiltshire, C.W., Symondson, W.O.C., 2006. Molecular quantification of slug density in the soil using monoclonal antibodies. Soil Biol. Biochem. 38, 2903–2909.
Miller, L.J., Allsopp, P.G., 2005. Phylogeography of the scarab beetle *Antitrogus parvulus* Britton (Coleoptera: Scarabaeidae) in south-eastern Queensland, Australia. Aust. J. Entomol. 44, 29–36.
Miller, L.J., Allsopp, P.G., Graham, G.C., Yeates, D.K., 1999. Identification of morphologically similar canegrubs (Coleoptera: Scarabaeidae: Melolonthini) using a molecular diagnostic technique. Aust. J. Entomol. 38, 189–196.
Miller, N.J., Birley, A.J., Tatchell, G.M., 2000. Polymorphic microsatellite loci from the lettuce root aphid, *Pemphigus bursarius*. Mol. Ecol. 9, 1951–1952.
Miller, N.J., Birley, A.J., Overall, A.D., Tatchell, G.M., 2003. Population genetic structure of the lettuce root aphid, *Pemphigus bursarius* (L.), in relation to geographic distance, gene flow and host plant usage. Heredity 91, 217–223.
Moran, N.A., McCutcheon, J.P., Nakabachi, A., 2008. Genomics and evolution of heritable bacterial symbionts. Annu. Rev. Genet. 42, 165–190.
Murray, P.J., Gregory, P.J., Granger, S.J., Headon, D.M., Johnson, S.N., 2010. Dispersal of soil-dwelling clover root weevil (*Sitona lepidus* Gyllenhal, Coleoptera: Curculionidae) larvae in mixed plant communities. Appl. Soil Ecol. 46, 422–425.
Okimoto, R., Chamberlin, H.M., Macfarlane, J.L., Wolstenholme, D.R., 1991. Repeated sequence sets in mitochondrial DNA molecules of root knot nematodes (Meloidogyne): nucleotide sequences, genome location and potential for host-race identification. Nucleic Acids Res. 19, 1619–1626.
Ortiz-Rivas, B., Martínez-Torres, D., Pérez Hidalgo, N., 2009. Molecular phylogeny of Iberian Fordini (Aphididae: Eriosomatinae): implications for the taxonomy of genera *Forda* and *Paracletus*. Syst. Entomol. 34, 293–306.
Perry, J.C., Rowe, L., 2011. Rapid microsatellite development for water striders by next-generation sequencing. J. Hered. 102, 125–129.
Pieterse, C.M.J., Dicke, M., 2007. Plant interactions with microbes and insects: from molecular mechanisms to ecology. Trends Plant Sci. 12, 564–569.
Pinzon-Navarro, S., Barrios, H., Murria, C., Lyal, C.H., Vogler, A.P., 2010. DNA-based taxonomy of larval stages reveals huge unknown species diversity in neotropical seed weevils (genus *Conotrachelus*): relevance to evolutionary ecology. Mol. Phylogenet. Evol. 56, 281–293.
Pompanon, F., Deagle, B.E., Symondson, W.O.C., Brown, D.S., Jarman, S.N., Taberlet, P., 2012. Who is eating what: diet assessment using next generation sequencing. Mol. Ecol. 21, 1931–1950.
Porazinska, D.L., Giblin-Davis, R.M., Faller, L., Farmerie, W., Kanzaki, N., Morris, K., Powers, T.O., Tucker, A.E., Sung, W., Thomas, W.K., 2009. Evaluating

high-throughput sequencing as a method for metagenomic analysis of nematode diversity. Mol. Ecol. Resour. 9, 1439–1450.

Porazinska, D.L., Sung, W., Giblin-Davis, R.M., Thomas, W.K., 2010. Reproducibility of read numbers in high-throughput sequencing analysis of nematode community composition and structure. Mol. Ecol. Resour. 10, 666–676.

Porco, D., Bedos, A., Greenslade, P., Janion, C., Skarżyński, D., Stevens, M.I., Jansen van Vuuren, B., Deharveng, L., 2012. Challenging species delimitation in Collembola: cryptic diversity among common springtails unveiled by DNA barcoding. Invertebr. Syst. 26, 470–477.

Powers, T., Harris, T., Higgins, R., Mullin, P., Sutton, L., Powers, K., 2011. MOTUs, morphology, and biodiversity estimation: a case study using nematodes of the suborder Criconematina and a conserved 18S DNA barcode. J. Nematol. 43, 35–48.

Rao, S., Liston, A., Crampton, L., Takeyasu, J., 2006. Identification of larvae of exotic *Tipula paludosa* (Diptera: Tipulidae) and *T. oleracea* in North America using mitochondrial cytB sequences. Ann. Entomol. Soc. Am. 99, 33–40.

Rasmann, S., Agrawal, A.A., 2011. Evolution of specialization: a phylogenetic study of host range in the red milkweed beetle (*Tetraopes tetraophthalmus*). Am. Nat. 177, 728–737.

Ritland, C., Ritland, A., 2000. DNA fragment markers in plants. In: Baker, A.J. (Ed.), Molecular Methods in Ecology. Blackwell Scientific, Oxford, pp. 208–234.

Roehrdanz, R.L., Levine, E., 2007. *Wolbachia* bacterial infections linked to mitochondrial DNA reproductive isolation among populations of northern corn rootworm (Coleoptera: Chrysomelidae). Ann. Entomol. Soc. Am. 100, 522–531.

Roehrdanz, R.L., Szalanski, A.L., Levine, E., 2003. Mitochondrial DNA and ITS1 differentiation in geographical populations of northern corn rootworm, *Diabrotica barberi* (Coleoptera: Chrysomelidae): identification of distinct genetic populations. Ann. Entomol. Soc. Am. 96, 901–913.

Schallhart, N., Tusch, M.J., Staudacher, K., Wallinger, C., Traugott, M., 2011. Stable isotope analysis reveals whether soil-living elaterid larvae move between agricultural crops. Soil Biol. Biochem. 43, 1612–1614.

Sheppard, S.K., Harwood, J.D., 2005. Advances in molecular ecology: tracking trophic links through predator–prey food-webs. Funct. Ecol. 19, 751–762.

Shi, W., Xie, S., Chen, X., Sun, S., Zhou, X., Liu, L., Gao, P., Kyrpides, N.C., No, E.-G., Yuan, J.S., 2013. Comparative genomic analysis of the endosymbionts of herbivorous insects reveals eco-environmental adaptations: biotechnology applications. PLoS Genet. 9, e1003131.

Staudacher, K., Pitterl, P., Furlan, L., Cate, P.C., Traugott, M., 2011a. PCR-based species identification of *Agriotes* larvae. Bull. Entomol. Res. 101, 201–210.

Staudacher, K., Wallinger, C., Schallhart, N., Traugott, M., 2011b. Detecting ingested plant DNA in soil-living insect larvae. Soil Biol. Biochem. 43, 346–350.

Staudacher, K., Schallhart, N., Pitterl, P., Wallinger, C., Brunner, N., Landl, M., Kromp, B., Glauninger, J., Traugott, M., 2013a. Occurrence of *Agriotes* wireworms in Austrian agricultural land. J. Pest. Sci. 86, 33–39.

Staudacher, K., Schallhart, N., Thalinger, B., Wallinger, C., Juen, A., Traugott, M., 2013b. Plant diversity affects behavior of generalist root herbivores, reduces crop damage, and enhances crop yield. Ecol. Appl. 23, 1135–1145.

Szalanski, A.L., Powers, T.O., 1996. Molecular diagnostics of three *Diabrotica* (Coleoptera: Chrysomelidae) pest species. J. Kans. Entomol. Soc. 69, 260–266.

Szalanski, A.L., Roehrdanz, R.L., Taylor, D.B., Chandler, L., 1999. Genetic variation in geographical populations of western and Mexican corn rootworm. Insect Mol. Biol. 8, 519–525.

Szalanski, A.L., Roehrdanz, R.L., Taylor, D.B., 2000. Genetic relationship among *Diabrotica* species (Coleoptera: Chrysomelidae) based on rDNA and mtDNA sequences. Fla. Entomol. 83, 262–267.

Taylor, H.R., Harris, W.E., 2012. An emergent science on the brink of irrelevance: a review of the past 8 years of DNA barcoding. Mol. Ecol. Resour. 12, 377–388.

Traugott, M., 2003. The prey spectrum of larval and adult *Cantharis* species in arable land: an electrophoretic approach. Pedobiologia 47, 161–169.

Vernon, R.S., Kabaluk, T., Behringer, A., 2000. Movement of *Agriotes obscurus* (Coleoptera: Elateridae) in Strawberry (Rosaceae) plantings with wheat (Gramineae) as a trap crop. Can. Entomol. 132, 231–241.

Vos, P., Hogers, R., Bleeker, M., Reijans, M., Van De Lee, T., Hornes, M., Frijters, A., Pot, J., Peleman, J., Kuiper, M., et al., 1995. AFLP: a new technique for DNA fingerprinting. Nucleic Acids Res. 23, 4407–4414.

Vuts, J.Z., Tolasch, T., Furlan, L., Bálintné Csonka, Ã.V., Felföldi, T.S., Márialigeti, K.R., Toshova, T., Subchev, M., Xavier, A.L., Tóth, M.S., 2012. *Agriotes proximus* and *A. lineatus* (Coleoptera: Elateridae): a comparative study on the pheromone composition and cytochrome *c* oxidase subunit I gene sequence. Chemoecology 22, 23–28.

Waite, I.S., O'Donnell, A.G., Harrison, A., Davies, J.T., Colvan, S.R., Ekschmitt, K., Dogan, H., Wolters, V., Bongers, T., Bongers, M., Bakonyi, G., Nagy, P., Papatheodorou, E.M., Stamou, G.P., Boström, S., 2003. Design and evaluation of nematode 18S rDNA primers for PCR and denaturing gradient gel electrophoresis (DGGE) of soil community DNA. Soil Biol. Biochem. 35, 1165–1173.

Waits, E.R., Stolz, U., 2008. Polymorphic microsatellite loci from northern and Mexican corn rootworms (Insecta: Coleoptera: Chrysomelidae) and cross-amplification with other *Diabrotica* spp. Mol. Ecol. Resour. 8, 707–709.

Waldner, T., Sint, D., Juen, A., Traugott, M., 2013. The effect of predator identity on post-feeding prey DNA detection success in soil-dwelling macro-invertebrates. Soil Biol. Biochem. 63, 116–123.

Wallinger, C., Juen, A., Staudacher, K., Schallhart, N., Mitterrutzner, E., Steiner, E.-M., Thalinger, B., Traugott, M., 2012. Rapid plant identification using species- and group-specific primers targeting chloroplast DNA. PLoS One 7, e29473.

Wallinger, C., Staudacher, K., Schallhart, N., Peter, E., Dresch, P., Juen, A., Traugott, M., 2013. The effect of plant identity and the level of plant decay on molecular gut content analysis in a herbivorous soil insect. Mol. Ecol. Resour. 13, 75–83.

Welsh, J., McClelland, M., 1990. Fingerprinting genomes using PCR with arbitrary primers. Nucleic Acids Res. 18, 7213–7218.

Whiteman, N.K., Jander, G., 2010. Genome-enabled research on the ecology of plant–insect interactions. Plant Physiol. 154, 475–478.

Williams, J.K., 2012. Mesofaunal recolonisation of degraded soils (Ph.D.). Plymouth University, Plymouth.

Williams, J.G.K., Kubelik, A.R., Livak, K.J., Rafalski, J.A., Tingey, S.V., 1990. DNA polymorphisms amplified by arbitrary primers are useful as genetic markers. Nucleic Acids Res. 18, 6531–6535.

Yu, D.W., Ji, Y., Emerson, B.C., Wang, X., Ye, C., Yang, C., Ding, Z., 2012. Biodiversity soup: metabarcoding of arthropods for rapid biodiversity assessment and biomonitoring. Methods Ecol. Evol. 3, 613–623.

Zhang, H.C., Qiao, G.X., 2007. Molecular phylogeny of Fordini (Hemiptera: Aphididae: Pemphiginae) inferred from nuclear gene EF-1 α and mitochondrial gene COI. Bull. Entomol. Res. 97, 379–386.

Zhang, M., Crocker, R.L., Mankin, R.W., Flanders, K.L., Brandhorst-Hubbard, J.L., 2003. Acoustic identification and measurement of activity patterns of white grubs in soil. J. Econ. Entomol. 96, 1704–1710.

Zhang, H.-C., Zhang, D., Qiao, G.-X., 2008. Association of aphid life stages using DNA sequences: a case study of tribe Eriosomatini (Hemiptera: Aphididae: Pemphiginae). Insect Sci. 15, 545–551.

INDEX

Note: Page numbers followed by "*f*" indicate figures and "*t*" indicate tables.

D

E

F

G

H

I

L

M

N

S

V

W

X

Printed and bound by CPI Group (UK) Ltd, Croydon, CR0 4YY
08/05/2025
01864962-0001